AF245497

Transportation for the Nuclear Industry

Transportation for the Nuclear Industry

Edited by
D. G. Walton
University of Manchester
Manchester, United Kingdom

and

S. M. Blackburn
The Institution of Nuclear Engineers
London, United Kingdom

PLENUM PRESS • NEW YORK AND LONDON

Library of Congress Cataloging in Publication Data

International Conference on Transportation for the Nuclear Industry (1st: 1988: Stratford-upon-Avon, England)
Transportation for the nuclear industry / edited by D. G. Walton and S. M. Blackburn.
p.   cm.
"Proceedings of the First International Conference on Transportation for the Nuclear Industry, held May 23–25, 1988, in Stratford-upon Avon, United Kingdom"—T.p. verso.
Includes bibliographical references.
ISBN 0-306-43379-6
1. Radioactive substances—Transportation—Congresses. 2. Radioactive substances—Packing—Congresses. I. Walton, D. G. II. Blackburn, S. M. III. Title.
TK9400.I62   1988   89-39478
621.48′38—dc20   CIP

Proceedings of the First International Conference on
Transportation for the Nuclear Industry, held May 23–25, 1988,
in Stratford-upon-Avon, United Kingdom

© 1989 Plenum Press, New York
A Division of Plenum Publishing Corporation
233 Spring Street, New York, N.Y. 10013

Printed in the United States of America

STEERING COMMITTEE

Mr.W.C.BIRKETT (Chairman)
CEGB

Mrs.S.BLACKBURN (Secretary)
The Institution of Nuclear Engineers

Mr.P.ABBOTT
British Rail

Mr.L.BAKER
Project Management Services Limited

Mr.J.BANKS
Pickfords Removals Limited, Industrial
Division

Mr.P.BEAVER
Health & Safety Executive, Nuclear
Installations Inspectorte

Mr.F.J.L.BINDON
The Institution of Nuclear Engineers

Mr.D.BLACKMAN
Department of Transport

Cpt.J.BRISTOW
James Fisher & Sons plc

Mr.J.E.MIDDLETON
British Nuclear Fuels plc

Mr.T.PETERS
CEGB

Mrs.J.SHEPPARD
Merseyside & North Wales Electricity
Board

Mr.V.L.SNAREY
Ministry of Defence

Mrs.J.WALES
Costain Engineering Limited

Co-sponsors:
The Institution of Civil Engineers
The European Nuclear Society
The Chartered Institute of Transport

PREFACE

The transport requirements of the nuclear industry are unique in many respects. This is not because cargoes are particularly large or hazardous by comparison with other industries but because standards of performance required in every aspect of the activity are so much greater than those required for any other industry.

Transport of nuclear materials is subject to existing statutory regulations applied not only nationally but internationally. In addition to this, users of transport demand the highest standards of performance for their own purposes particularly in the area of quality assurance. Similar considerations also apply to the transport of non-nuclear materials where the transport link often has to tie in with project management and quality assurance requirements. Safety and security of nuclear materials are of paramount importance but even when these aspects are of a completely acceptable standard public attitudes to the transport activities have to be addressed adequately.

The transport system itself consists of many components. The route, the vehicles, the containers, and the individual packages. The performance of each component determines the performance of the total system: all these factors were presented in the 1988 Conference on Transportation for the Nuclear Industry, giving a broad over-view of current practice together with wide ranging consideration of future requirements and developments.

D.G.Walton
13th June 1989

CONTENTS

SESSION 1: CASK DESIGN ANALYSIS AND TESTING

SESSION 2: PACKAGING SYSTEMS AND STRATEGIES

SESSION 4: OPERATIONS

ROAD CASK FOR THE TRANSPORTATION OF CANDU IRRADIATED FUEL

D.J.Ribbans

Ontario Hydro
700 University Avenue
Toronto
Ontario
Canada

ABSTRACT

Ontario Hydro, the electric utility for the Province of Ontario, Canada, currently stores all irradiated fuel from its nuclear generating stations at the station site. As part of its commitment to the Canada/Ontario Nuclear Fuel Waste Management Program, Ontario Hydro has designed, licensed and manufactured Canada's first large-scale type B(u) transportation cask for the road transportation of irradiated fuel.

The construction of the Road Cask is monolithic 304 stainless steel with an impact limiter constructed from redwood encased in stainless steel. The Road Cask is licensed to carry 192 fuel bundles of 10 year cooled fuel. A capability to carry a smaller number of Recently Discharged fuel bundles is being developed and a licensing submission is currently being produced for review by the Atomic Energy Control Board (AECB) of Canada. To support the licensing applications, engineering evaluation, computer analyses supported by benchmark tests and scale model testing were used to demonstrate that the Road Cask will comply with all the tested and accident conditions as required by the AECB Transport Packaging of Radioactive Materials Regulations.

This paper describes the Irradiated Fuel Transportation Road Cask design and its transport trailer. It will discuss compliance with the regulations, the model testing and computer code benchmarking. It will also include a discussion on testing and analysis for conditions beyond the regulatory requirements. Finally the paper discusses the manufacturing of the prototype Road Cask, the commissioning/trials program and the public communications program.

INTRODUCTION

Ontario Hydro, the electric utility for the Province of Ontario in Canada, is generating 46 percent of its electricity through the use of CANDU (CANadian Deuterium Uranium) pressurized heavy water reactor (PHWR) systems at two major sites, Pickering and Bruce. The third site, Darlington Generating Station, consisting of four PHWR's, is now under construction. When completed in 1992, Ontario Hydro will have a nuclear generating capacity of 136000 MWe, from 20 CANDU PHWR units.

By the year 1993, the fuel arising from Ontario Hydro's stations are expected to reach 92000 fuel bundles (1900 Mg) per annum. At each of the Nuclear Generating Stations (NGS), water-filled Irradiated Fuel Bays (IFB) are built to store this fuel.

A very modest system for irradiated fuel transportation has met our needs in the past. Three casks, of 2-bundle (44 kg) and 72-bundle (1.6 Mg. Pegase cask) capacity have been used infrequently. They have been used mainly for shipping fuel from the NGS to the Atomic Energy of Canada Limited (AECL) research laboratories at Whiteshell, Manitoba and Chalk River, Ontario. Also an on-site flask is used for transportation of IF among IFB's (from primary storage to auxiliary storage bay) at the Pickering NGS.

As part of its commitment to the Canada/Ontario Nuclear Fuel Waste Management Program, Ontario Hydro has developed a cask for the large scale transportation of irradiated CANDU fuel. The prototype cask has been licensed and constructed (Feb 1988). It will be used to provide a demonstration that CANDU IF can be transported safely and economically. Although the cask is specifically designed for road transportation it could be used for rail or water modes, both of which are currently being studied.

CASK DESCRIPTION

Figure 1 shows a pictorial view of the road cask design. In this configuration the cask is designed to carry 192 fuel bundles (approximately 4000kg U) housed in two transportation/storage modules, see Figure 2. These modules have been specially designed to both maximize storage in the water pools at the stations, and to meet the transportation structural requirements. Irradiated fuel is currently being stored in these modules at the Pickering NGS and they will be used at the Darlington NGS when it becomes operational in 1988.

The construction of the cask is in monolithic stainless steel, type 304. Stainless steel was chosen to eliminate any concerns regarding brittle fracture and to facilitate decontamination. Cost evaluations were undertaken which showed that solid stainless steel was no more expensive than a ferritic steel cask with a stainless steel cladding. The licence allows the cask to be manufactured as a welded fabrication, a one piece forging or a one piece casting. The prototype cask is constructed as a one piece forging whereas the 1/2 scale model used in the drop test program was a welded fabrication.

The cask lid is secured by 32 Nitronic 60 stainless steel bolts, each 48 mm diameter, and sealed with two elastomeric '0' rings. The '0' rings are housed in two parallel dovetail section grooves in the cask lid. The fit of the lid in the cask body is such that any lateral impact forces are transmitted through the cask body and not as shear on the lid bolts. An impact limiter, constructed from redwood encased in a thin stainless steel shell, is attached to the lid of the cask by eight Nitronic 60 bolts. The purpose of the impact limiter is to both protect the closure from impact and from the heat of a fire. Nitronic 60 was chosen as the bolt material following galling tests undertaken with different combinations of stainless steel.

The cask has vent and drain connections which are also sealed by elastomeric '0' rings. Seal selection is discussed later.

Two trunnions, bolted to the cask body, are provided for lifting,

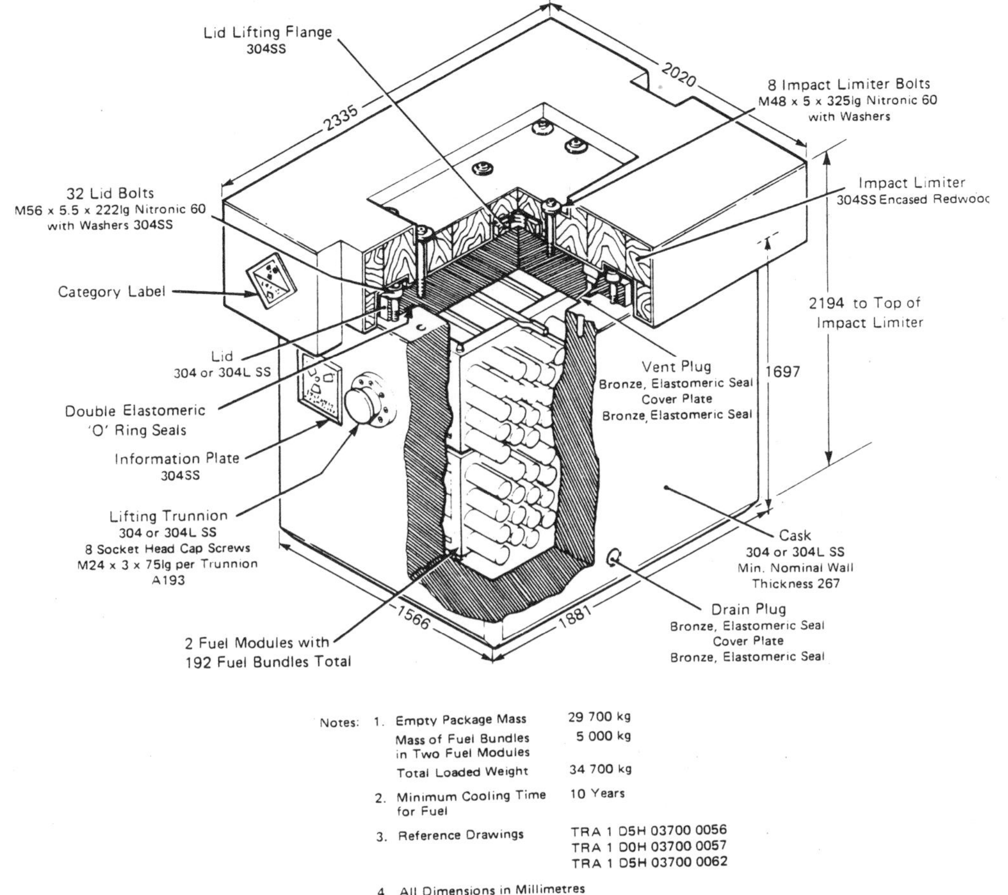

Fig. 1. CANDU irradiated fuel road cask.

using the dedicated lifting beam, and to enable the cask to be attached to the trailer via the tie down system.

The cask is licensed to carry 10 year cooled fuel and in this configuration is designed to be transported dry with an internal pressure of one atmosphere of air. Dry transportation has been adopted because it facilitates interfacing with dry loading and unloading, minimizing turn round time, decontamination problems and operator dose.

The cask can also be loaded and unloaded with fuel in a water pool. The cask is designed so that simple draining is sufficient to reduce the residual water to a safe level following wet loading. This eliminates the need for time consuming vacuum drying procedures, which would increase the turn-around time.

TRACTOR TRAILER

Figure 3 shows the cask mounted on a four axle flat bed trailer. The cask Lifting beam is carried at the rear of the trailer. The total payload, including the tractor/trailer weight is 59,000kg. With the design axle spacing, the Ontario Provincial regulations allow the cask to

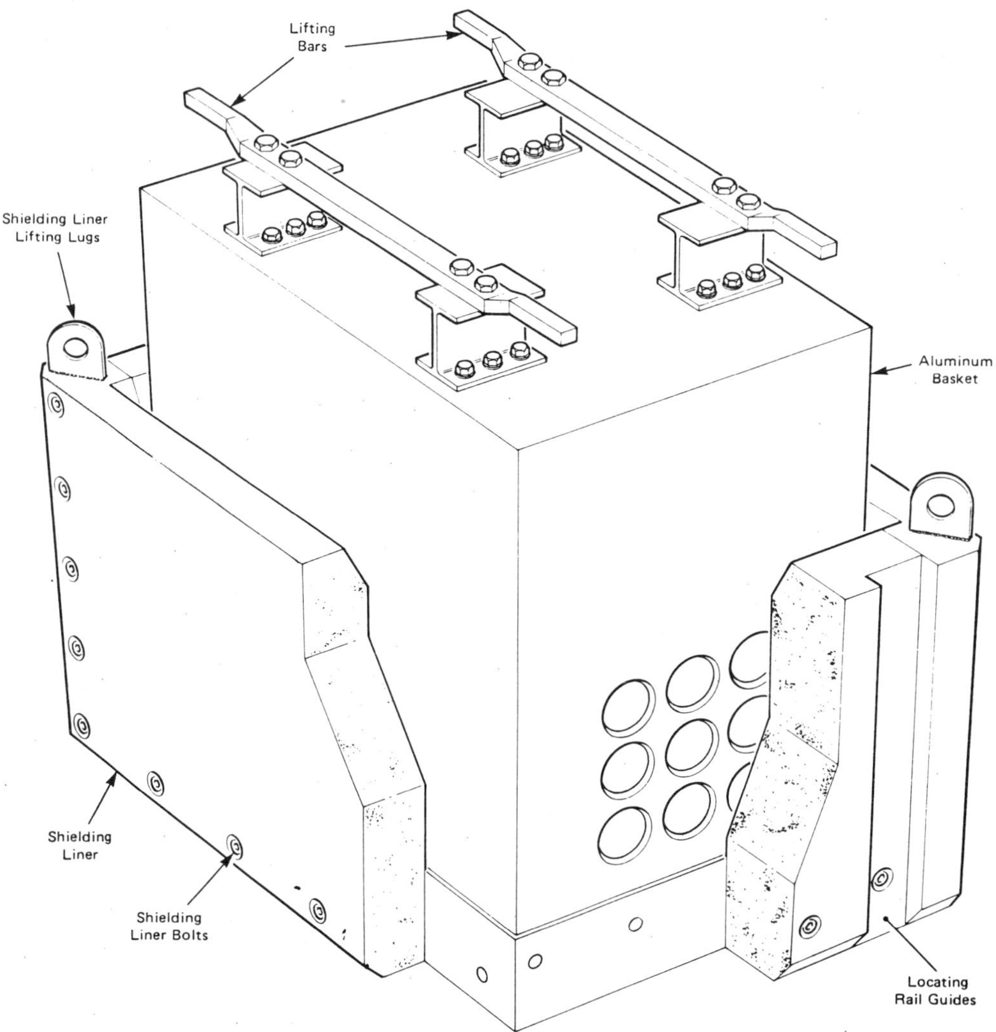

Fig. 1A. Recently discharged fuel transportation basket
and shielding liner.

be transported without special permits. The bed of the trailer is
sheathed with stainless steel to facilitate decontamination. The tie down
system, attached to the cask trunnions, is designed to withstand
accelerations of 4g in all three directions. During transportation the
cask is covered by an aluminum weather cover.

DESIGN CONSIDERATIONS

Different concepts were evaluated using the following criteria:

(a) The cask must be licensable as a type B(U) package.
(b) The cask must be easily transportable, i.e. no special permit
     required for travelling on the highway
(c) Loading and unloading of irradiated fuel must be easily accomplished
     and occupational doses for the operation must be minimized.

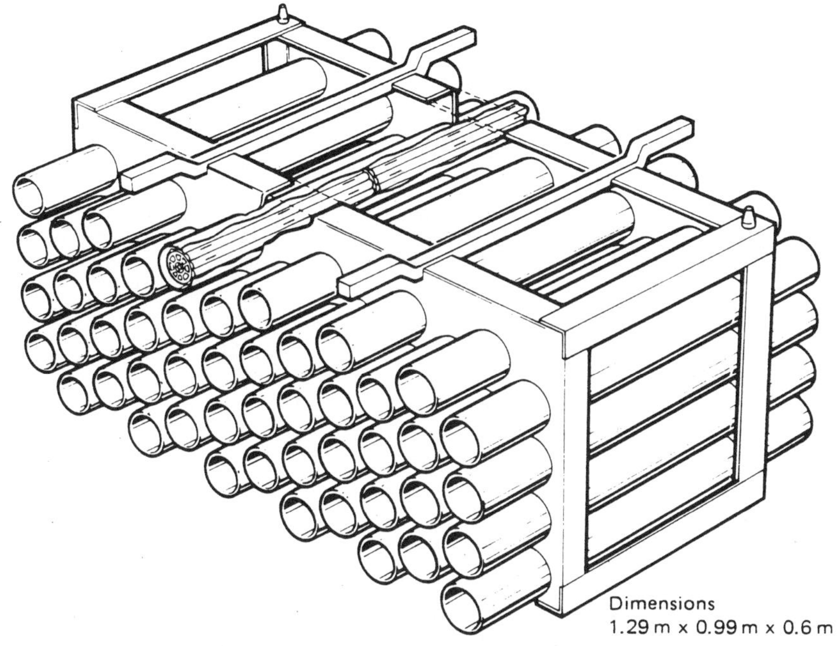

Fig. 2. Irradiated fuel shipping and storage module

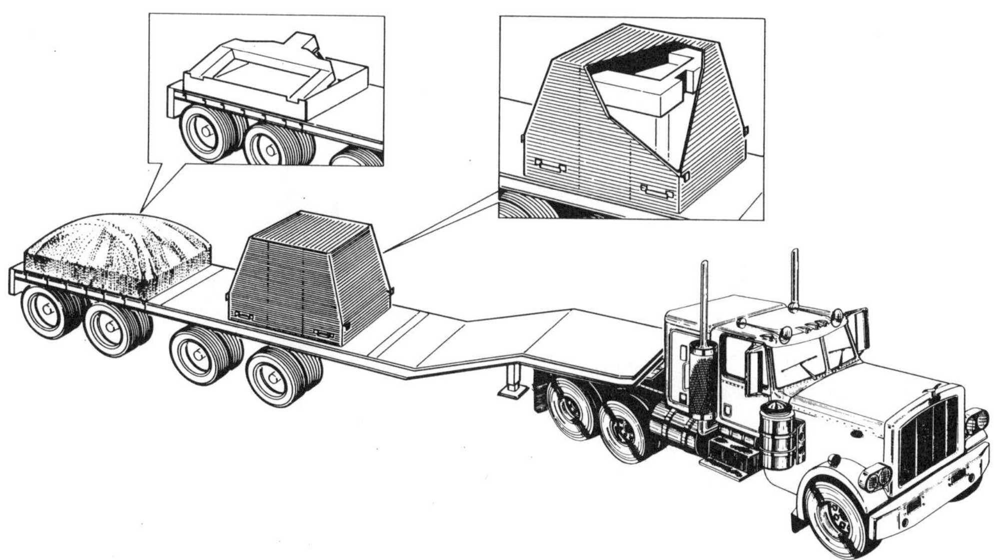

Fig. 3. CANDU irradiated fuel road cask/tractor trailer

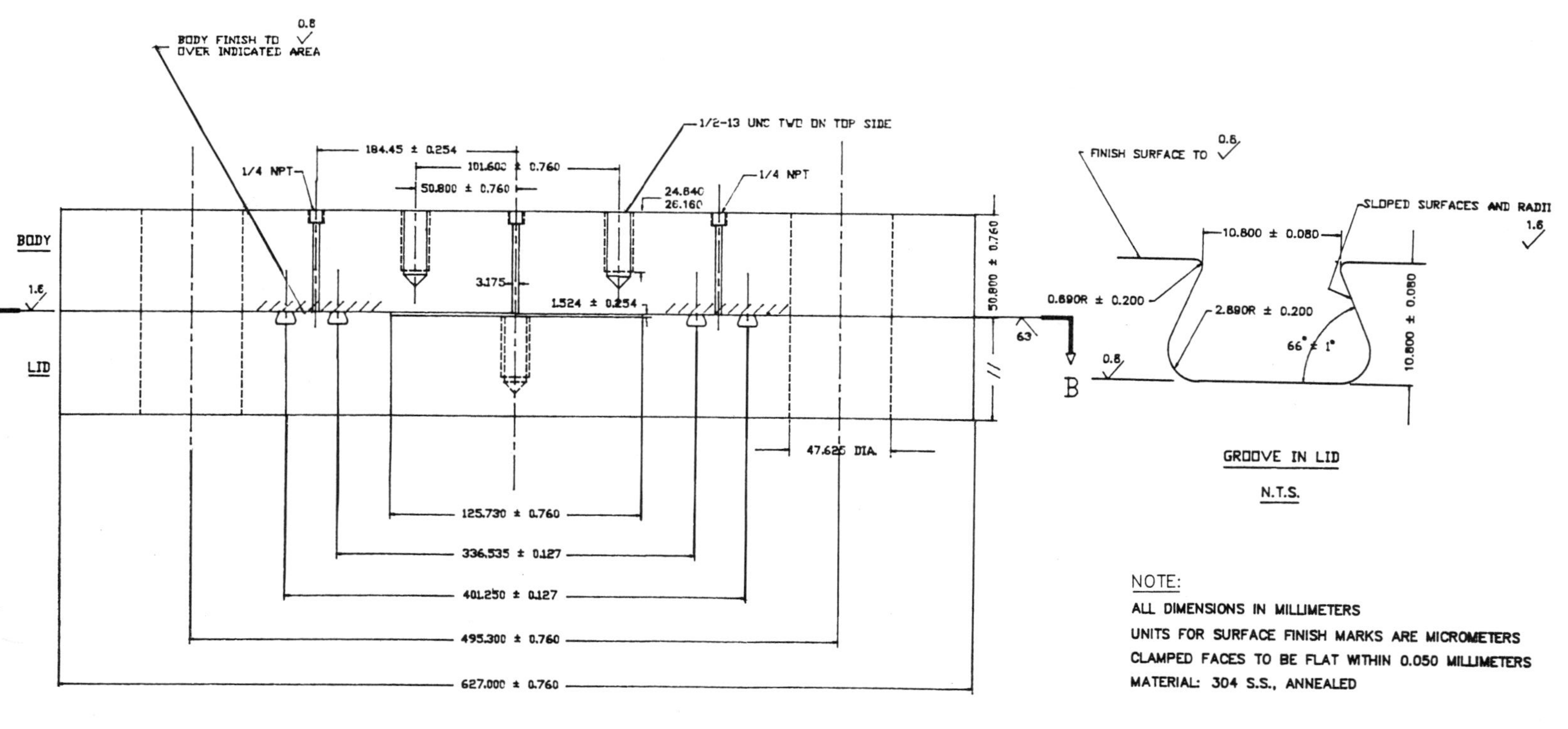

Fig. 4. Irradiated fuel road cask seal test rig.

After evaluation of various design concepts, the present concept was chosen for further development. Detailed design, engineering analyses and testing have been followed by the manufacture of a prototype cask.

CASK PERFORMANCE DEVELOPMENT PROGRAM

The prototype cask design has resulted from an extensive development program. Prior to any design work being undertaken, the relationship between real accident conditions and the IAEA/AECB transportation regulations was reviewed. It was determined that compliance with the regulatory requirements provided for adequate safety, however it was decided that where economically possible, an increased safety margin would be provided.

Ontario Hydro has developed an in-house capability for analysing cask performance. This was initially used to compare a wide range of alternative design concepts, prior to the selection of the final concept. Where appropriate, the performance of the selected design has been verified by testing.

The following is a general outline of the development program. In addition some of the key results obtained are presented.

Fuel Performance

In order that the fuel bundles may be automatically removed form the transportation/storage module following transportation, dimensional stability and structural integrity of the assembly must be guaranteed. Shock and vibration analysis and testing of fuel bundles and UO_2 oxidation experiments were undertaken. In addition the performance of the fuel with respect to fission product release was also investigated experimentally.

Shock and vibration. To assess the performance of the CANDU fuel bundle under normal transportation conditions, a combination of tests and analysis was undertaken. These include:

(1) Field studies to determine typical frequencies and accelerations experienced during road (and rail) operations;

(2) Shaker table and impact testing of both irradiated and unirradiated CANDU fuel bundles.

(3) Vibration analysis of typical tractor trailer configurations and of the cask, storage module and fuel bundle assembly.

Initially it was thought that it may be necessary to axially clamp the fuel bundles during transportation to avoid damage. However the shock and vibration program proved this to be unnecessary, and that:

(a) CANDU fuel bundles can withstand greater than 32 g without failure

(b) CANDU fuel bundles are adequately resistant to fatigue cracking to ensure that they will not break apart during normal transportation by road (or rail).

UO_2 Oxidation. In order to simplify the operating procedures, and reduce the turn-around time, the cask will not be purged of air before transport. To avoid any significant oxidation of exposed UO_2, such as undetected defective fuel or fuel damaged during transport, it is necessary to ensure that fuel temperatures are controlled at a safe level. Experiments on oxidation of irradiated UO_2 were performed in the temperature range $175^{\circ}C$ to $250^{\circ}C$, and a safe transportation temperature of $200^{\circ}C$ was established.

Fission Product Release. Severe transportation impact accidents could result in fuel cladding failure. Deliberately defected irradiated fuel pencils were tested at $200^{\circ}C$ and $400^{\circ}C$. The fission product release was measured on initial cladding puncture and subsequently during a 24 hour sweep test in air. The only significant fission product released was found to be Kr-85.

The allowable leakage from the cask permitted by the 1985 revision of the IAEA regulations is 100 TBq in a period of one week. Even at 400° the quantity released in the above tests was only 1.07 percent of the inventory. Based on the figure, a total of 2.4 TBq would be released from the fuel if all 7104 fuel pencils failed in an accident, (with the cask carrying 192 fuel bundles of 10 year cooled fuel). The cask seals would also have to fail for the 2.4 TBq to be released from the cask.

Seal Performance. Each opening in the cask, the lid and the vent and drain, is sealed by two elastomeric 'O' rings. This allows the seal integrity to be verified and provides redundancy. A major reason for selecting elastomeric 'O' rings is their ability to maintain a seal in the event of a gap opening up between the lid and the cask body. This gap could occur during an impact accident due to bolt stretch, or as a result of thermal distortion in an extended fire accident.

An extensive seal test program has been undertaken to evaluate seal performance over the temperature range $-40^{\circ}C$ to $250^{\circ}C$ with a lid gap of up to 1.0 mm. These tests were conducted using a simple seal tests shown in Figure 4.

The tests showed that:

(1)   Leakage is limited to permeation through the seal material i.e. there is no significant by-pass leakage.
(2)   The 'O' rings are capable of sealing when the lid to body gap is 1.0 mm. This is well in excess of the maximum gap resulting from thermal distortion or bolt stretch.
(3)   The seals remain functional over the temperature range $- 40^{\circ}C$ to $250^{\circ}C$.

Viton, a fluorocarbon rubber, was chosen for the road cask 'O' ring seals because of its low permeability combined with an adequate working temperature range.

A test program to evaluate the performance of metallic seals has been inititated. Metallic seals would provide better high temperature performance in the event that the road cask is required to withstand a fire well in excess of the regulatory duration. The first phase of this program has been to determine the resiliancy of a variety of metallic seals at room temperature.

Impact Performance

The minimum requirement for impact resistance of the road cask is the regulatory nine metre drop onto an unyielding surface and a one metre drop onto a steel punch. The three main steps in designing the cask for impact resistance and assessing the impact performance were (i) the design of an impact limiter, (ii) theoretical analysis and (iii) scale model drop testing.

Impact Limiter Design. The impact limiter reduces the forces experienced by the lid bolting arrangement during an impact accident. The impact limiter is constructed from redwood encased in stainless steel. Redwood has been selected because of its energy absorbing and thermal insulating properties and its ability to resist combustion (The section on Thermal Performance describes these last two properties).

Scale Model Impact Testing To provide confirmation of the cask performance under impact and as a basis for licensing, a series of six drop test were performed on a half scale model cask. A series of one-seventh scale tests was first carried out to identify any design problems and to perfect the testing technique. The six licensing drops performed on the half scale model comprised four nine metre and two one metre punch tests. The tests demonstrated that the cask meets the regulatory requirements with a large margin of safety, a summary of the results being:

Leakage: For all drops leakage before and after the test was by permeation only. The maximum gap which developed between the lid and the body was 0.2 mm, a gap in excess of 1.0 mm is required to initiate seal failure.

Shielding: No loss of shielding occurred other than minor deformation of the bottom corner of the cask.

Impact Limiter Attachment/Damage: The impact limiter remained attached to the cask. The corner of the impact limiter was damaged on the top corner drop, however the loss of thermal protection was not significant.

Thermal Performance

The decay heat of the fuel must be dissipated from the cask so that fuel sheath, seal and cask wall temperatures do not become excessive under normal conditions of operation. The cask must also be capabale of surviving fire accident conditions without excessive leakage.

Heat Dissipation. Ten year cooled fuel has a very low decay heat, for the reference case the decay heat rating is only 1.25 kW. This allows some very important design and operating simplifications.

(1)   No cooling fins are required which reduces the construction costs and makes decontamination easier:
(2)   Fuel temperatures are maintained at a level which makes it Unnecessary to purge the cask of air to avoid oxidation of exposed UO_2 or the zircalloy sheathing:
3)   Vacuum drying of the cask interior is not required because over pressurization will not occur even with the trace quantities of water remaining after draining.

There were three stages in the thermal analysis and testing addressing heat dissipation.

The first stage was an experimental program to determine the relationship between fuel sheath and module tube assembly. Tests were performed using electric heaters simulating the fuel bundles placed inside a module tube. Temperature profiles were established for a range of decay heat ratings.

The second stage was the development of a three dimensional finite difference program to model heat dissipation from the module tube to atmosphere. The major component of the program was the modelling of convection and radiation inside the cask cavity.

The third stage was verification of the computer code by full scale heat dissipation testing. Two production fuel modules, with electric heaters simulating the fuel bundles and a full size simulated cask were used.

The results of the testing and analysis program show that the temperature distribution throughout the cask is well within acceptable limits under extreme normal operating conditions (38°C ambient plus insolation as required by the IAEA regulations). See Table 1.

A similar thermal analysis and testing program was undertaken for the Recently Discharged Fuel case, the results of this program are shown in Table 2.

Fire Accident. The finite difference code developed for heat dissipation was adapted for use in the fire accident analysis by adding convective and radiant heat inputs. The results of this analysis for the 10 year cooled fuels summarized in Table 3, showed that the cask design has a large margin of safety when subjected to the regulatory 30 minute, 800°C pool fire.

Table 1. Heat Dissipation Under Normal Conditions for 10 Year Cooled Fuel

Maximum Temp/Pressure		Limit
Cask Wall Temp:	59°C	82°C (IAEA regulations, for readily accessible surface)
Fuel Temp:	109°C	200°C (for oxidation of exposed UO_2)
Seal Temp:	58°C	250°C (for viton)
Cask Pressure:	84 kPag	700 kPag (IAEA regulations)

Table 2. Heat Dissipation Under Normal Conditions for Recently Discharged Fuel

Maximum Temp/Pressure		Limit
Fuel Temp:	151°C	200°C
Seal Temp:	130°C	250°C
Cask Pressure:	393kPag	700 KPag

Table 3. Response to IAEA Fire for 10 Year Cooled Fuel

Maximum Temp/Pressure		Limit	
Fuel Temp:	$164^{\circ}C$	$400^{\circ}C$	(for oxidation of exposed UO_2 for short duration).
Seal Temp:	$114^{\circ}C$	$250^{\circ}C$	(for viton seal)
Cask Pressure:	410 KPag	700 KPag	(before seal leakage occurs)

Fire accident analysis for Recently Dishcharged Fuel has not been completed and therefore results for this case are not available.

Scale Model Fire Testing

Although the Road Cask licence was obtained on the basis of the fire accident analysis, a subsequent fire test program was performed on the half scale model cask. Fire testing was carried out to provide information on real fires and to confirm cask performance under real fire conditions.

The pool fire was achieved by burning kerosene. Thermocouples were introduced into the flame characterization and the model was fully instrumented to provide internal, external and seal temperatures. A pressure tranducer was used to provide a record of the cask internal pressure throughout the fire.

The half scale model cask was subjected to the equivalent of the regulatory fire. The test demonstrated that the fire accident analysis appears to be conservative, with the predicted seal temperature being higher than obtained during the test.

Subsequent to the regulatory fire the half scale model cask was subjected to a fire of extended duration, the equivalent of a 1 hour fire at $800^{\circ}C$. This test provided very useful information on cask performance in real fires beyond the regulatory requirements. The cask cavity was provided with a positive pressure prior to the test and no pressure loss was recorded either during or subsequent to the fire.

Impact Limiter/Thermal Insulator

One of the functions of the impact limiter is to provide thermal insulation to the cask closure during a fire accident. A series of fire test was carried out on the redwood used in the impact limiter. Redwood blocks were directly exposed to open flames consistent with the regulatory fire of $800^{\circ}C$ for 30 minutes. The redwood only burned momentarily and on completion of the test had only charred to a depth of 20 mm, with thermal effect being insignificant 50 mm below the surface.

The impact limiter used in the half scale model fire test program had a damaged sheathing, resulting from the impact testing. The surface of the exposed redwood burned during the fire and subsequent examination confirmed the above results.

<u>Public Communications Program</u>

One reason for developing the Road Cask at this time is to demonstrate the safety of the system and to assure the public that this technology is available to Ontario Hydro well in time to meet the transportation requirements. An important part of the development program has been to address potential public concerns and to effectively communicate the safety of the system. To these ends the following actions have been undertaken:

(1) Focus group sessions at the design stage to receive input on the public's perception of safety relating to the cask appearance.

(2) Inviting members of the public and media to the drop tests to demonstrate an open approach to safety.

(3) Development of a public communication program. This program covers the development and presentation of communication material. This material includes video on the test programs, a video showing the development of the Road Cask, a colour brochure and audio/visual material for use at presentations.

(4) Public surveys to assess public attitudes to the transportation of radioactive materials.

A province-wide information program has been in place since 1979 to inform emergency officials such as police, fire and ambulance groups of Ontario Hydro's radioactive materials transportation programs. This program, originally conceived as an information program for emergency officials along the transportation routes, now includes a variety of features such as seminars, Ontario Hydro's 24 hour emergency response system, quick reference field guides and detailed training programs. These programs are being further enhanced in support of irradiated fuel transportation starting with the road cask demonstration program in 1988.

<u>Trials and Commissioning Program</u>

A trials and commissioning program is planned, commencing in the spring of 1988.

The program comprises four stages, each stage becoming progressively more involved.

Stage 1 Trial Loading and Unloading at the NGS without fuel or with dummy fuel bundles. This will identify any problem with the proposed operating procedure and familiarize operators.

Stage 2 Trial Loading and Unloading at the NGS with irradiated fuel. This will confirm radiation shielding.

Stage 3 Trial Transportation between sites with dummy fuel bundles. This will identify any off-site/transportation problems.

Stage 4 Trial Transportation between sites with irradiated fuel. This will set a precedent for irradiated fuel transportation and provide proof of the durability of the CANDU fuel.

THE DESIGN OF TRANSPORT CONTAINERS
FOR RADIOACTIVE WASTE MATERIALS

R.F. Keene,[1] P. Donelan,[1] J.C. Miles,[1] B. Marlow[2]

[1]Ove Arup and Partners
13 Fitzroy Street
London W1P 6BQ

[2]UK Nirex Ltd
Curie Avenue
Harwell
Didcot
Oxfordshire OX11 ORH

ABSTRACT

This paper describes the design process related to the development of a family of transport containers for intermediate level radioactive waste materials. These steel containers, each large enough to carry four 500 litre drums full of immobilised waste, all have similar internal cavities, but have wall thicknesses chosen for the activity levels of the waste streams to be transported. Four different wall thicknesses have been chosen, ranging from 70mm at the lightest end of the spectrum to 285mm at the heaviest end.

Three principal steps are described in the paper. First, a series of simple calculations is used to identify concepts which have the potential to satisfy the design aims. Following this, a concept appraisal identifies the preferred scheme and more elaborate computer-aided calculations are used to confirm its adequacy and develop the details of the design. Finally, some scale model prototypes are manufactured and tested to prove their performance under extreme load conditions and to validate the predictive modelling.

The paper outlines some of the methods which have been used for the development of components such as shock-absorbers and fasteners and shows how relatively simple calculations can be used to obtain preliminary sizings. Some attention is also paid to more advanced analytical methods (such as impact prediction using finite element techniques) and examples are also given of the use of computers to produce 3-D visualisation of the concepts in the early stages of a project.

INTRODUCTION

This paper describes the design process and techniques used to develop a family of transport containers for intermediate level waste

(ILW). The paper is based upon a project which lasted about 18 months and was carried out by Ove Arup and Partners on behalf of UK Nirex Ltd.

Intermediate level waste will normally be packaged in 500 litre steel drums and immobilised with cement as a solid matrix. An earlier investigation by Nirex indicated that an array of 4 drums conveyed within a stillage inside a cuboid, shielded, and re-usable transport container offered many advantages. This information formed the starting point for developing transport container designs.

The transport containers are required to carry a large variety of different types of waste and therefore need to be capable of conveying drums having a wide range of strength and activity.

To accommodate the different levels of activity a family of transport containers was conceived (Nirex 1984; Atom 1985) with shielding thicknesses of 70mm, 145mm, 210mm and 285mm of steel. The optimum transport configuration is shown in Figure 1 and is a square array of four drums held in a stillage, both to ease handling and to protect the drums from damage and abrasion whilst in transit. The stillage and drums then go into a container. The design concepts for this will be discussed later.

A project to prove the viability of this concept was formulated consisting of three main parts.

a)   An investigation into the dynamic strength of the waste material.

b)   An analysis of the combined strength of the waste confined inside the drum.

c)   The development of the transport container concepts used to convey the filled drums.

This paper deals mainly with the work covering part (c), namely the design of the transport containers to resist impact. The result of parts (a) and (b) may be found in Donelan et al (1986). Although the performance of the containers in fires and other requirements are outside the scope of this paper, some of these aspects have been discussed elsewhere (Smith & Marlow, 1987). The objective of the work was to develop a fundamental understanding of transport container impact behaviour, and then to use this knowledge to produce concept designs of viable containers.

In the following sections, the various aspects of the design process are discussed, using the container designs that were developed as examples.

DESIGN OBJECTIVES

The transport containers have to offer a high level of protection to the drums which they carry to ensure that there is no release of activity during normal transport and the minimum allowable release even in very severe accidents.

It is expected that ILW transport operations will not begin until early in the next century, some 12-15 years away, although plants are already being built to condition the waste. This presents a constraint in that design options have to be kept open to accommodate any changes which might be thought necessary in the interim period. The general approach has been:

14

- throughout the study, to develop a thorough generic understanding of the underlying engineering behind the transport containers. This means that the concepts are well understood in terms of how they work and what the key aspects of their designs are. This will allow design changes to be more easily incorporated as the designs develop.
- a design target drop height of 36 metres (60 mph impact) was deliberately chosen in order to inflict measurable damage on the package. This has allowed different design concepts to be compared, enabled the mechanisms of failure to be closely examined and identified any inherent weaknesses.

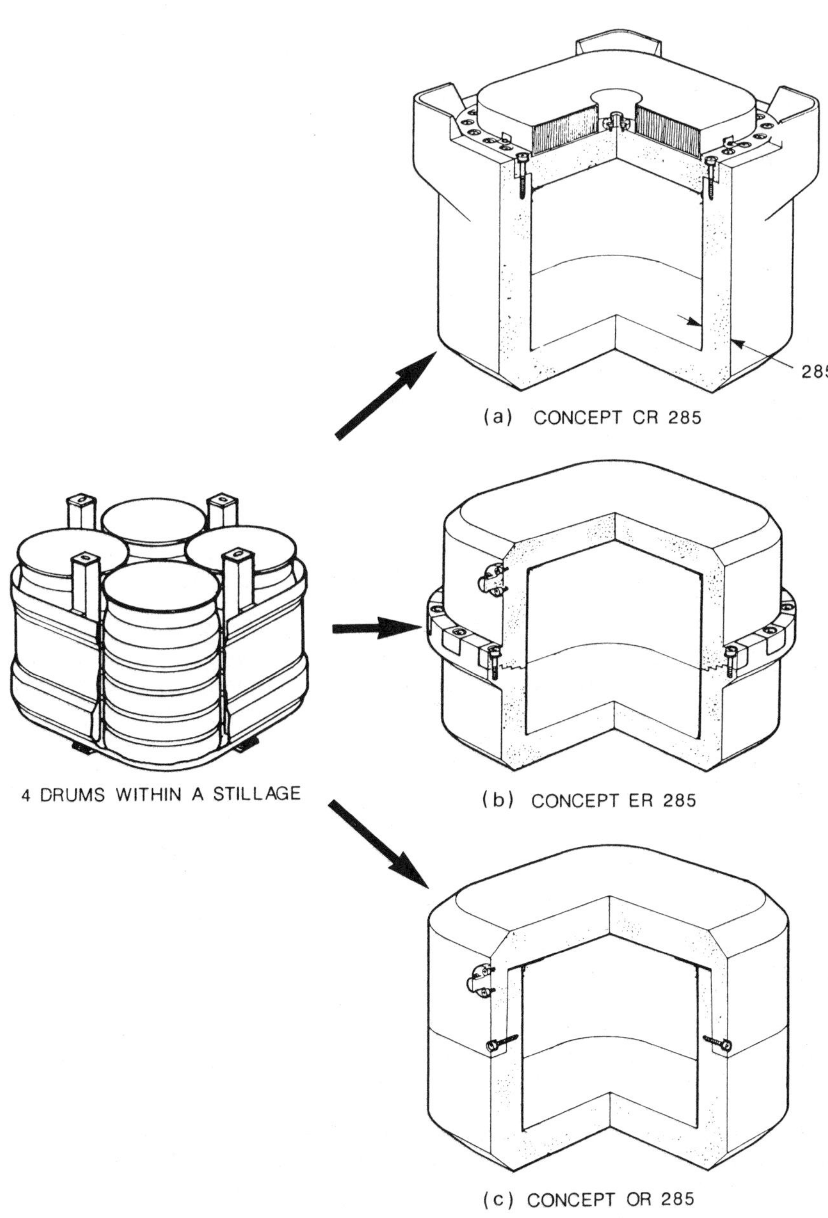

Fig. 1. Transport container concepts.

This study concentrated on designing the containers against impact, although factors arising from the day to day transport operation and the manufacture of the containers were also considered.

The main design requirements formulated in compliance with this philosophy were as follows:

- the drum payload should be virtually undamaged in minor incidents. This was represented as a 2m drop onto a rigid target

- the loaded containers must pass the IAEA regulations (IAEA, 1985) for type B packages. These include a 1m drop onto a punch and a 9m drop (30mph impact) onto a rigid target

- the loaded containers must retain their shielding in a 36m drop (60mph impact) onto an unyielding target.

DESIGN METHOD

The procedure used to design the containers can be split into three main stages:

1) Scheme calculations - these allow a large number of concepts to be considered in a fairly short space of time and allow an understanding of container behaviour to be built up.

2) Detailed calculations - having identified a number of likely concepts, more sophisticated computer analysis is used to gain a greater understanding of the behaviour of the containers and to detail their design. This stage may include computer simulations of various impacts to ensure the highest confidence that the containers will perform as anticipated and that no unexpected modes of behaviour exist.

3) Impact testing - finally the preferred concept is impact tested to prove its capability and to validate the design techniques.

Inevitably there is considerable feedback between these various stages, particularly from stage 2 to stage 1. Ideally stage 3 is merely a confirmation that the design works, but, if this is not the case then further cycles will be necessary. In practice, even if the test is satisfactory, so much is learnt from the production and testing of prototypes that a host of improvements often come to light resulting in significant design changes.

The cuboid transport container proposed by Nirex, although new to the carriage of radioactive waste, closely resembles the shape of the CEGB Magnox spent fuel flask. Superficially both have a lot in common and have to satisfy many of the same requirements. Many flasks have been successfully developed and used over the last 30 years to transport irradiated fuel and as such can provide a starting point for this study.

There are however a number of key differences between the constraints governing the design of irradiated fuel flasks and those governing the design of ILW transport containers:

a) The contents of an irradiated fuel flask represent about 5% of the total mass of the flask, whereas in the thinner-walled IWL containers (those with 70mm shielding) the ratio is almost 50%. Since the objective of the container is to keep the contents contained within it, this is a much more formidable task.

b) Irradiated fuel elements conveyed inside skips or multi-element bottles absorb a large proportion of their own energy in an impact. Damage sustained to the contents in transit is less important as the fuel has completed its reactor life, but the flask contents must be capable of being satisfactorily handled at the reprocessing site.

c) Drums of immobilised waste on the other hand have to arrive undamaged fit for disposal. The solidified waste can vary widely in strength. Drums containing strong waste will not deform on impact and therefore will absorb virtually none of their own kinetic energy. This energy must therefore be absorbed elsewhere. Drums containing weak waste will be prone to damage, and to prevent this the stillage conveying the drums was given shock absorbing capacity.

d) Finally the immobilised waste is less dispersible than irradiated fuel presenting a less severe requirement for containment on the containers than would be necessary in an irradiated flask.

For these reasons designing an ILW transport container is significantly different and in some ways more difficult than designing an irradiated fuel flask. However a large amount of expertise has been developed over the years for the fuel flasks and much of this was used to develop the concepts. In particular the calculation methods used to assess the various schemes were an extension of those previously used to investigate the behaviour of Magnox irradiated fuel transport flasks (Dallard, 1985).

SCHEME CALCULATIONS

In order to consider as wide a range of schemes as possible in the search for those with potential, it is important to use simple calculations. These can be carried out quickly and, because they are often expressed algebraically, rather than in absolute numbers, allow a greater appreciation of the underlying physics of the problem.

The sequence of calculations used to assess the impact performance of various schemes and to size key components involves the determination of the following parameters:

a) The amount of energy absorbed by the container in an impact

b) The impact load acting on the container

c) The forces which are set up within the container

d) The size of the components to withstand these forces.

These steps are shown diagrammatically in Figure 2 for a vertical impact and are now covered in more detail. The methodology is applicable to any impact situation.

Energy Absorbed

In a vertical impact a proportion of the kinetic energy of the container will be absorbed by deformation of the container itself while the remainder goes in the translational and rotational energy of the container as it rebounds. By treating the container as a rigid body the

amount of rebound energy can be estimated for any attitude and graphs
similar to that of Figure 2a derived showing how much of the initial
energy is absorbed by the container.

The largest amount of energy has to be absorbed when the container
impacts with its centre of gravity directly above the point of impact.
This energy can be absorbed in a number of areas, by deformation of the
contents, elastic and plastic distortion of the walls, base and lid of
the container, and by plastic distortion of the container close to the
point of impact.

It has already been stated that at least some of the IWL to be
transported is very strong and very little deformation of the drums will
occur. Body distortion of the container itself is not an ideal way to
absorb energy since the large distortions involved place severe
requirements on the fastening system used to hold the lid to the body.

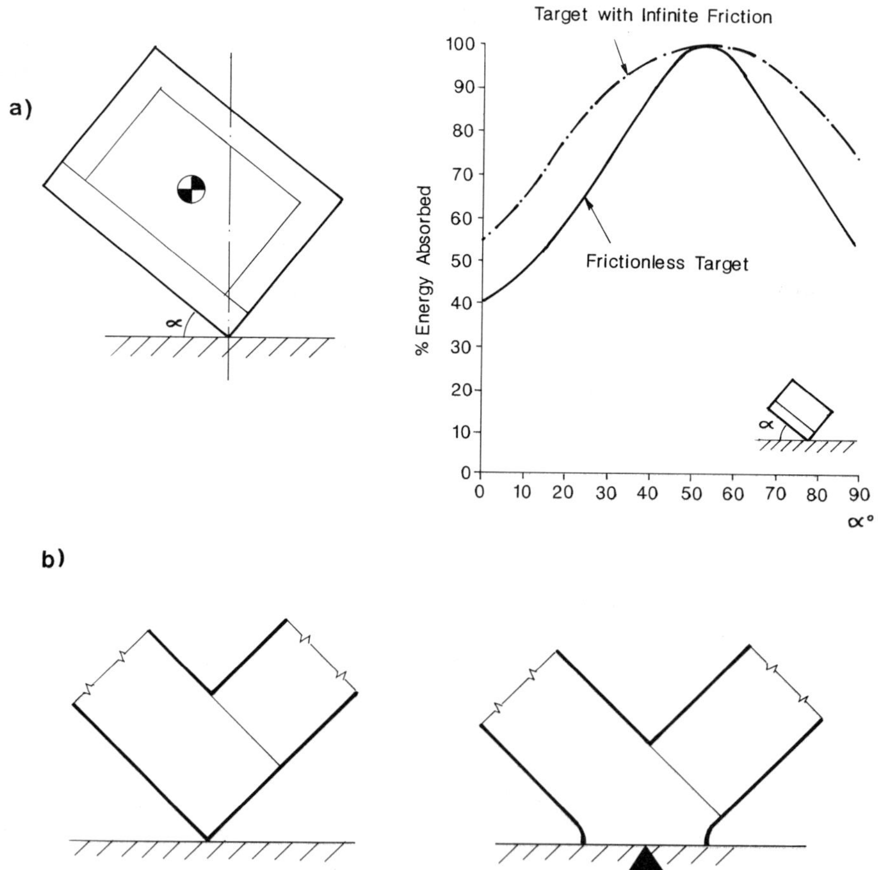

Fig. 2. Scheme calculations. (a) Using rigid body momentum calculations,
the variation of energy to be absorbed with impact attitude is
estimated; (b) Using plastic work calculations, the distortions
local to the point of impact caused by the absorption of this energy
are determined. The peak force acting on the container can then be
calculated.

18

The containers were therefore designed to absorb the impact energy close to the impact point, either in specially provided shock absorbers, or in local distortion of the container itself.

Impact Load

The amount of deformation produced and the impact load can be estimated by assuming a failure mechanism, estimating the amount of plastic work associated with the mechanism and then equating this to the energy that has to be absorbed. Figure 2b shows an example where the edge of the container hits the targets. For complex geometries the determination of the volume of material displaced becomes difficult and solid modelling techniques were used to provide this information quickly and accurately.

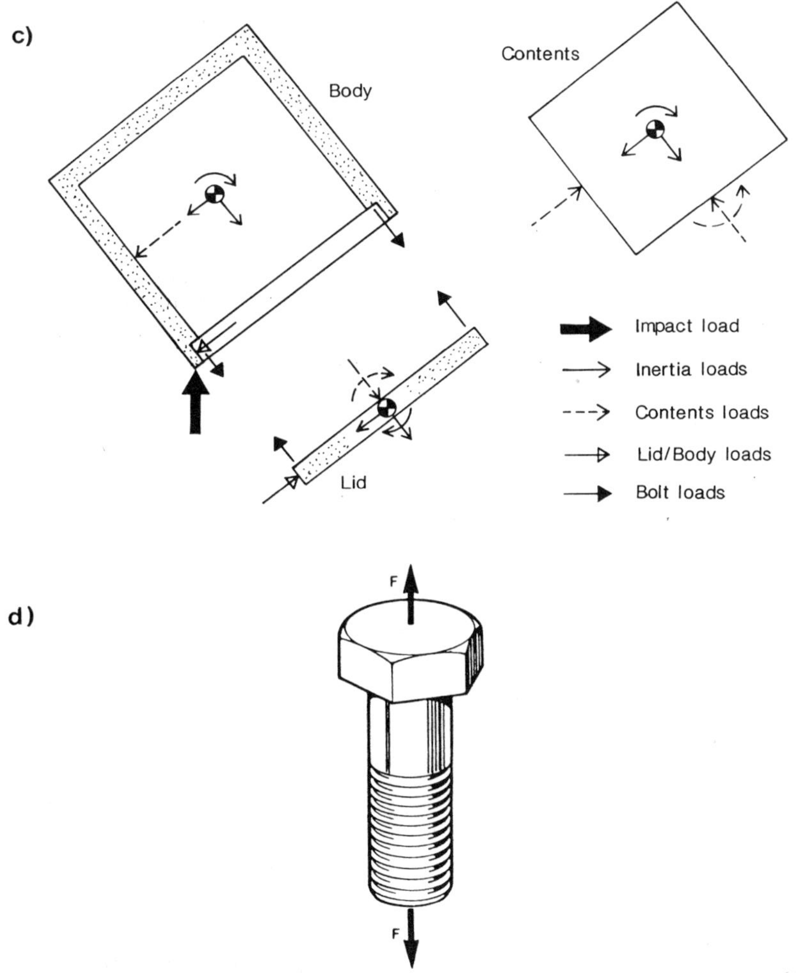

Fig. 2 (contd.) Scheme calculations. (c) The system of internal loads acting between the components, and the inertia loads caused by the decelerations, can then be calculated since all the parts must be in equilibrium. (d) Knowing the forces acting in the various components, such as the bolts, they can be sized to carry these loads.

Internal Forces

Since the container must remain intact during the event, this external force must be in equilibrium with the internal forces acting between the various parts and the inertia (D'Alembert) forces caused by the acceleration of the various parts. (Figure 2c). By considering equilibrium of the various parts, the magnitude of these forces can be calculated.

By altering the geometry of the container, and the mass and inertia of the various components, the force actions will change and their magnitudes will alter.

Component Design

The forces shown in Figure 2c can be used to design the various components. The simplest example of the use of these forces is the bolt loads necessary to keep the lid attached to the body (Figure 2d). Bolts of sufficient number, size and strength must be provided so they are capable of carrying the required load.

It should be remembered that these loads are often very large and sizing the components is not an easy task. Many iterations are necessary before a viable scheme is achieved.

Scheme Visualisation

As various schemes were identified, computer generated solid models of them were produced. This allowed the concepts to be visualised at an early stage, easing the communication of new ideas. Scheme visualisation also provided more accurate geometric data, such as the containers' mass and inertia, for use in the hand calculations.

SCHEME EVALUATION

Using calculations of the type described above three schemes of container design were identified as worthy of further considerations (Figure 1) labelled 'CR', 'ER' and 'OR' respectively.

Concept CR

This concept is shown in Figure 1a and draws most heavily on irradiated fuel flask technology, although in practice it is significantly different from any fuel flask. It does, however, retain certain features, such as the relatively small top fitting lid attached to a heavy body using a large number of high strength bolts.

The force diagram for this container in a lid edge impact is shown in Figure 2c. Of particular importance are the very large bolt and lid-spigot loads which result as a direct consequence of the need to hold the contents in the container.

Figure 3 shows the maximum load that can be carried by the container in various lid edge attitudes, assuming the lid is held in place with 32 50mm diameter high strength bolts. Also shown is the peak load on the container if no shock absorbers are provided. For many attitudes the load generated exceeds the capacity of the bolts and the bolts would break. The third curve shows the peak load on the container if it is fitted with special shock absorbers. For this case the capacity of the container exceeds the generated loads for all attitudes except those where the container is almost upside down.

To cope with these attitudes an external wooden shock absorber is fitted to the lid itself. This pushes the lid onto the body on impact producing considerable strength as it does so and changes the mode of failure of the container. For this case the capacity of the container exceeds the generated loads in all attitudes as shown by the fourth curve on Figure 3.

The main features of the CR concept are:

a) Specially shaped steel shock absorbers are provided at the top corners to reduce the impact load.

b) The top of the container has a square flange to support the shock absorbers and provide greater resistance to distortion of the lid aperture.

c) The lid has a wooden shock absorber to protect it in lid-down impacts.

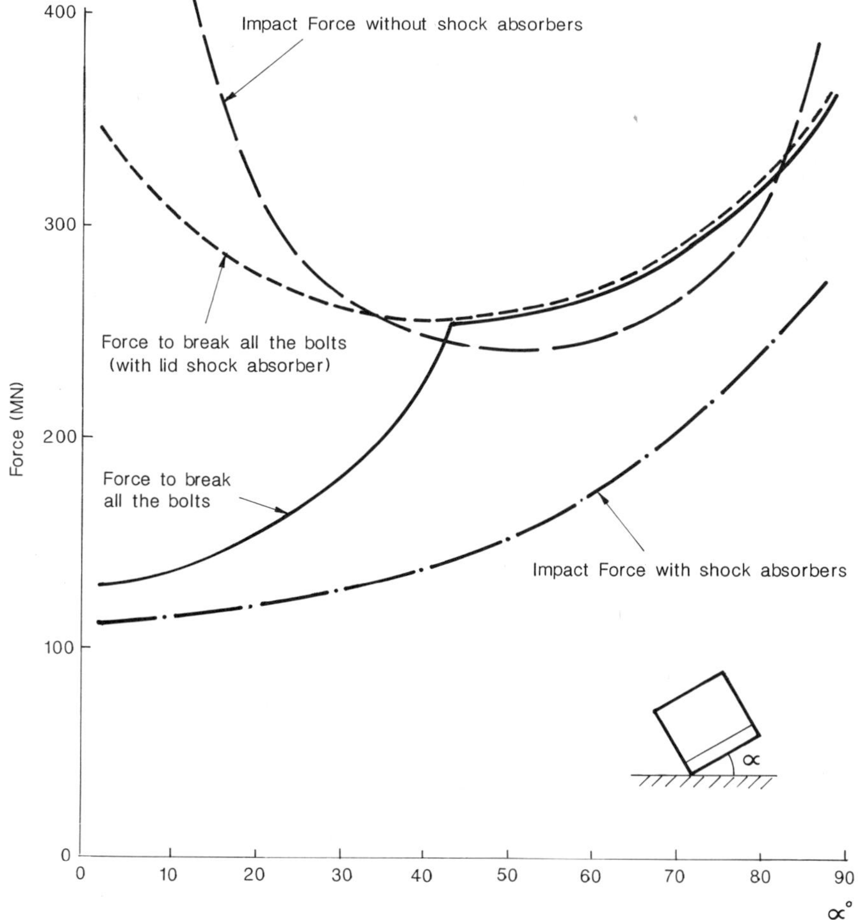

Fig. 3. Concept CR — forces.

d)   The inside of the lid has an aluminium honeycomb shock absorber to
     reduce the load caused by the contents hitting the lid.

e)   Thirty two, 50mm diameter, high strength bolts attach the lid to the
     body.

Concept 'ER'

     Many of the difficulties in designing the 'CR' container arise
because of the very large loads that must be transferred across the
interface between the lid and body. These loads can be dramatically
reduced by shifting the split between the lid and body away from the
point impact.

     The 'ER' container (Figure 1b) uses this principle to reduce the
number of bolts to 16 and reduce their size to 35mm. Figure 4 shows the
force diagram for this container which is very similar to that for the CR
container shown in Figure 2c. However the changes in geometry and masses
of the components are significant.

     Figure 5. shows the increase in strength of a container with a
central split over one with a top split. For some attitudes (35 to 80
degrees) the bolts carry no load at all - the net force across the
interface is always compressive.

     The most important features of the ER concepts are:

a)   It is lighter and stronger than the CR concept.

b)   Only sixteen, 35mm diameter high strength bolts are needed to
     secure the lid.

c)   A castellated joint is incorporated in the bolt flange to provide
     shear resistance between the lid and body.

d)   No shock absorbers are required.

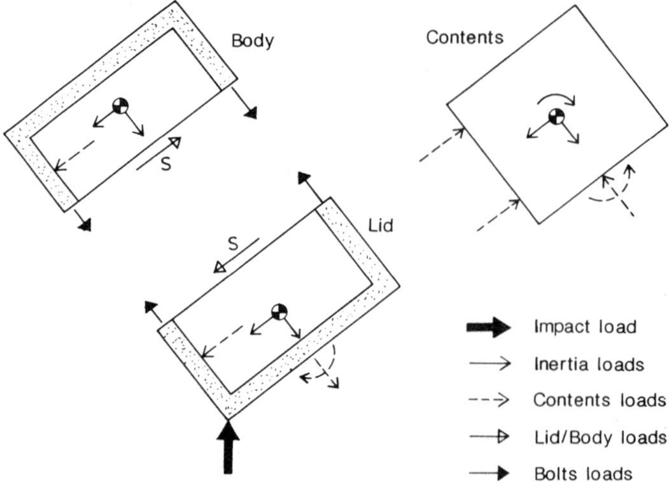

Fig. 4. Concept ER - internal loads.

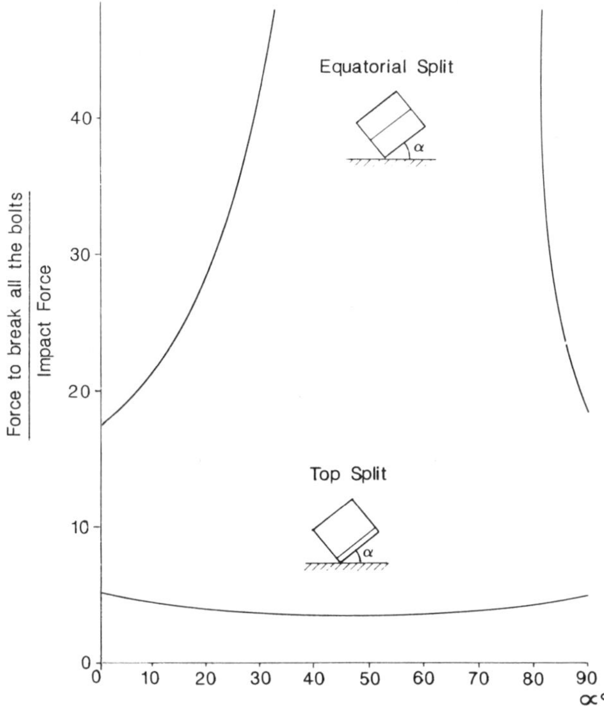

Fig. 5. Benefits of a central split.

Concept 'OR'

Concepts 'CR' and 'ER' both use bolts to prevent separation of the lid and body during the impact. By providing the lid with a deep skirt that fits over the body (Figure 1c) then, as the lid and body rotate relative to each other on impact, they will interact and prevent separation. Figure 6 shows the force diagram and it can be seen that no 'bolt' load is now necessaray to maintain equilibrium of the various components. In practice some form of fastening is necessary for everyday operation and security, but this can be minimal since this is not required for impact protection.

The integrity of the container on impact is maintained by the lid and body interacting and preventing separation. Naturally some movement of these components is necessary before they can interact and so some gaps will be present after the impact. However this is acceptable since the radiation level does not exceed the allowable dose following accidents.

The key advantages of the 'OR' concept are:

a) It has the absolute minimum size and weight. No more material is needed other than that required for shielding.

b) The skirts of the lid and the body have to be tapered for casting which also facilitates fitting and removal of the lid.

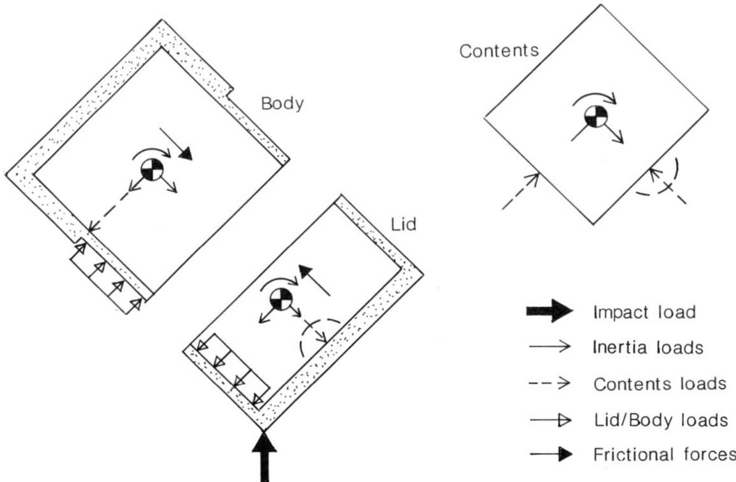

Fig. 6. Concept OR - internal loads.

c) No shock absorbers are required.

d) Only a few small bolts are needed to fasten the container, resulting in

e) Much reduced operational exposure and ideally suitable for remote handling.

DETAILED CALCULATIONS

Having identified the schemes worthy of further consideration, more detailed calculations are necessary for three reasons.

a)   To gain a greater understanding of how the schemes work.

b)   To gain greater confidence that the concepts will work in practice.

c)   To detail the designs to the stage where prototypes can be manufactured.

As the required level of accuracy of the calculations increase, so computer aided calculations become increasingly necessary. Of these three types are particularly useful;

a)   Lumped parameter models

b)   Finite element analysis

c)   Nonlinear impact simulation.

Lumped Parameter Modelling

Lumped parameter modelling is particularly useful for looking at the dynamics of the various components during the impact. The components are modelled as rigid parts, with appropriate mass and inertia, connected by springs. The force/displacement characteristics of the springs are representative of the components being modelled. Figure 7 shows a typical model of a transport container and its contents.

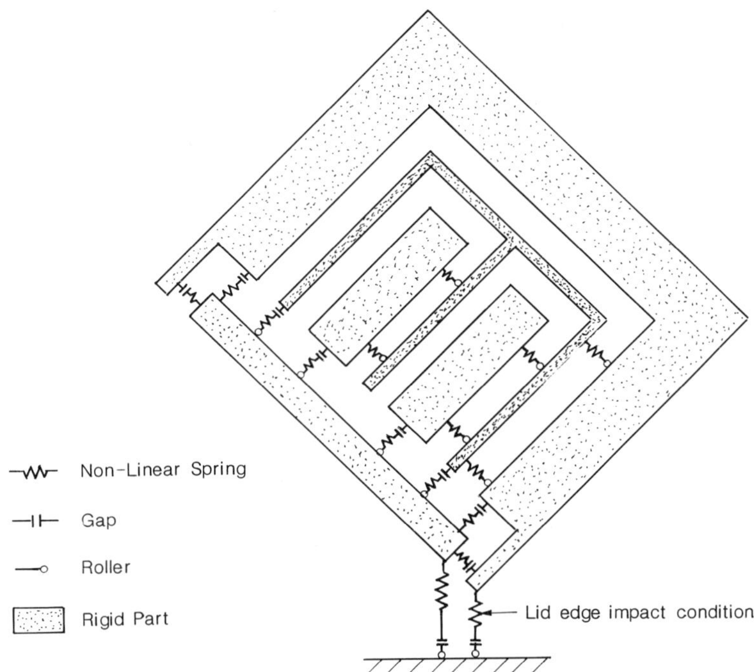

Fig. 7. Lumped element model.

This type of analysis allows the areas where energy is absorbed in the package to be determined. It is particularly useful when deciding the best location and strength of shock absorbing material. It also provides information on the dynamic interaction of the various components that make up the container.

Finite Element Analysis

Finite element analysis was used extensively to design key components, to ensure they were capable of carrying the very large loads required of them. Simple components could be designed adequately by hand, but many had complex three dimensional geometry and, since the containers were being designed for a limit state condition, it was desired to reduce design margins above this limit state to an absolute minimum.

An example is shown in Figure 8 which illustrates the results of an analysis used to size the castellations of the ER container. The calculation is a linear elastic analysis, using the loads derived from the earlier scheme calculations.

The peak stress in the castellations is 1000 N/mm^2 and, even allowing for dynamic enhancement of yield stress due to strain rate effects, some yielding would be expected. In the subsequent test of the ER container, the castellations were found to suffer permanent deformation and locked together after the impact.

Nonlinear Impact Simulation

The previous design stages have used relatively simple techniques to look at various aspects of the problem in isolation. Many of them are quasi-static in their approach, neglecting dynamic effects, and all of them contain inherent simplifying assumptions.

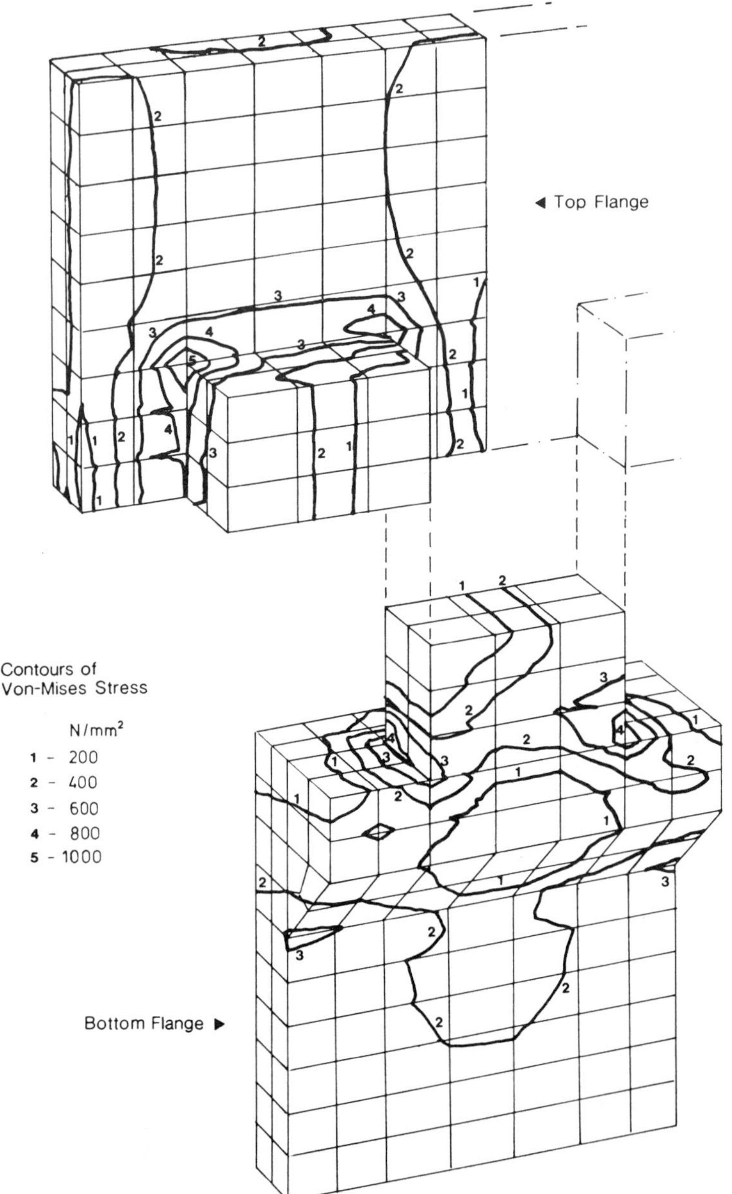

Fig. 8. Concept ER - analysis of castellations.

To gain greater understanding of the way the containers behave, particularly regarding the dynamics of the components themselves and the way they interact during the event, a computer simulation of the impact was performed. This was done using a specialised finite element program, DYNA3D (Hallquist 1987), which has many features making it suitable for impact analysis. In particular it has large displacement and large strain formulation and highly developed contact algorithms to model the interaction of components, including frictional effects. The program had previously been correlated with test data (Dallard 1985) and has been used successfully for a wide range of impact analysis.

The finite element mesh used for the CR container is shown in Figure 9. The mesh is cut away to show the contents. The key features of the container were included; the drum, stillage, bolts and lid shock absorber, as well as the lid, body and body shock absorbers. The lid edge impact was considered one of the most severe impacts for this container and was subsequently chosen as one of the attitudes to be tested. This allowed a direct comparison between the computer simulation and test.

Details of these tests are left to the section on Testing, but Figure 10 shows a comparison of the velocity/time characteristic obtained from an accelerometer mounted at the back of the test container compared with that predicted by the DYNA3D simulation. They can be seen to be in good agreement, but more importantly the measured behaviour of the container agreed with that predicted by both the simulation and the design calculations giving confidence in the techniques used to design them.

The tests on the ER and OR containers were carried out in the same attitudes as the CR to provide a direct comparison between the three concepts. However this attitude is not necessarily the worst for these designs. Impact simulations were therefore carried out using DYNA3D to check that performance in the other more critical attitudes would be satisfactory. The ER container has a projecting bolt flange which, if the container were dropped directly on it, would be pushed back into the container, distorting the sidewalls and prying open the bolted joint. Figure 11 shows the results from DYNA3D analysis of this situation illustrating that the flange has indeed been pushed into the container. However the level of distortion indicates that the container will remain intact in this situation.

The mesh used for this analysis is relatively simple. When running this type of analysis the mesh density should be chosen to be commensurate with the level of accuracy of the results that are required. For this analysis it was not necessary to calculate the stresses or distortions to a great degree of accuracy. What was sought were the modes of deformation and indications of the effects on the bolts. Where a higher degree of accuracy is needed, then a more detailed mesh should be used. An example is the analysis of the CR container, which was compared with test data.

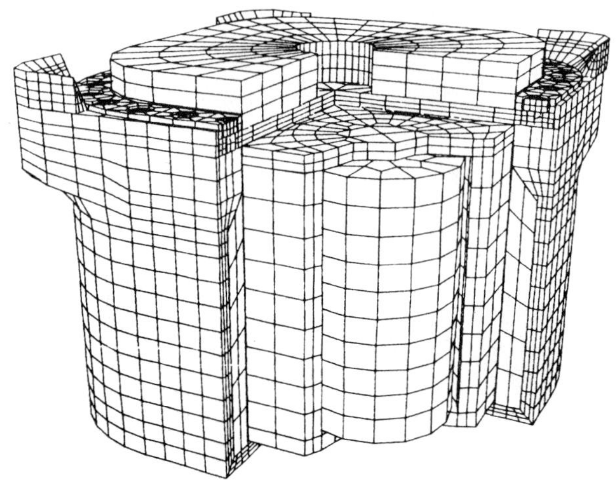

Fig. 9. Concept CR - DYNA3D finite element mesh.

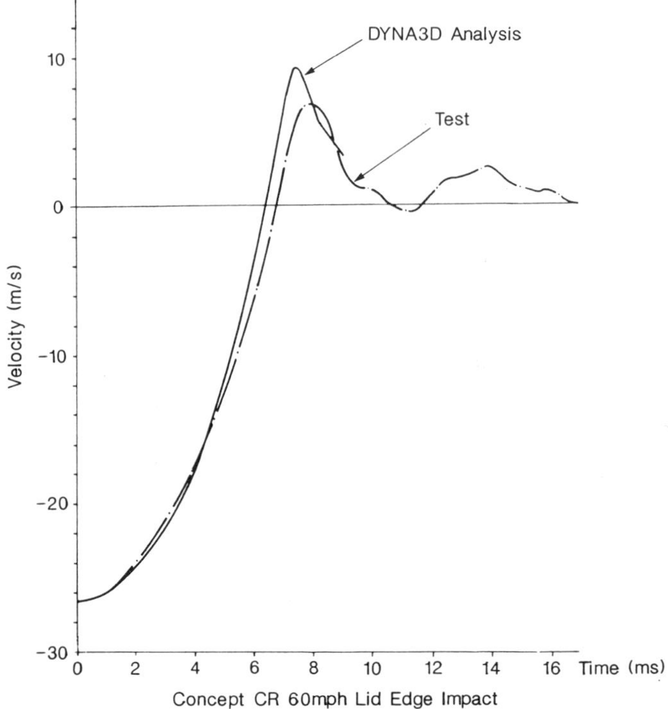

Fig. 10. Predicted and measured velocities.

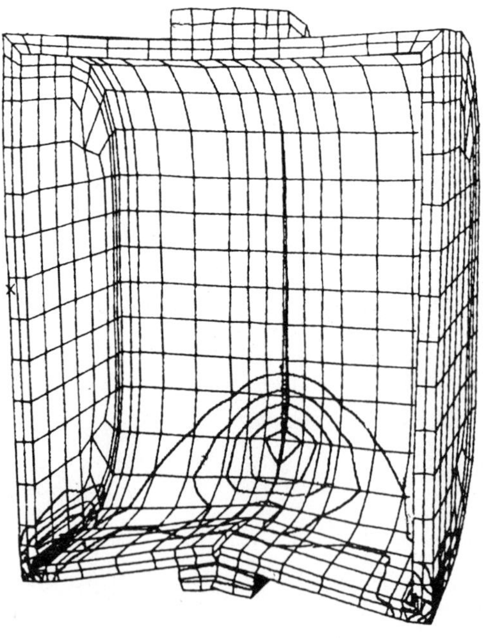

Fig. 11. Concept ER - Flange impact analysis.

TESTING

Testing is essential, both to prove the container concepts themselves and to validate the techiques used in their design. However the use of more detailed computer simulations can reduce the number of tests that have to be carried out in two ways. Firstly they allow the prototypes to be developed to greater extent before committing to testing, since the designs can be better assessed. This gives greater confidence that the container will perform as expected and so fewer iterations requiring testing are needed before a final design is achieved. Secondly, once the design calculations are verified by a number of tests, other attitudes and minor design changes can be investigated by calculation and only the most critical tested.

Four containers were built and tested:

1) Concept CR - 20m lid corner impact.
2) Concept CR - 36m lid edge impact.
3) Concept ER - 36m lid edge impact.
4) Concept OR - 36m lid edge impact.

All the tests were carried out on 1/4 scale models. Previous work on irradiated fuel flasks (eg Donelan & Dowling 1985) has proved the validity of scale model testing and the similarity of fuel flasks and ILW transport containers means that this work is directly relevant.

The testing was carried out by the CEGB Structural Test Centre in Cheddar using a guided rig to ensure a high degree of accuracy in the impact attitude. The containers were instrumented with accelerometers and strain gauged bolts to record transient information during the impact, and the high speed cameras filmed the impact from three directions. Detailed measurements were taken after impact to determine the amount of distortion that had occurred.

After each test the results were analysed and compared with the behaviour expected from the calculations.

In all cases the containers passed the test successfully, retaining their contents inside the container but the results were very close to the limit. In the case of the CR and ER containers no bolts were broken, the lid was firmly held in place and the gaps between the lid and body were very small. The OR container broke its small fastenings, and the desired interaction of lid and body occurred preventing separation of the halves.

Most importantly all three containers behaved in very much the expected manner with no major surprises. This was particularly gratifying since this was the first test on three designs containing many novel and innovative features. It demonstrated the effectiveness of the methods used to design them.

CONCLUSIONS

A method of designing transport containers for ILW has been presented that allows an understanding of the design to be developed and places a lower requirement on testing than used previously.

Three different concepts of container have been identified which incorporate many innovative features and yet offer a very high level of impact resistance with little penalty in size and weight.

All three designs have been drop tested at 60mph and have been shown to meet the demanding requirements placed upon them. The final choice of container concept will be made following a consideration of all the transport and handling aspects.

REFERENCES

Atom Number 344 June 1985 pp 25-29.

Dallard, P.R.B., 1985, Flask Analytical Studies in "The Resistance to Impact of Spent Magnox Fuel Transport Flasks", Mechanical Engineering Publications.

Donelan, P.J. and Dowling, A.R. 1985, The use of Scale Models in Impact Testing in "The Resistance to Impact of Spent Magnox Fuel Transport Flasks", Mechanical Engineering Publications.

Donelan, P., Milloy, C., Miles, J.C., Marlow, B., 1986, The Mechanical Properties of Immobilised Intermediate Level Waste, in "Proceedings of 8th International Symposium on The Packaging and Transportation of Radioactive Materials" (PATRAM 86), Davos, Switzerland.

International Atomic Energy Agency Safety Series 6, 1985, "Regulations for the Safe Transport of Radioactive Materials".

Hallquist, J.O., 1987, "DYNA3D Users Manual". UCID-19592, Revision 3. Lawrence Livermore National Laboratory, California.

Nirex Press Release November 1984.

Smith, M.J.S., and Marlow, B., 1987, The Packaging and Transport of Radioactive Waste in "IMechE Seminar on the Management & Disposal of Intermediate and Low Level Radioactive Waste", Mechanical Engineering Publications.

FLASKS FOR TRANSPORTING IRRADIATED NUCLEAR FUEL

I.J. Hunter[1], J. Emmison[2]

[1]Nuclear Transport Ltd
Risley
Warrington WA3 6AS

[2]British Nuclear Fuels plc
Risley
Warrington WA3 6AS

ABSTRACT

The design of flasks for the transport of future nuclear fuels is complicated by the high neutron emissions from 'High burn-up' and 'MOX' fuels.

The paper concludes that it is possible to design suitable flasks for future fuels without departing fundamentally from existing designs. It is acknowledged that well founded computer programs are available for calculating the main technical and engineering inputs and performance data, but this does not eliminate the need for demonstration testing to prove the flask designs.

INTRODUCTION

Irradiated nuclear fuels have been transported from reactor sites to reprocessing or storage plants for some 30 years. In this period there has been a steady evolution in the design of flasks used for transporting this fuel safely. Flasks designs have had to cope with increased heat loadings, higher levels of radioactivity, larger payloads and more stringent regulations as the size and sophistication of reactors has changed over this period.

Nuclear Transport Ltd (NTL) and Pacific Nuclear Transport Ltd (PNTL) are carriers of nuclear fuels in the European and Japanese areas respectively; transporting irradiated nuclear fuels to the reprocessing plants at La Hague (France) and Sellafield (UK). For the most part this involves the transport of Uranium Oxide fuels from light water thermal reactors. The EXCELLOX range of flasks is used extensively in these operations - Figure 1.

The trends in reactor operations are toward fuels of higher initial enrichment and increased burn-up. These trends, coupled with the potential adoption of mixed Uranium/Plutonium oxide fuels, all have implications for future designs. The main design problems are those of criticality and nuclear radiation; either of these considerations could have an effect on the payload carried by the flask and on the internal

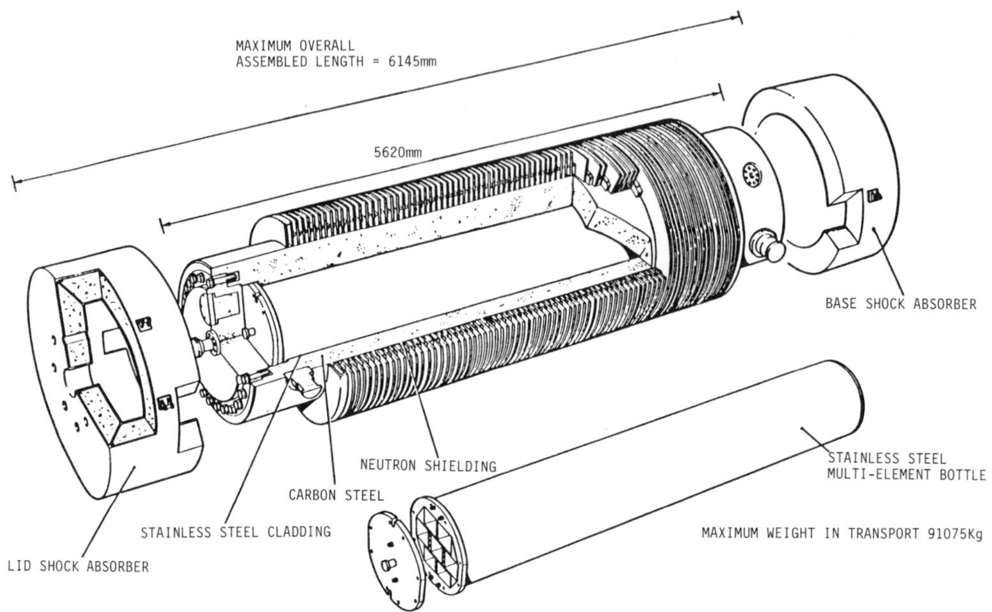

Fig. 1. Excellox spent fuel flask.

and surface temperatures of the flask. For these reasons NTL and BNFL have been examining new designs of flasks.

IRRADIATED FUELS

Electrical power generating utilities have vested interests in using present nuclear fuel resources as effectively as possible. To these ends research programmes are yielding results which will permit future reactor operations using richer fuels for longer operational periods - ie higher fuel 'burn-up'. For the purposes of the current design studies, the following fuels are being considered (Table 1).

Table 1. Design Fuel Types

Reference Fuel	High burn-up Fuel		Mixed Oxide Fuel
Rating	40 MW/Te(U)	40 MW/Te(U)	40 MW/Te(Ueq)
Irradiation	40 GWD/Te(U)	55 GWD/Te(U)	40 GWD/Te(Ueq)
Initial Enrichment	4^w/o U_{235} in U	4^w/o U_{235} in U	4^w/o Pu

These fuels are likely to be reprocessed at (or after) a period of 5 years has elapsed from the time of removal from the reactor - i.e. 'Cooling' time. However, they are likely to be shipped from the reactor site to the reprocessing plant at some time during this period - probably within the first two years.

The characteristics of these fuels have been calculated using industry standard computer codes, viz:

FISPIN   : to produce radiation source and thermal data.
McBEND   : to estimate neutron dose rates and secondary gamma emissions.
RANKERN : to estimate primary gamma emissions.

The main features of these calculations are shown in Table 2. These approximate results depend upon some initial assumptions; consequently the values are quoted only for illustrative purposes.

Table 2. Comparison of Fuel Characteristics
(Values relative to Reference Fuel)

	High burn-up Fuel (Ratios)	Mixed Oxide Fuel (Ratios)
Heat output	1.2 - 1.5	1.5 - 2.0
Neutron Emission	4	15
Gamma Emmission		
Ru - 106	1.4	1.7
Ce - 144	1.0	0.8

These results indicate that neutron shielding is the most significant problem to be tackled in the design of new flasks since the calculated flux is about 10^{10} neutrons per second per tonne of uranium. This problem will be exacerbated if current proposals to double the quality factor for neutrons in the conversion to dose equivalent rates are adopted. Thus, it may be necessary to improve the neutron shielding of flasks by a factor exceeding 30 over current practice.

NEUTRON SHIELDING

The overall sizes and weights of flasks are dictated by the capacity of handling equipment at the despatch and receipt locations, at intermediate trans-shipping points - if appropriate, and by the loading restrictions for road, rail and ship transporters. Flasks are designed within these constraints for carrying the maximum payload of irradiated fuel.

Within the existing range of EXCELLOX flasks the water content provides some moderation of the neutron flux. Also, the multi-element bottle which fixes the spacing and relative locations of the fuel elements is fabricated from either Boral - which is a commercial designation for boron carbide (B_4C) in an aluminium matrix - or boronated stainless steel, and provides a measure of neutron absorption. These effects are small and are insufficient for the estimated increases in neutron fluxes. The problem, therefore, is to incorporate sufficient neutron absorbing material, without compromising payload or the ability to dissipate heat significantly, within the restricted dimensions and weights.

A number of neutron absorbing materials have been investigated to verify their shielding performance and to define the manufacturing route for fabricating and placing the materials. A boronated silicone rubber material, manufactured by Fulmer Yarsley Limited of Trowers Way, Redhill, Surrey, has been identified because it can be cast readily into a cavity created between fins. The steel capping piece required to form the cavity also acts to provide some attenuation of the secondary gamma emission arising from the interaction with the boron loading in the silicone rubber (Typically $1.0^w/o$).

CRITICALITY

In the design of multi-element bottles the spacing and relative locations of the fuel elements are defined by the worst case reactivity of the assembly. In general this implies a full loading of non-irradiated elements with a reactivity limit $k < 0.95$. The typical layouts used in the EXCELLOX range of flasks is shown in Figure 2.

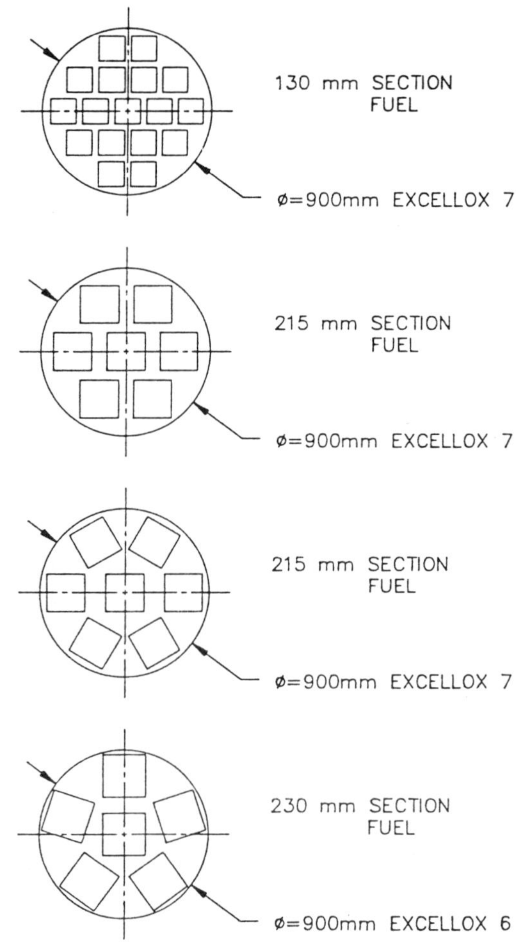

Fig. 2. Multi-element bottle arrangement.

These arrangements currently limit the maximum initial enrichment to about 3.5 ^{W}/o U_{235} in U. It is possible to load fuels of 4% initial enrichment when they are spent (or partially spent) because the reactivity decreases with burn-up. Therefore, a pre-condition for the transport of these fuels is the demonstration of an appropriate burn-up. Alternatively this problem is amenable to redesign using a smaller payload or more neutron absorbent internal materials.

IMPACT ENERGY ABSORPTION

Bolt-on shock absorbing assemblies are fitted at each end of EXCELLOX flasks in order to limit the forces applied to the flask components and contents under accident impact conditions. Under IAEA regulations (IAEA 1985) the flask is required to sustain a 9m drop test without breaching shielding or allowing the leakage (or dispersal) of radioactive material. Tests are required with the flask falling flat on its end, and falling with the centre of gravity over the rim of the flask ends.

The EXCELLOX absorbers are steel enclosures filled with balsa wood. The performance of these units has been calculated using a commercially available computer code (DYNA 2D) using data derived from model testing. The results indicate a need for material having multi-directional energy absorbing properties similar to those of end-grain balsa wood.

FLASK CONSTRUCTION

The current range of EXCELLOX flasks has fabricated steel bodies and is fitted with lead liners which absorb the major part of the gamma emissions. This arrangement has worked well in practice but it complicates and extends the periodic maintenance of the flasks. For future flasks, especially those for carrying high burn-up and MOX fuels, it is anticipated that the flasks will be made from forged or cast steel bodies clad internally with stainless steel. This arrangement will eliminate the handling of lead liners and the removal and replacement of internal paint during maintenance.

CONCLUSIONS

We conclude:-

a) That it is possible to design flasks for transporting 'High burn-up' and 'MOX' fuels without departing fundamentally from current flask designs.

b) It is possible to calculate the main technical and engineering inputs and performance data using well founded computer programmes.
[This of course does not eliminate the need for demonstration testing to satisfy licensing authorities and to verify the computations].

c) Continuing effort is required to design flasks economically to transport irradiated nuclear fuel safely and also to ensure their continuing reliability by simplifying maintenance techniques.

ACKNOWLEDGEMENTS

Nuclear Transport Ltd is an international company incorporated in

England with shares held by French, German and British organisations. It provides services for the transport of irradiated nuclear fuels.

Pacific Nuclear Transport Ltd is an international company incorporated in England with shares held by British, Japanese and French organisations. It provides services for the transport of irradiated nuclear fuels from Japan to Europe.

British Nuclear Fuels plc provides a comprehensive service in the nuclear fuel cycle field.

The authors gratefully acknowledge the permission of these organisations to publish this paper.

REFERENCES

International Atomic Energy Agency (IAEA) 1985, Safety Series No. 6, Regulations for the Safe Transport of Radioactive Materials.

FURTHER DEVELOPMENTS IN HIGH-LEVEL WASTE TRANSPORT TECHNOLOGY

R.Gowing[1], R.D.Cheshire[1], R.J.Sills[1],
I.J.Hunter[2], L.G.James[2], A.R.Cory[2]

[1]British Nuclear Fuels plc
 Risley, Warrington, Cheshire WA3 6AS

[2]Nuclear Transport Limited
 Risley, Warrington, Cheshire WA3 6AS

ABSTRACT

British Nuclear Fuels plc, with the assistance of Nuclear Transport
Limited as contractor, are developing and designing a new transport flask
in order to be ready for the return of vitrified high level waste to
customers when the Windscale Vitrification Plant and Vitrified Product
Store come on stream early in the 1990s.

The development and operation of the vitrification plant and store
are described briefly. The nature of the vitrified residue is discussed,
with particular reference to how it is affected by the different reactor
types from which irradiated fuel is reprocessed at Sellafield, and the
properties of the resulting residue packages or filled containers in
relation to transport are described.

The requirements and constraints for transport are discussed and the
reasons given for designing a new flask rather than adapting an existing
irradiated fuel flask. The design concept for a flask which maximizes
the payload is described, paying special attention to the novel
arrangement of the internal support structure, and the handling
arrangements, which will allow the flask to be transported using existing
rail wagons and ships, are outlined. Harmonization of design features
with the equivalent French flask, to enable both types to be unloaded at
the same facilities, is discussed.

The factors considered in the preparation of the concept design are
listed and a series of development tests is described which have proved
the basic design concept of the internal arrangements and demonstrated
its ability to meet the requirements for heat transfer and handling of
the waste containers and internal support structure.

Other aspects of detail design are mentioned and an indicataion is
given of the time schedule for proving the impact resistance and
leaktightness of the package, preparing the Design Safety Report and
obtaining competent authority approval as a Type B(U) Package Design, and
construction of the flasks ready for the commissioning of the despatch
facility at the Windscale Vitrified Product Store.

HIGHLY ACTIVE LIQUID WASTE AT SELLAFIELD AND ITS TREATMENT

The highly active liquid waste resulting principally from evaporation of the reject stream of fission products and actinides from the first stage of the separation process has been safely stored at British Nuclear Fuels' Windscale Works ever since reprocessing began there in the early 1950s. Although the storage tanks incorporate comprehensive multiple systems for removing decay heat, keeping any deposited solids in suspension and dealing with the unlikely event of leaks, it has long been recognised that this waste can be more easily and safely stored, and eventually disposed of, if it is converted into a massive solid form.

Vitrification processes have been studied at Windscale and Harwell over the years and the feasibility of a one-step batch process was demonstrated. Meanwhile, the French developed a two-stage continuous vitrification process and by 1978 had constructed a full-scale active demonstration plant, Atelier de Vitrification de Marcoule (AVM). Adoption of this process, combining the French plant design with British developments in the glassification of the Windscale waste stream, has resulted in a saving of several years in the projected introduction of industrial-scale vitrification in Britain in the WVP (Windscale Vitrification Plant).

The process and its development have been described in detail elsewhere (see References). Briefly (Figure 1) the liquid waste with calcination additives is fed to a calciner (a rotating inclined tube inside a furnace). The waste travels down the tube and is evaporated, dried and thermally denitrated. The dry powder or calcine then flows into a melting pot.

Glass-making materials are fed in small batches (a ratio of about 25% waste to 75% glass) to the melting pot which is heated to about $1150^{o}C$ by an induction furnace. The calcine and glass-making materials fuse to form vitrified residue which is periodically allowed to run through a freeze valve into a container which is large enough to hold at least two pours form the melter.

The temperature of the container is kept at about $1150^{o}C$ before and during pouring two minimise thermal shock on the glass and ensure that the separate pours fuse together. Once the container has received its quota of vitrified residue, a lid is fitted and sealed by automatic fusion welding, thus completely enclosing the contents.

The welding is checked and the container decontaminated before it is transferred in a shielded product flask to a natural circulation air-cooled store, the Vitrified Product Store (VPS), shown in Figure 2. Here the containers are stacked up to 10 high in stainless steel thimble tubes. These tubes are sealed at the bottom to avoid the possibility of contamination of the air stream by the containers, and are provided with shock absorbing devices to limit the damage to the containers if these are dropped from the level of the operating area.

The containers of vitified residue can be kept in the air-cooled store for a long period,until such time as deep disposal is available in the U.K. This is still in the early stages of investigation. However, such delay is unnecessary for waste of overseas origin and it is desirable for commercial and political reasons to return this residue at an early stage after vitrification. Accordingly the store is being constructed from the outset with a fully operational despatch facility for loading the containers into transport flasks ready for despatch on

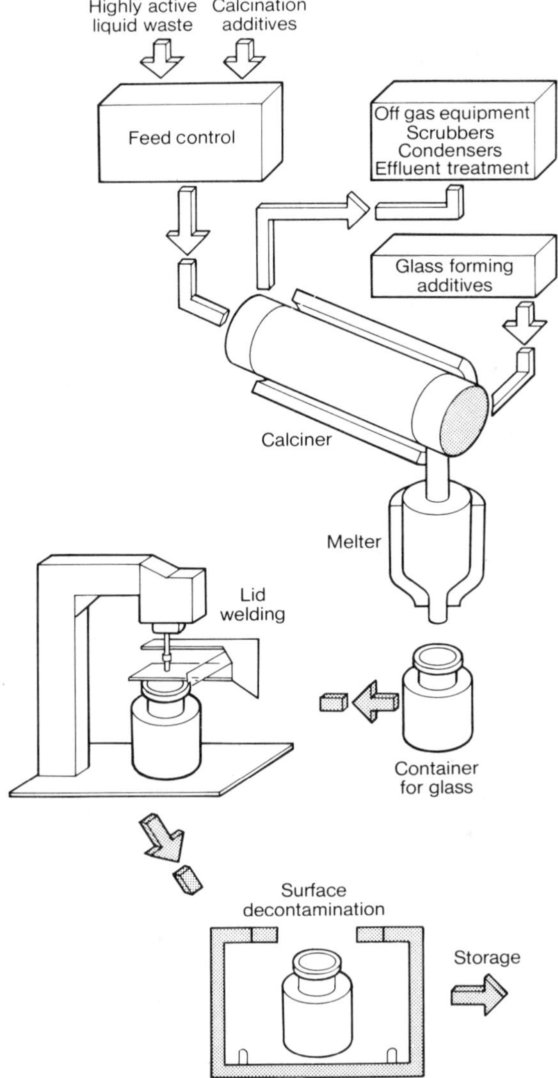

Fig. 1. Vitrification process.

rail wagons from the works. The development and design of a flask for this purpose forms the main content of this paper.

TWO WASTE STREAMS: THEIR EFFECT ON THE VITRIFIED RESIDEUE

One feature of the WVP is that it has to handle HAL (highly active liquid) wastes arising from irradiated metal fuel from Magnox reactors as well as oxide fuel from LWRs and AGRs. The specification of the glass sets a limit to the material quantity of the HLW incorporated, which is the same for both types of fuel. The low burn-up (4.8 GWd/t) and rating (21.3 Mw/t) for Magnox fuel compared with oxide (33 GWd/t burnup, 33 MW/t rating) result in 4.5 tonnes of Magnox fuel being reprocessed to produce the same amount of HLW as 1 tonne of oxide fuel. The susceptibility of

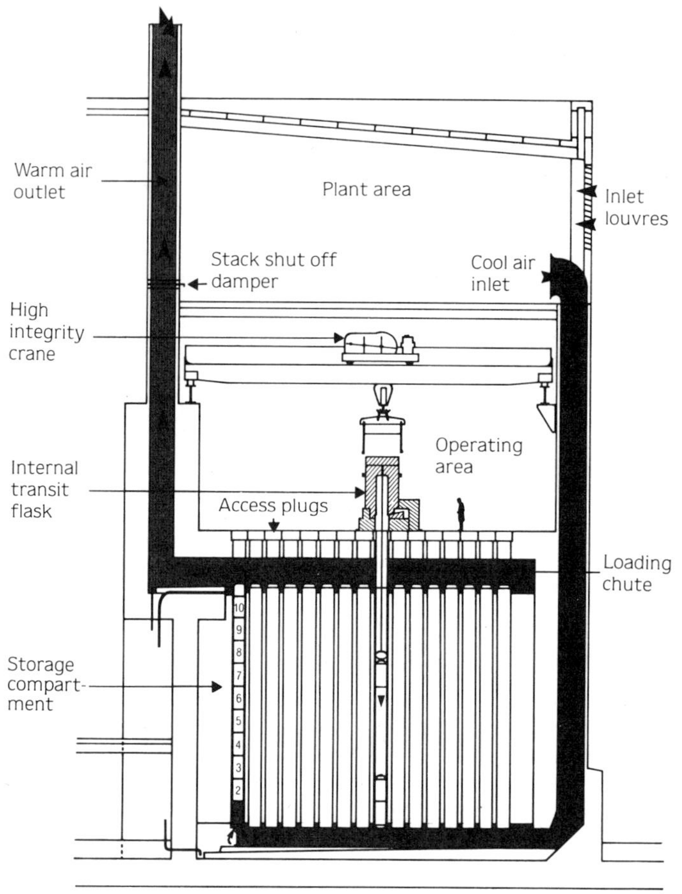

Fig. 2. Typical section through vitrified product store.

Magnox fuel to corrosion means that it must be reprocessed early, within 3 years of discharge, whereas oxide fuel is free of the constraint and the cooling time is set by economic considerations at 7 years. These specifications are also geared to the same maximum heat output of 2.5 KW per canister.

The shorter cooling time of the Magnox HLW gives rise to a higher gamma activity than the oxide HLW, by a factor of 2 in the dose-rate from the container. Conversely the high burnup of the oxide fuel results in a higher concentration of actinides, in which spontaneous fission gives rise to neutron dose rates 10 times higher from the oxide than from the Magnox derived containers.

For reasons of operating flexibility and economy, HAL from the reprocessing of oxide and Magnox fuels may be held in the same tanks, so that the composition can range from that arising from the reprocessing of Magnox fuels to that arising from oxide fuels, and the plant and transport system must be capable of dealing with the radioactivity and neutron and gamma radiation from any combination of fuel inputs. This could give rise to problems in collaboration in flask design with reprocessors who only have oxide fuel material to consider.

40

GLASS AND CONTAINER

The HLW matrix will be a lithia-soda-borosilicate glass incorporating about 25wt% of waste oxides, of which about 20% will be fission product and actinide oxides and about 15% will be inactive additions (process additives, fuel additives, burnable poisons, corrosion products, etc.) The choice of lithia-soda-borosilicate glasses for the incorporation of HAL wastes is based on the development programmes referred to above, in which the capability of such glasses to accommodate successfully the wide range of HAL waste compositions arising from the different reactor fuels has been demonstrated.

The transformation temperature of the glass, at which it undergoes transition from a liquid to a solid, is $495 - 520^{o}C$ and the liquidus temperature, above which all components are liquid, lies in the range $975-1050^{o}C$. The container is fabricated in stainless steel, in accordance with rigid quality assurance procedures, to the design shown in Figure 3. It is designed for easy lid placement and welding, convenient, reliable and safe handling with nesting ends to facilitate stacking, and ability to withstand the stresses due to handling and storage conditions. It weighs 85 Kg empty and has a capacity of 169 litres, a diameter of 430 mm and a stacking height of 1263 mm. The container is filled with 400 Kg of residue (92 Kg of waste with 308 Kg of glass-formers) occupying 151 litres, arising from 1.9 t U of LWR fuel or 8.65 t U of Magnox fuel to form a "residue package". The residue package has the strength to withstand the streses of prolonged storage in a 10 high stack in the VPS and of accidental release from the top of a VPS thimble tube. This has been proved by impact tests in a simulated thimble tube in which an inactive container was dropped the full height of 17 m onto the thimble tube shock absorber, which deformed 420 mm to give a deceleration of 40g. In a further test a container was placed at the bottom of the tube and another dropped the full remaining height onto it. In both instances, although there was some minor deformation of the container, the lid seal weld retained intact and containment of the glass was maintained.

The maximum activity of a residue package is 29.6 PBq beta-gamma, 111 TBq alpha, and maximum surface dose rate will be $3x10^3$ Gy/h$\beta\gamma$ and 25 mSv/h neutrons.

Non-fixed surface contamination, averaged over the surface of the container, will not exceed:

$3.7 \text{ Bq/cm}^2 \beta\gamma$

$0.37 \text{ Bq/cm}^2 \alpha$

The maximum heat output at transport will be 2.5 kW and the average heat output for containers to be loaded to a flask will be below 2 kW.

From the transport point of view the residue packages must be treated as "radioactive contents", but in many aspects they are effectively "special form radioactive material". To avoid confusion with the IAEA concept of "package", the residue packages will henceforth be referred to as "containers".

THE TRANSPORT REQUIREMENT

Since 1977, all new overseas reprocessing contracts for Sellafield include an option for the return of residues and it is anticipated that the vitrified waste will be returned to its country of origin. The VPS export facility is expected to be available from 1994 and it is estimated that, during the first 10 years, up to 2,000 containers will need to be shipped overseas, of which approximately 60% will be to Japan and the remainder to Europe. The mode of transport will be by rail and sea.

Curtis et al (1986) described how existing irradiated fuel flasks were considered for the transport of HLW containers.

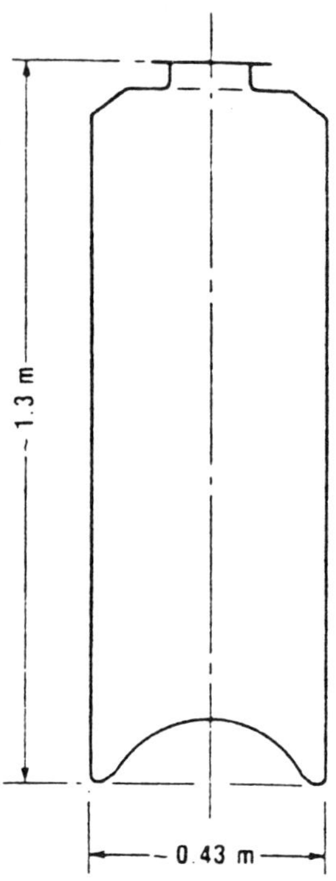

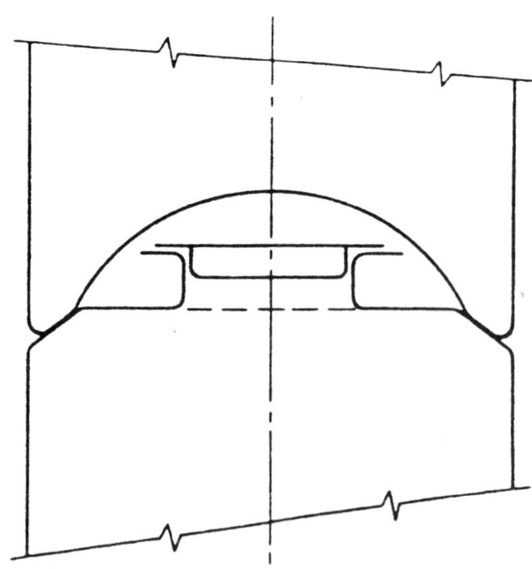

Fig. 3. Waste container.

It was found:

1)   The fuel flasks could not carry an economic number of containers

2)   The neutron shielding of the fuel flasks, being matched to the
     emission from the fuel which peaks axially at the centre, is
     inadequate at the ends

3)   That the use of water as heat transfer and shielding medium in the
     fuel flasks is incompatible with the dry handling of the containers
     in storage and despatch

4)   That the resulting packages would have to be licensed against the
     1985 IAEA Regulations.

A new, purpose designed transport flask was therefore required.
This would be designed to carry the maximum payload within the existing
dimensional and weight limits of the BNFL/BR rail wagon, the PNTL ships
and the handling facilities, and to meet the requirements of the 1985
IAEA Regulations and the operating rules (with special regard to dose
uptake) at the despatch and receipt sites.

DESCRIPTION OF FLASK - THE CONCEPT DESIGN

The flask concept design proposed for the transport of HLW
containers was introduced by Curtis et al (1986) at PATRAM 86, and the
main features are shown in Figure 4.

The flask carrying 21 HLW containers constitutes a Type B(U) Package
with a gross weight of about 110 tonnes in an envelope of approximately
2.4 metres diameter by 5.5 metres long. It consists principally of a
one-piece cylindrical body and base with a lid attached by about 40
high-tensile steel bolts, the lid being set into the body to provide
protection for the bolts against sideways impact. The flask and lid are
made from a medium strength carbon steel, specially produced to give good
low temperature toughness.

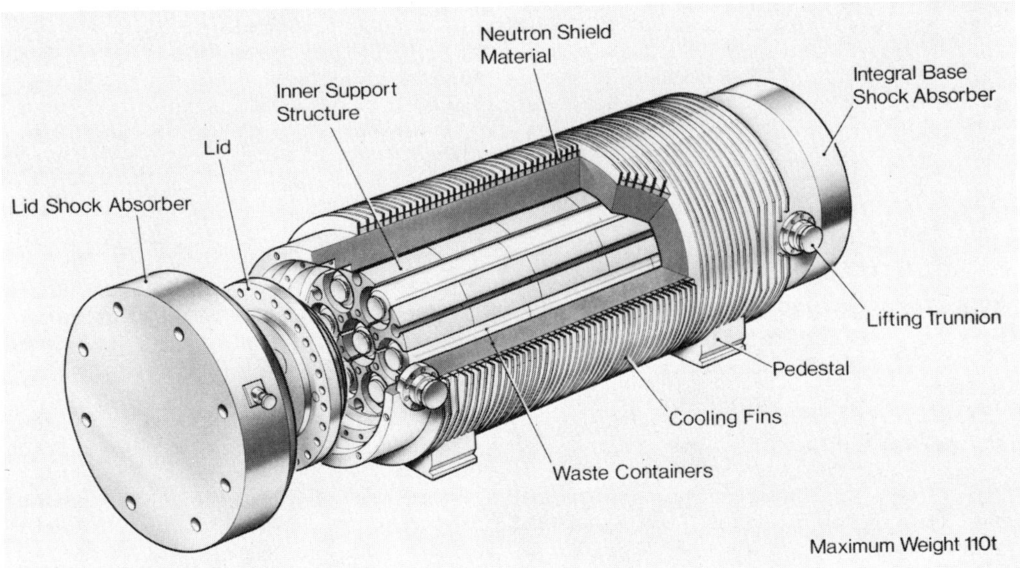

Fig. 4. Vitrified waste transport flask: part section

The cylindrical surface of the body is surrounded by longitudinal steel compartments which encapsulate neutron shielding material; these are embraced by circumferential fins which dissipate the 42 kW maximum heat load. The neutron shielding material reduces the radiation level by a factor of about 10; its loss in a fire accident would be acceptable since the IAEA regulations permit a 100 x increase in radiation level as the result of an accident. Further neutron shielding is encapsulated on the outer surface of the lid and inside the base shock absorber.

Protection of the flask against accidental impact damage during handling and transport is provided by lid and base shock absorbers, the latter being integral with the flask body, and by the cooling fins. The lid shock absorber also protects the lid and its seals against the effects of a fire accident. For lifting and tie down the flask has two pairs of trunnions, and is supported and secured during transport by two pedestals attached to the body. The cavity is accurately machined to provide close contact with the internal support structure and protected against corrosion by a metal-sprayed coating. The exterior of the flask is treated with a high-quality paint system specially developed to give a hardwearing and readily decontaminable surface, while surfaces that are vulnerable to damage by handling equipment are protected locally by stainless steel overlays.

The lid has double elastomer facing seals and the interspaces between these can be tested via test points in the upper surface of the lid. A shielded orifice in the lid gives access to the flask interior for sampling the cavity atmosphere during site operations.

The waste containers are carried in an internal support structure which is built up from some 30 individual aluminum segments to form seven longitudinal channels, each carrying three stacked containers. This support structure, which is a special feature of the design, has a number of advantages over a more conventional monolithic structure. Each segment is firmly fixed to the wall of the flask cavity, thus maximising the flow of heat from the contents into the flask wall. The assembly clearances between each segment minimise differential thermal expansion effects and the relatively small size of the segments simplifies manufacturing and handling. The final shape and surface treatment of the segments have now been chosen, based on the experiments described below, to minimise wear and avoid snagging during loading and unloading. The active support system can be remotely assembled and dismantled; this also was confirmed in the tests described below.

HANDLING ARRANGEMENTS AND TRANSPORT

The flask as presented for transport is shown in Figure 5. The method proposed for attaching the flask to the transport frame is by captive bolts in the frame which engage in grooves in the flask pedestal. This will provide a firm tiedown while minimising the working time spent close to the flask. The frame has standard bolt holes to fit the decks of BNFL/PNTL wagons and ships.

Lifting arrangements are similar to those in established use for BNFL fuel flasks. The flask with its transport frame ia lifted horizontally by the main trunnions at either end. Vertical lifts take place only at the loading and unloading sites and are carried out without the lid shock absorber, which is unbolted from the rim of the flask and lifted off using its own trunnions. The flask can then be lifted off using the lid end trunnions, pivoting on the base end trunnions before being lifted free and transferred to a bogie for remote handling into the vitrified product store or repository.

44

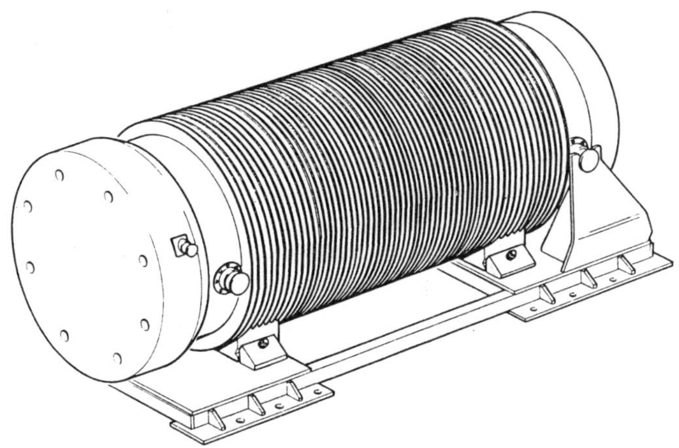

Fig. 5. Flask as presented for transport

Transport to the British port or to continental repositories will take place on the BNFL rail wagons which are constructed to European rail and ferry standards. Transport to other European destinations and Japan will be effected using BNFL/PNTL's fleet of specially constructed nuclear fuel carriers.

HARMONIZATION OF THE DESIGN

Reprocessing for power utilities in Europe and Japan is divided between the two leaders in the field, BNFL and its French counterpart CEGEMA. While these two concerns are using the same vitrification process and producing waste containers to the same mechanical specification, for reasons discussed above two different flask designs are emerging.

It is desirable however that these two flask designs can both be handled in customers' receipt facilities, and for this reason agreement has been reached between the two companies on a list of dimensions and features interfacing with receipt facilities which must be common to both flask designs. These include overall outside diameter, lid aperture and base dimensions, flask and channel internal diameter, and design of lifting trunnions, lid and shock absorber bolts, sampling orifice with protective cover, and leak tightness test orifice.

DEVELOPMENT OF THE DESIGN

The preparation of the concept design has entailed consideration, (involving calculation where necessary) of thermal behaviour, structural behaviour including impact analysis, shielding, activity leakage, manufacturing, maintenance, and handling. This has been supported by a series of experiments to develop and confirm the behaviour of the mechanical clamping system, to confirm the thermal properties of the segment and to demonstrate the thermal, loading and dismantling characteristics of a significant portion of the flask assembly. This three phase programme is illustrated in Figure 6.

Test rig No. 1 was intended to investigate the clamping arrangement between the aluminium segment and the body in order to determine the size of any gap between the segment and the body and how it would be affected by operational temperatures and temperature cycling.

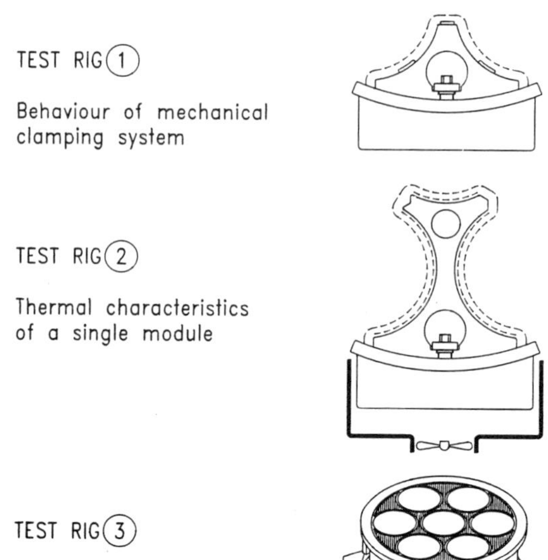

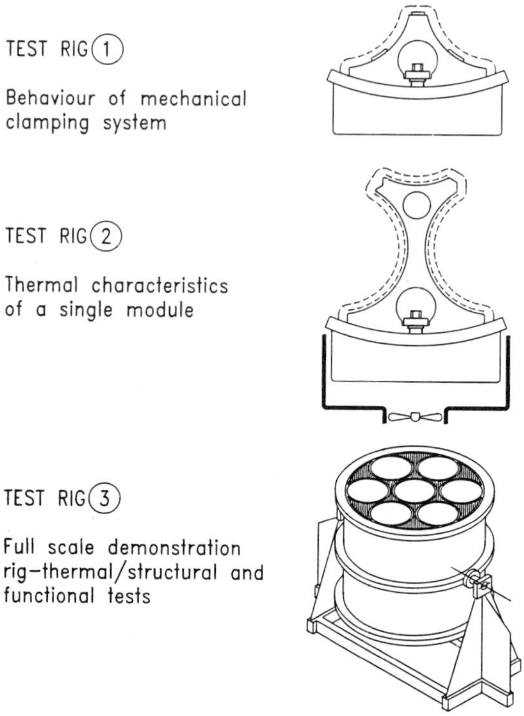

Fig. 6. Development test rigs.

The experimental rig, shown in Figure 7, consisted of the outer part of one of the aluminium segments clamped onto a steel base plate representing the flask shell. The heat load of the central and outer waste containers was represented by electric heating pads applied to the topmost surface (heater A in Figure 7) to simulate the central container, and to the concave surfaces (heater B) to simulate the outer containers. The rig was generously fitted with thermocouples, strain gauges and dimensional reference marks, and was lagged on all surfaces except the bottom which represented the heat-dissipating surface of the flask.

Heat was applied in cycles to represent loading, transport and unloading. The tests showed that:

1) The torque required on the clamping bolts to maintain good thermal contact between the aluminium segments and base plate is of a value readily achievable in practice by remote handling

2) The temperature rise across the gap between the segment and the base plate is not high enough to interfere significantly with heat transfer

3) As the aluminium heats up and expands, the gap between the segment and the base plate reduces until they come into intimate contact. This obviously contributes to good thermal contact between the segment and the base plate and is a good design feature

4) Thermal cycling of the rig did not appear to produce any 'creep' or 'ratcheting' effects.

Test Rig No. 2 was intended to determine what temperatures would be expected in a single segment and in the body shell, under typical operating conditions and in the event of poor thermal contact between the segment and the body.

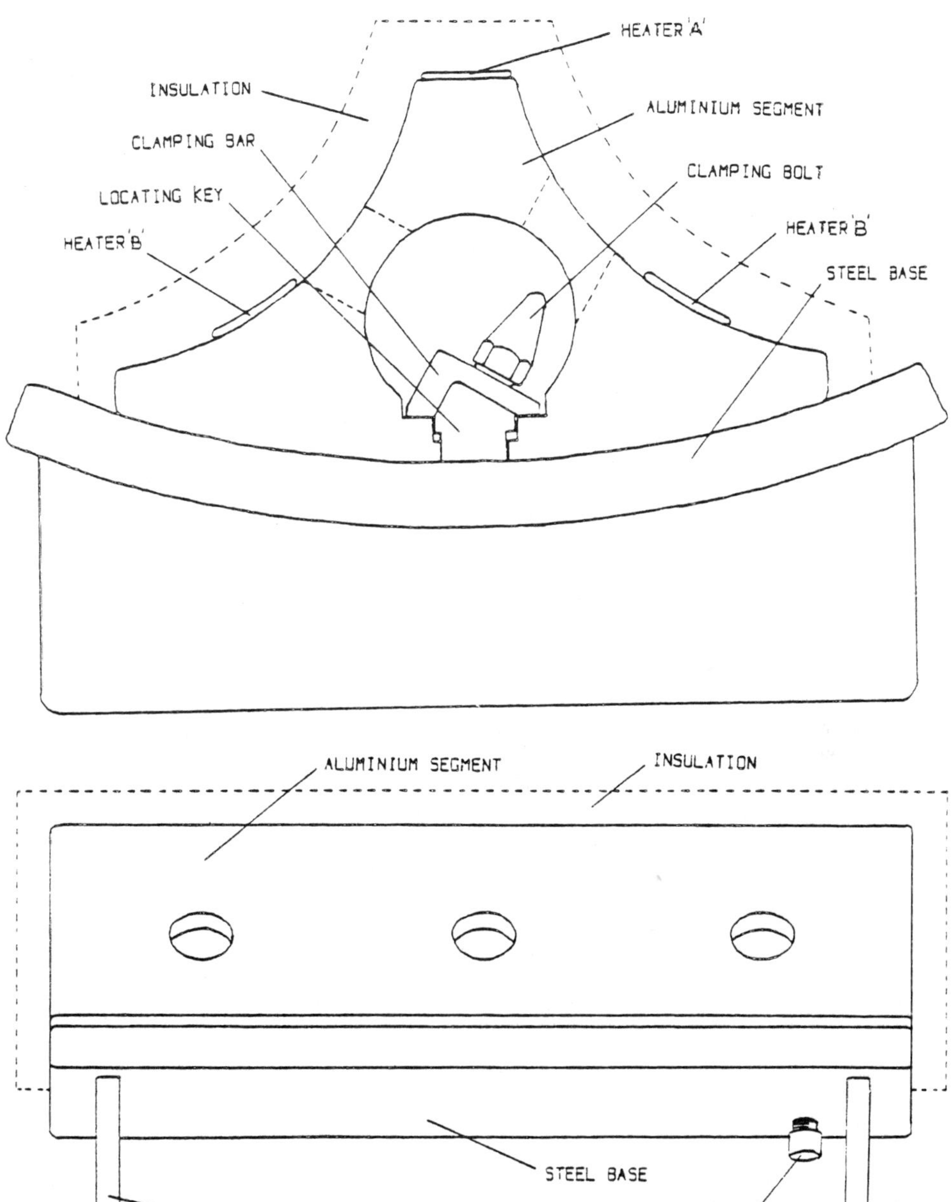

Fig. 7. Test rig No.1: clamping rig.

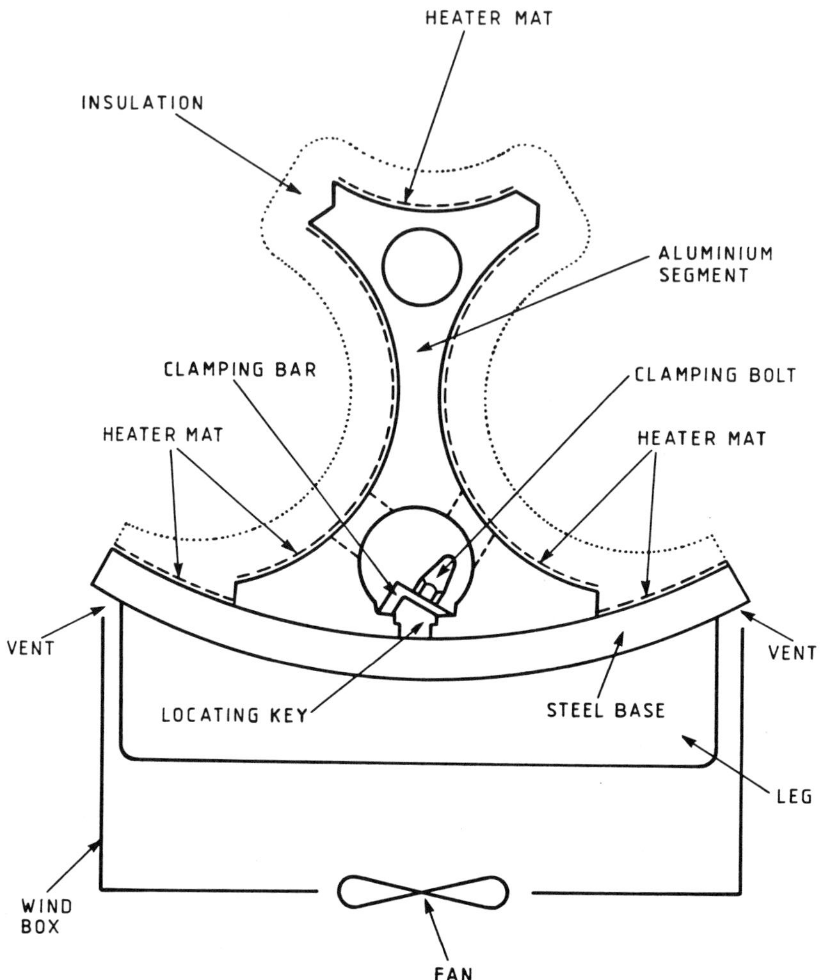

Fig. 8. Test rig No.2: thermal rig.

The experimental rig, shown in Figure 8, was similar to Rig No 1, comprising one whole full-size aluminium segment clamped onto a steel plate representing the flask shell. The heat load of the waste containers was represented by flexible rubber mat heaters bonded to the curved surfaces of the segment and, in order to avoid edge effects, to the bare topside of the flask shell. Accurate representation of the heat input was ensured by separate variable power supplies to the three areas (top, side and shell), and the whole of the heated part of the rig was insulated by a 50mm thickness of ceramic fibre blanket. The heat loss from the flask outer surface by natural convection was simulated by a wind box with two fans blowing air over the exterior. The speed of these fans could be varied, and the flow of air through each could be controlled by iris shutters. The rig was fitted with 33 thermocouples feeding to a data logger with visual display and printout, and reference marks enabled changes in dimensions due to thermal and mechanical stresses to be measured.

Dimensions were checked before and after clamping up the segment, and after each signficant stage of the heat input experiment. Heat was applied to simulate various loadings of the flask, and further tests were carried out with shims inserted betwewen segment and flask shell to investigate the effect of bad thermal contact.

Results from this and the earlier clamping rig, showing that a torque on the clamping bolts of 200 Nm was sufficient rather than the 400 Nm originally designed for, led to the idea that the bolts attaching the segment to the flask wall could be tightened up from within the hole at the bottom of the aluminium segment rather than through holes in its sides. This enabled the clamping bar and key way to be redesigned so that the bolts were now vertical rather than angled. This new simpler arrangement is shown in the diagrams of Test Rigs 1 and 2 on Figure 6. A further test was then carried out to confirm that changing to this new clamping bar did not affect the heat transfer between aluminium segment and flask wall.

The results of the tests confirmed that the temperatures encountered in normal flask operation would be within those predicted by theory and that minor imperfections in contact between aluminium segment and flask wall would not produce unacceptable temperatures. They also showed that a simpler arrangement could be used to clamp the segments to the flask wall.

Having developed and validated the segment clamping system and determined the heat transfer from the segments to the flask wall, it remained to carry out a third test series with a full cross-section of the flask assembly which would give a more exact replication of operating conditions and allow demonstration of handling of waste containers and components. The objectives of this third test series were:

a)   To establish the most effective assembly procedures for the flask's internal support structure
b)   To investigate the behaviour of the internal support structure under assembly, thermal cycling and dismantling conditions and demonstrate the suitability of the design under such conditions
c)   To demonstrate the ability to load and unload the flask under different operational conditions
d)   To investigate the ability of the internal support structure to transmit heat from the waste containers into the flask wall when subjected to symmetric and asymmetric heat loads using heaters to simulate the waste containers.

The rig for these tests, shown in Figure 9, comprised a cylindrical shell with a support stand, an assembly of 12 aluminium segments complete with clamping systems, and 7 simulated waste containers. The internal diameter of the shell was identical to that of the proposed flask and its length represented 40% of that of the flask. It was constructed from carbon steel plate, reinforced by hoops so as to be representative of the behaviour of the monolithic flask body.

The shell was supported on two central trunnions resting in a support frame, and could be rotated to aid assembly. The 12 aluminium segments were positioned in two layers of six, and when fully assembled formed 7 circular compartments. Each segment had a vertical slot which located onto a 'T' shaped key firmly attached to the wall of the shell; once located, the segment was forced against the wall by means of the clamping system.

Fig. 9. Test rig No.3: full scale demonstration rig.

The simulated waste containers were fabricated stainless steel cylinders, 20% greater in length than the container height to suit the length of the rig. They were fitted with internal electric heaters with individual variable supplies and the space inside was filled with vermiculite to minimise natural convection and make the temperatureuniform. The arrangement of the shell, segments and heaters is shown in Figure 10.

The rig was insulated at each end to minimise heat losses,the lagging being cut on the top to allow each cylinder to be withdrawn during hot handling trials. Dissipation of heat from the shell surface was assisted and controlled by a system of ducted cooling fans provided with individually variable power supplies and airflow controls.

The simulated waste containers, segments and shell were fitted with a comprehensive array of over 100 thermocouples feeding to a data logger as in the previous experiments.

The segments were assembled using the torque setting of 200 Nm derived from the earlier experiments, with a tool devised to work inside the clamping channels.

In the first thermal trial, uniform heating was applied at a range of power imputs and the equilibrioum temperatures recorded. At the maxium power input, each heater was disconnected and removed in turn and loading trials conducted in the vacant cavity using an inactive waste

50

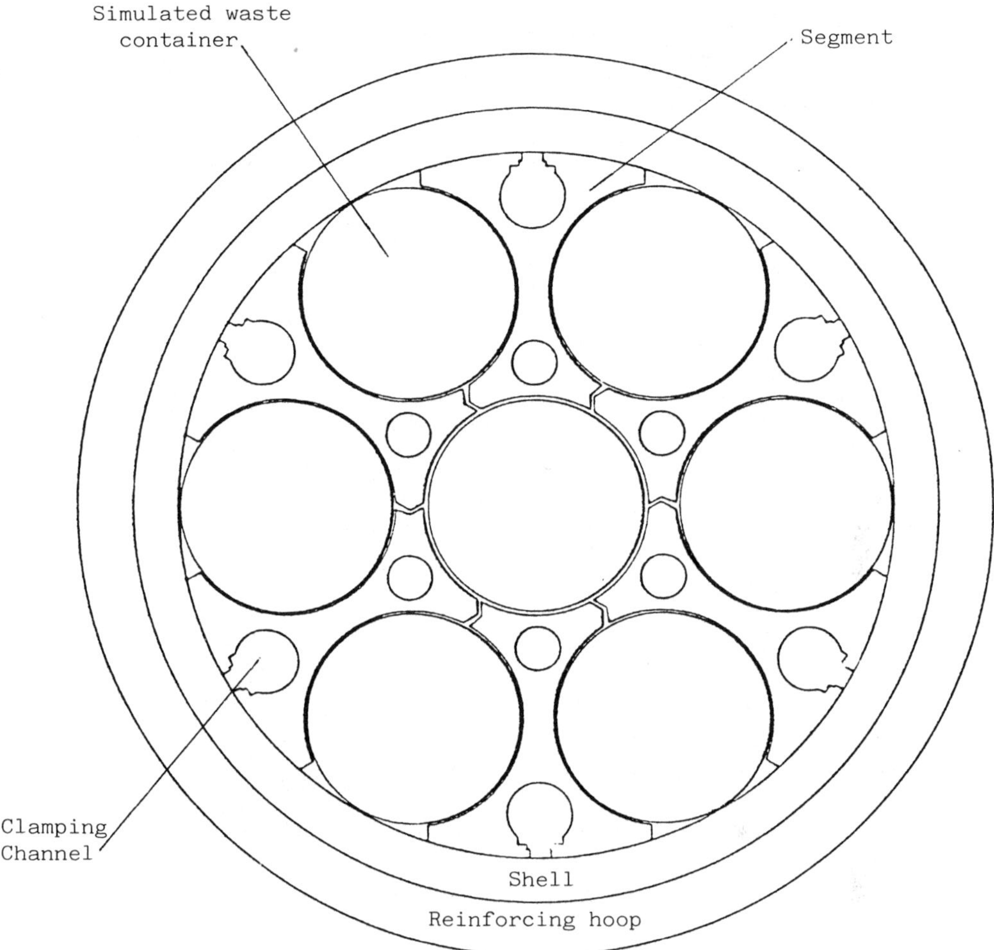

Fig. 10. Cross section of rig No.3.

container. In all cases, the container passed freely in and out of each compartment. Measurements of compartment diameters and relative segment displacements were noted.

In the second series of thermal trials, various asymmetric arrangements of heaters and empty channels were tested, with temperatures and dimensions being recorded and loading trials conducted.

At the end of the test series, the rig was allowed to cool down and the heaters removed. The segment clamps were released using the remote spanner devised for the purpose, noting the de-torque values.

Conclusions

1)   On all loading trials, the waste container passed freely into and out of the cavities in the flask rig. After completion of the thermal trials, the diameter of the waste container was increased to 440mm by temporary attachments to simulate a mandrel, which also passed freely into and out of the cavities.

2)   The diameter of the cavities was measured and shown to have remained
      stable; generally the diameter was repeated to within 1mm. The
      largest change was 2.5mm.

3)   The heater transfer of the rig was notably better than theoretical
      predictions. The maximum glass temperature in the waste containers
      is estimated not to exceed 470°C, relating to a heat load of 2.5kW
      per waste container.

4)   A small loss of preload was noted on some segment attachment bolts,
      but overall the de-torques were within 10% of the initial values and
      hence the clamping system was shown to behave satisfactorily.

The stability of the internal dimensions and the success of the
loading trials have now vindicated the design concept of the Vitrified
Waste Flask. The thermal performance was shown to be better than
predicted, and it is considered that a major factor in achieving this is
the higher than expected emissivity of the aluminium segments.

The detail design of the interior arrangements may now be regarded
as complete, and the next stage is to continue with the development of
the detail design of the containment and external features.

Matters to be addressed in detail design include the lid features –
sealing and bolting, the neutron shielding and its encapsulation, the
trunnions and the attachment to the transport frame. Although the
activity release limits of the IAEA regulations translate into minute
quantities of glassified waste, the high integrity of the containers
under normal and accident conditions, with rigorous quality control of
the welding of the lid and thorough decontamination of the exterior
surfaces of the container in the WVP, coupled with the low driving
pressure of the dry flask atmosphere, make it difficult to conceive a
mechanism for activity release from the flask.

The impact strength and leak tightness of the flask will be
demonstrated in a comprehensive series of impact tests on scale models.
Preparation of the models is scheduled to take place concurrently with
that of the Design Safety Report during 1988, and the tests are envisaged
as taking place sometime in 1989, at one of a number of British
establishments that are experienced in this type of work, such as AEE
Winfrith of the CEGB's Structural Test Centre at Cheddar. The complete
Design Safety Report would be in the UK competent authority's hands by
the end of 1989, with design approval in 1990 intime to build flasks
ready for commissioning in 1993 with a view to the commencement of HLW
transport in 1994.

CONCLUSION

BNFL, with the assistance of NTL as contractor, are designing and
developing an efficient transport flask which will be ready for the
return of vitrified high level waste to customers when the Windscale
Vitrification Plant and Vitrified Product Store come on stream early in
the 1990s.

REFERENCES

British Nuclear Fuels plc, Information Services (1984), Vitrification of
      Highly Active Liquid Waste,
Clelland, D.W., Corber, A.D.W., (August 1982) Vitrifying Britain's waste,
      in Nuclear Engineering International 33.

Curtis, H.W., et al, (1986), The transportation of high level waste, IAEA-SM-286/254, PATRAM 86, Davos

Larkin, M.J., (November 1986), Development of highly active waste conditioning at Sellaffield, _in_ Atom No 301 - 25.

Smith, W., (March 1985), Vitrification of Sellafield wastes, _in_ Atom No 341 - 3.

Wilkinson, W., (February 1985), High-level waste management in the UK, _in_ Nuclear Europe 13.

Woodall, A., Maillet, J., (August 1987), Solidifying Sellafield's high level waste, _in_ Nuclear Engineering International 44.

DISCUSSION FOLLOWING SESSION 1: Papers 1 - 4:

MR.R.GOWING, BRITISH NUCLEAR FUELS plc for Paper 1:1
The lid seals are stated to remain functional over the temperature range
-40° C to 250° C. Certain European competent authorities are
questioning the low temperature performance of cask seal materials -
which grade of Viton is used on the CANDU casks?

MR.D.J.RIBBANS, ONTARIO HYDRO, CANADA:
We are using a Grade B Viton, which has an operating temperature range
down to -40° C, and we are in the process of undertaking a test
programme into seal performance.

MR.A.HARRISON, FORGEMASTERS ENGINEERING LTD., SHEFFIELD for Paper 1:1
Given the large amount of successful experience with ferritic steel
forgings and cask manufacture, is the use of an entirely austenitic cask
likely to be taken up by other irradiated fuel transport companies?

MR.D.J.RIBBANS:
I can't answer that question, but at the concept stage we evaluated
various materials and compared costs; as solid stainless steel was of a
similar cost to ferritic steel with stainless steel cladding, we chose
the solid stainless steel to overcome the potential low-temperature
brittle fracture and this has considerably simplified the design testing
process.

DR.H.CODEE, COVRA N.V., THE NETHERLANDS for Paper 1:1
What was the reason for extending the duration of your fire tests beyond
the IAEA guidelines of 30 minutes, to a period of one hour?

MR.D.J.RIBBANS:
We are trying to determine the cask performance in conditions exceeding
the regulatory requirements both as part of our public communications
programme to prove the the public that we have a safe transportation
system, and in the event that sometime in the future the line the
regulations change. We believe that the thirty minute fire is a very
severe test but we are just trying to determine what the cask performance
is in conditions beyond the regulations - and the fire test is one
aspect.

DR.H.CODEE:
May I comment on that. Does it not hint that since you show so much
attention to informing the public about the safety of your cask, when you
go beyond the IAEA regulatory guidelines you show some slight disbelief
in these regulations, and this just makes your position so much weaker?

MR.D.J.RIBBANS:
No, I don't believe so. There is a recurring question that we receive

from the public. In Canada we had a fire on a rail line in Toronto and based on the media reports, that fire lasted anywhere from 1 hour to 6 hours. So it is in the public's mind that fires last longer than 30 minutes. When we speak to the public we always start by giving your point that the 30 minute 800° C fire is very severe, and that we believe that it is a severe enough test, but if our cask performance will go beyond that, why should we not demonstrate it?

MR.M.S.T.PRICE, ATOMIC ENERGY AUTHORITY, WINFRITH for Paper 1:1
In the slide you showed of the fire test, was the flask wholly immersed in the fire or were the ends sticking out?

MR.D.J.RIBBANS:
Yes, the fire tests that we did were not licensing tests. We license the test on the basis of analysis. Subsequent to that, with the half-scale model, we did the fire test and the fire did completely engulf the cask. We have also produced a video of the fire, for as long a duration as we could, which indicates that it was fully engulfed.

MR.D.BLACKMAN, DEPARTMENT OF TRANSPORT, LONDON for Paper 1:3
To what extent have you gone down the road towards adopting burn-up credit in your approach to criticality safety? As the United Kingdom licensing authority we are naturally very interested in how soon we will be faced with applications embodying this principle.

MR.I.J.HUNTER, NUCLEAR TRANSPORT LIMITED, RISLEY:
A very short answer is 'Quite soon'. We are now approaching the enrichment limits of some of the fuels which we are carrying. In the current designs of MEB's we are approaching 4% on some of our fuels. We are preparing concepts for new MEB's to introduce but I would expect within the next 12 months or so, approaches will be made to competent authorities with the use of burn-up credit and systems for verification of burn-up to back this up.

MR.M.S.T.PRICE: ATOMIC ENERGY AUTHORITY, WINFRITH for Paper 1:2
Could you comment on the radiological advantages/disadvantages of the various designs and what radiation doses might be experienced by workers at the Nirex site relative to the ionizing radiations regulations and ALARP?

DR.J.MILES, OVE ARUP AND PARTNERS, LONDON:
That depends entirely on the methods that are used to operate the flasks or transport containers, but if you assume that there is no high degree of automation then it is the concept 'O' that is the most attractive because it has the fewest bolts. So there are a very small number of bolts involved and therefore it would not take very long to remove and replace them. From that point of view it would be the most attractive design, but I think those issues are very much balanced by the methods you would use to deal with the flask and therefore the conclusion is not black and white.

MR.T.W.STEPHEN, JAMES FISHER AND SONS plc for Paper 1:2
First of all I would like to congratulate the authors on a very enlightening paper. Dr.Miles, could I ask if you have considered the sea transport of these containers, and if so, are the containers designed to withstand hydrostatic pressure, and what criterion was used?

DR.J.MILES:
The answer is that sea transport was not explicitly considered. The original brief centred around rail and road transport and the concepts were developed with those forms of transport in mind. However, I can't

see any reason why there should be any particular problems with over-pressure due to being submerged in the sea. If one is considering loss at sea with damage then the question of sealing may arise but I can't see that there will be any big issues there either, even if sea water did get in. So in concept I see no major problem. The size of the containers is such that I think the pressure due to sea water would not compromise their integrity unless they were already damaged before they went into the sea, which again I think is difficult to imagine.

DR.J.MILES:
I would like to just pick up on the point that Mr. Price raised and which has to do with the advantages of the different concepts put forward. I believe there is no clear-cut winner, but there are three different concepts, each one of those concepts has a different advantage, and it would be the circumstances which doctate the one that would be chosen for the task in hand.

The concept C has the advantage, first of all, of being conventional, there is little emotional resistance to it, but it also has some advantages in terms of protection round the lid body area and being able to guarantee a seal. The Concept E on the other hand is by far the strongest design because inherently it is very well locked together and the forces across the intersection are minimized which is very attractive, except for the fact that having quite a wide waist band it gets in the way of the rail gauge, so there is a limit on the maximum wall thickness you can have and that isn't quite as high as you might like. That could be modified but it would require some work. Concept O at the end of the day is very attractive because it is the lightest possible design and the only amount of material which is there is that which is required for shielding, but alternatively it is rather difficult to seal because it works fundamentally by jamming the two halves together and there is relative motion. So there are different advantages with the different concepts and it depends on a great number of facts as to which would be chosen.

MR.R.GOWING, BRITISH NUCLEAR FUELS PLC:
I would like to take up a point that came up in the discussion of the Ove Arup paper about the radiation exposure in sealing the flasks. One thing that I have become aware of recently on a visit to the new receipt facilities at Sellafield is that we have the capability of remote bolting and unbolting of flask lids and when this is coupled with the facilities for loading the cylinders in the vitrified product despatch facility, it means that direct operator exposure to the flasks would be very limited.

MR.I.J.HUNTER, NUCLEAR TRANSPORT LIMITED:
I would like to broaden the presentation slightly. The paper which I presented was looking at one answer to the question of what is the problem with high burn-up fuels for the future. As BNFL NTL has a large fleet of existing flasks we will of course continue to operate these and be able to modify them to cope with the future. That is, only a small proportion of the future fuel will need to have new flasks specifically developed.

MR.I.A.WOOD, GEC ENERGY SYSTEMS LIMITED for Paper 1:2
A question for Mr. John Miles. With the long skirt engagement that you have on your final type of flask, have you been able to demonstrate that operators in fact can insert one on top of the other with very little problem?

DR.J.MILES, OVE ARUP AND PARTNERS:
Yes, we have done some work in that direction but not an exhaustive

amount, as it did not fall within the original brief although of course
common sense says one has to pay some attention to practicalities. In the
event, if the thing is produced by a cast method, which would be quite a
suitable way of doing it, there would necessarily be a taper between the
top and bottom halves, which provides quite a convenient lead in, and so
as long as you can crane the top half over and lower it down, there
wouldn't be any serious interference during the time of engagement. We
satisfied ourselves as to the practicality of doing that, the sort of
clearances that would be involved and the level of equipment that might
be required. I don't think that one would require anything out of the
ordinary. Superficially it looks as if you need a large clearance in
order to get the top half of the container off and take it across to one
side. In fact when you add up all the clearances, that are necessary for
the other two designs as well, you do not require any greater lift height
to get that lid of than you do for any of the others. So, it is not as
disadvantageous as it might appear. You don't have to lift it as high as
one of the other designs and in fact there is quite a convenient way of
getting it in and out with the taper that would be on the interface.

DR.H.GEISER, GNS, ESSEN, F.R.GERMANY for Paper 1:2
I have a question on your very impressive computer codes - is it DYNA 3D,
the Lawrence Livermore Code?

DR.J.MILES: Yes.

DR.H.GEISER:
What kind of computer are your running?

DR.J.MILES:
It is a VAX, 11785. It is not a very powerful machine but I think you may
have noticed from the models I was showing that they are not very
complex. We have done a lot of work in trying to assess just how complex
a model has to be in order to give useful results, and a lot of our work
centres around the use of quite simple models. They may look complex on
the screen, but when you compare them to the many thousands of elements
you sometimes see in finite element models for impact analysis, they are
really quite crude. So a lot of our effort as designers and engineers has
gone in to developing sufficient background in the use of the code to be
able to use it with a simple model but still get useful results out. The
key to that development programme I described was being able to go
through very quickly and assess various different concepts, understand
whether there are any fundamental problems that would stop the thing from
working, and from a large number of initial ideas, hone them down to just
1, 2 or 3 that would take them through to the end of the programme. So
the emphasis was on speed of use and quickness of turnround.

DR.H.GEISER:
How did you execute the benchmark of the code?

DR.J.MILES:
This is a wider question. We have been involved in using the code for 5
to 6 years now, and we have used it in a wide number of different
applications. In fact we distribute the code, so there are a lot of other
aspects to this question. We have quite a large number of theoretical and
experimental benchmarks against the code itself, specifically in
connection with the design of transport containers. There was a lot of
work done prior to the project I have described in which we were looking
at the Magnox fuel flasks for the CEGB. A lot of scale model testing and
component testing was done during that programme and one for one
comparisons were made between the computer codes and the experimental
results and of course there was a degree of correlation done during this
programme too.

MR.M.L.McKENNA, ROLLS-ROYCE & ASSOCIATES LIMITED for Paper 1:3
Does Mr.Hunter find conflict between "on-site" radiological levels for
the work force and those required by IAEA Regulations? I am thinking of
influences of ALARP and NII requirements.

MR.I.J.HUNTER:
With fuel flasks the IAEA regulations stipulate certain dose limits at
distances from the package and the derivation of these dose limits is to
limit the total dose uptake to members of the public. The dose limits are
set typically at 2m from the package and of course, for operators, in a
totally different environment, it is more of a hands on operation. What
we have found with irradiated fuel flasks is that if we work to
regulatory limits imposed by the Japanese which are 10 milirem at 1m,
then the sort of surface dose rates which we get on the flasks are not
too onerous for operators. Of course the dose uptake is a function of the
dose rate outside the package and the time involved in operation. Most of
the operators of flasks, in particular BNFL, are designing systems to
minimise the hands on operation time on the packages with the use of
robotics, so we don't see a great dose conflict in fuel flasks.

MR.R.F.KEENE, OVE ARUP AND PARTNERS:
That was specifically a concept that came up after these flask concepts
were first developed, and some discussions took place as to what
additional shielding could be provided over and above the IAEA
regulations because questions were asked as to the suitability of the
IAEA regulations when in close contact, for handling, particularly at a
repository site, perhaps, where transport containers are coming in all
the time. So one of the design exercises we went through after this
project was complete was to look at each one of the concepts and see to
what extent the walls could be fattened up. When I spoke about the
concept E being limited by the waist band that goes around the centre. It
was in that respect because all of those three concepts meet the basic
IAEA wall thickness requirements, but thereafter there are fairly serious
limitations as to how far each one of the concepts can be extended. It is
an issue that excited some interest and I think possibly will continue to
excite some interest and may indeed influence the final design.

MR.I.A.WOOD, GEC ENERGY SYSTEMS LIMITED for Paper 1:1
Could you amplify what contributes to the 0.2mm gap (lid/body). Is the
figure due to impact only or is it a summation of both impact and thermal
excursion for the one hour or half hour fire?

MR.D.J.RIBBANS, ONTARIO HYDRO, CANADA:
The gap is produced by a combination of the two – bolt stretch in the
impact and thermal distortions during the fire.

MR.I.A.WOOD:
Was that with your regulatory fire or the one-hour fire?

MR.D.J.RIBBANS:
The gap has been determined by analysis, we didn't measure the gap during
the fire, we measured the sealing of the cask, but the maximum gap due to
thermal distortions occurs quite early on in the fire when you have a
large temperature differential across the wall with our shape of cask. As
the fire proceeds and the temperature is equalized internally and
externally, the distortion reduces. So the maximum gap would be in the
shorter fire. The gap doesn't increase as the duration of the fire
increases.

MR.G.LUCK, CEGB, GDCD for Paper 1:1
My question follows on from the last one; you made some comments in your

paper about the use of redwood in the impact limiter and its good
resistance to fire, could you comment on whether, in your fire test, (a)
was it tested with a damaged impact limiter and (b) how you would expect
this impact limiter to stand up to the extended fire?

MR.D.J.RIBBANS:
(a) The half scale model cask and impact limiter were the same cask that
went through the impact testing, so the impact limiter was damaged on one
corner from a top corner drop. It had split in several places, not
severely. You saw the slide of the sectioned impact limiter and that
proved some earlier testing we did on redwood blocks that - when
subjected to a fire - the redwood chars on the outside and then is self
limiting. We get a maximum depth of two inches on the redwood and that is
what we saw with the impact limiter.
(b) We used the same impact limiter for a 30 minute fire and then pulled
everything down and then did the one hour fire and the same impact
limiter was used for both fires so it actually saw the equivalent of one
and a half hours.

MR.P.DONELAN, OVE ARUP AND PARTNERS for Paper 1:1
I was interested to hear that he is investigating the use of metallic
seals. How are the results coming out - is it looking feasible?

MR.D.J.RIBBANS:
We have only done some preliminary testing - some resiliency testing - on
a number of metallic seals. They lack the resilience of the elastomeric
seals and in our rectangular seal configuration we found that with the
small radii at the corners this reduced the resilience still further. We
are now homing in on the best of those for further testing. They do lack
the resilience and we are not sure whether they are going to be adequate
for the gaps that we see in the impact testing. But of course the beauty
is that they can withstand the high temperatures that we would see in an
extended fire.

MR.R.GOWING:
I would like to add to that comment on the use of metallic seals. In our
early use of EXCELLOX flasks we had a a metallic asbestos lid seal and it
needed a very high torque on the lid bolts to achieve a good seal. When,
through the use of an insulator and shock absorber, we became able to use
Viton seals we were able to reduce the lid bolt torque considerably; the
elastomeric seals we have found by experience to be far more satisfactory
than metallic seals - especially for the lid.

EXPERIMENTS CONDUCTED IN SUPPORT OF
TRANSPORT FLASK HEAT TRANSFER ASSESSMENTS

E.Livesey, W.E.Swindlehurst

British Nuclear Fuels plc
Risley
Warrington
Cheshire WA3 6AS

ABSTRACT

Water filled and gas filled flasks are used to transport irradiated
LWR fuel. A programme of experimental work has been conducted to provide
data for thermal modelling. Operational, regulatory and extra regulatory
conditions have been covered.

In some wet flask designs the fuel is carried with an inner
container separated from the flask body by a narrow water filled annulus.
Experiments have been conducted to correlate the heat transfer across
the annulus. In dry flask designs however a significant temperature drop
occurs between the fuel array and its surroundings and a series of tests
on a simulated fuel element in a gas environment has been carried out.
Fins are generally fitted to the outside of flasks carrying high heat
output fuel in order to improve the convectional heat transfer to
ambient. Design data have been derived from tests on large diameter
finned sections with various combinations of pitch and fin height. Other
work has included an experimental survey of flask emissivity data and
thermal trials on a full size flask.

During a fire, heat is transferred to the flask surface by radiation
and convection from the flames. To provide for heat input data, tests
have been carried out to determine the heat flux to a full size finned
section of a flask surface when subjected to a hydrocarbon pool fire.
Although not a regulatory requirement, BNFL have also investigated the
thermal response of a wet flask in the extremely unlikely event of loss
of water. Tests have been conducted on a wet flask operated dry,
containing simulated fuel elements. Transient tests were carried out
over a range of fuel powers, filling gases, internal pressures and
partial refill with water.

INTRODUCTION

LWR fuel, transported by BNFL and its subsidiary company Pacific
Nuclear Transport Ltd, is carried within water filled (wet) or gas filled
(dry) cylindrical flasks.

Assessment of the thermal performance of a flask under normal

operating and accident conditions can be made by mathematical modelling.
However in order to provide appropriate physical data for modelling, BNFL
have conducted a programme of experimental work on flask heat transfer.
This paper presents a summary of experiments carried out during the
period 1978 to 1987 and their application to design assessments.

TRANSPORT FLASKS

LWR Fuel transport flasks, as used by BNFL, may be conveniently
described in terms of four main design features. These are:

- lead shielded: where a separate lead liner provides the main
  radiological shielding;
- monolithic: where the steel body of the flask is sufficiently thick
  to meet shielding requirements, and either
- water filled or
- gas filled.

The flasks are cylindrical in construction (typicallly 2m diameter
and 6m long) with a thick base and bolted lid, protected against impact
by stainless steel encased balsa wood shock absorbers. The outer surface
is finned to dissipate heat to the surroundings.

Lead Shielded Flasks

Lead shielded flasks are water filled in order to transfer
effectively the fuel decay heat from the fuel to the flask body.

Figure 1 shows the cross section of a flask typical of the BNF/PNTL
Excellox 3 and 4 flasks. The fuel is carried within a sealed inner
container, known as a Multi Element Bottle (MEB). Both the MEB and
surrounding flask cavity are water filled, with an ullage to allow for
water expansion. Water convection is the dominant mode of heat transfer
between the fuel and flask body. Conduction through the flask body, and
convection and radiation form the external fins to the surroundings
completes the heat transfer path and the fuel is kept cool (typically
$50-100^{o}C$).

In the event of a surrounding fire the heat from the fire raises the
flask/MEB water temperature. It must be shown that during consequent
expansion of the water the flask does not leak.

If a loss of flask/MEB coolant water were to occur the heat transfer
within the flask would be mainly by thermal radiation with some
contribution from convection. This situation results in higher fuel and
flask component temperatures.

Monolithic Flasks

Cross sections of both water filled and gas filled monolithic flasks
are shown in Figure 2. Water convection heat transfer dominates within
the wet flask (as for the lead shielded type). However the dry flask
internal heat transfer is much more dependent upon thermal radiation and
conduction through the fuel support basket. Dry flask fuel temperatures
are higher than for a wet flask.

The response of a monolithic wet flask to a fire or loss of coolant
is similar to that of a lead shielded type. However a monolithic dry
flask exhibits a relativley small rise in internal pressure during a fire
(in the absence of water expansion). Also in the case of the dry flask,

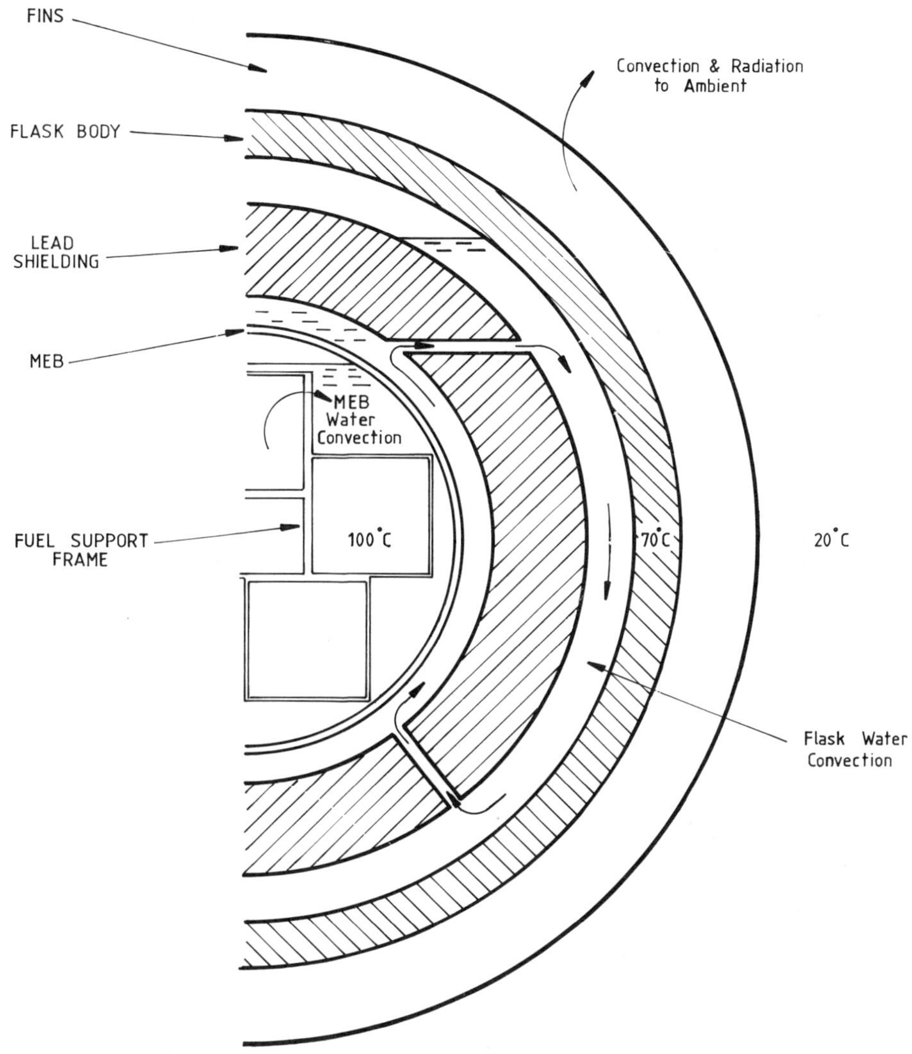

Fig. 1. EXCELLOX flask.

loss of coolant can only mean loss of gas pressure, resulting in a minimal rise in fuel temperature.

THE EXPERIMENTAL PROGRAMME

Background

In any given situation, whereas numerical models are used to establish the typical temperature profiles described above, experimental work plays an essential role in validating and underpinning the mathematical modelling work. More specificallly, within the transport programme experiments have been undertaken with one or more of the following objectives.

(i) to provide an improved understanding of complex heat-transfer processes,

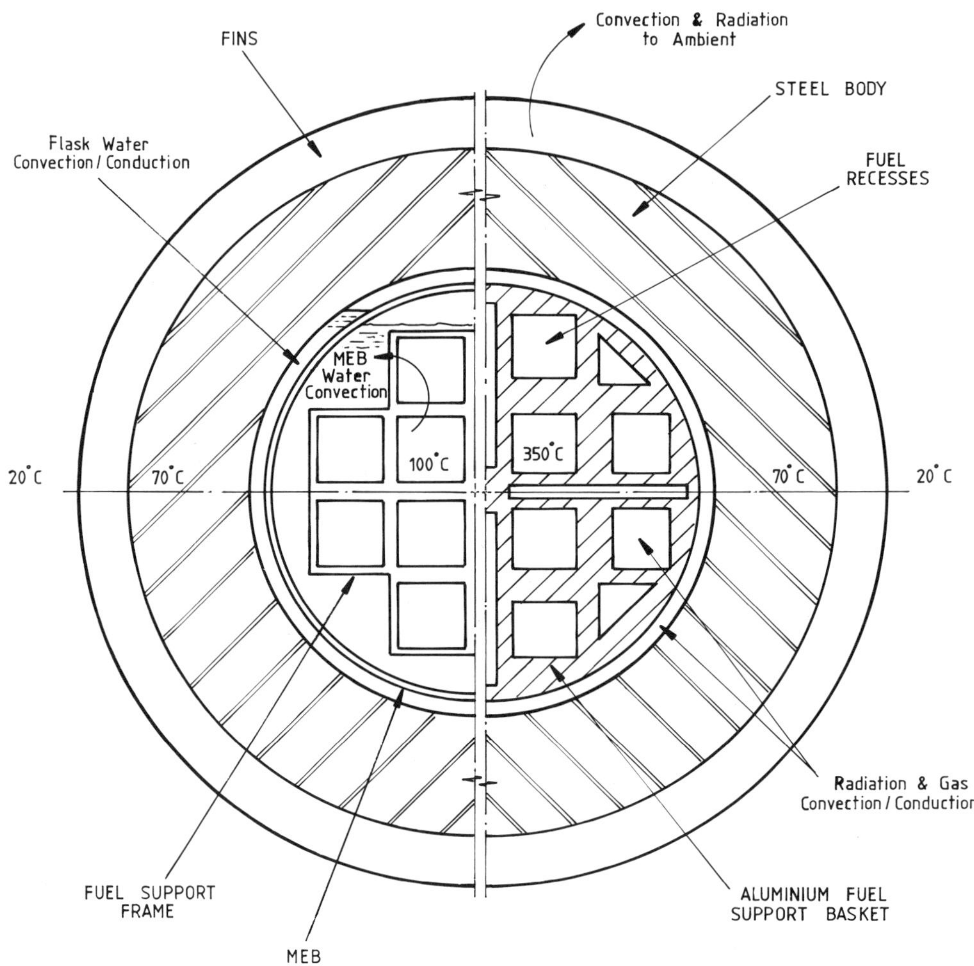

Fig. 2. Monolithic flasks - water filled/gas filled.

(ii)   to establish and quantify heat-transfer correlations relating to specific physical situations

(iii) to validate and establish the accuracy of numerical code production,

(iv) to determine thermal properties of flask component materials and surfaces,

(v)   to demonstrate overall compliance with regulatory requirements.

The series of experiments described briefly below covers both dry and wet transport and addressed the major areas of uncertainty which have been encountered in the definition of flask temperatures. With reference to Figures 1 and 2, more specifically these areas are:

(i)   External heat transfer to and from the outer finned flask surface, for both normal operation and fire accident conditions.

(ii) heat transfer between the central fuel element, MEB and the flask body via the narrow water filled annulus of Figure 2 or the

interconnected concentric MEB/liner/flask annuli of Figure 1.
(iii) heat transfer within the fuel pins and elements under dry transport
     or loss of coolant conditions.

Experiments Supporting Normal Operation

<u>Heat Transfer in a Single Fuel Element in a Gas Environment</u> A major constraint on dry fuel element transport is the maximum temperature of the fuel cladding itself. It is therefore essential that accurate modelling of heat transfer mechanisms within a fuel element assembly can be undertaken to enable element clad temperatures to be determined to an accuracy of $\pm 20^{\circ}C$. This particular experimental programme was devised to both confirm the understanding of the controlling heat transfer mechanisms and to provide a reference set of experimental data against which a computer code validation could be undertaken (Fry et al, 1983).

The experimental assembly consisted of a 16 x 16 horizontal array of Zircaloy-4 tubes held together by 9 spacer grids. Heaters were placed within 236 of the tubes with 20 vacant tubes representing control element positions. The axial flux was arranged to represent an irradiated PWR assembly.

The assembly was located within an aluminium box as shown in Figure 3. The box was surrounded by a wooden enclosure through which air was blown at a controlled rate to obtain a constant, pre-determined box temperature. Instrumentation comprised 224 thermocouples located as indicated in the Figure. As temperature distributions were assumed to be symmetrical about the vertical centre-line, the greater majority of the thermocouples were positioned on one side of the assembly.

The test programme comprised 17 tests. The first eleven covered three different powers and various box temperatures with the box air-filled. The next three tests were conducted at different powers with the box filled with helium to investigate the role of gaseous conduction. In the final three tests the inside of the aluminium box was painted matt-black to check the effect of emissivity.

The temperatures and powers were measured only after equilibrium had been achieved; no transient measurements were recorded.

The experiments showed that natural convection inside the air-filled assembly had a significant effect upon the temperature profile. The temperatures in the lower half of the assembly were less than those in the upper half. These effects were reproduced in a simple numerical model demonstrating that these convection currents also reduce the peak pin temperatures slightly. Filling the assembly with helium caused a large reduction in peak pin temperature, due to the increased gas conductivity. The temperatures in the lower and upper halves of the assembly were almost identical, indicating an absence of convection – attributed to the low density of helium.

The agreement between measured and calculated peak pin temperatures was in general very good, especially when the effects of convection were modelled. For an assumed reduction of about $5^{\circ}C$ due to convection for the tests in which the assembly was air filled, then the agreement over all the tests between the measured and calculated peak pin temperatures was $\pm 7^{\circ}C$. A significant part of this error is attributed to random errors in experimental power measurement and material properties.

The results of the experiment provided a validation of numerical

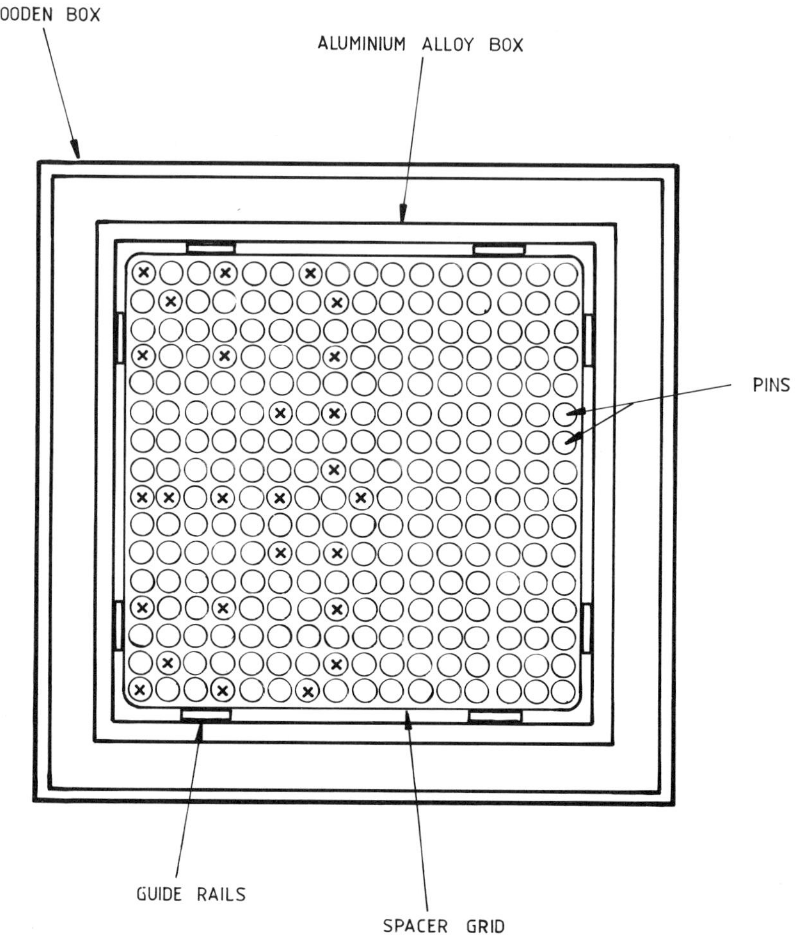

Fig. 3. Experimental assembly: single fuel element experiments.

modelling techniques, enabling temperature distributions in large arrays of heated pins in a stagnant gas environment, to be predicted with confidence. In particular it can be shown that peak temperatures can be predicted accurately.

Annular Gap Heat Transfer Experiments   For wet monolithic flask designs, heat transfer between the central MEB containing the fuel element assembly and the steel flask body is sensitively dependent on the intervening annular gap heat transfer coefficient. The geometry of the flask design results in a very low annular gap to diameter aspect ratio (of the order of 1.10^{-2}) for which little relevant experimental heat transfer data is available. Consequently an experimental programme was devised to assess heat transfer behaviour for this particular geometry and evaluate the applicability of existing data (Drake, 1988).

The experimental arrangement chosen was a full scale section of a short length of flask, which eliminated the need for scaling factors and

allowed tests to be performed near the actual flask operating
temperatures. A simplified view of the experimental arrangement is shown
in Figure 4. Two concentric steel cylinders rigidly fitted together with
a leak-tight water-filled annulus were completely immersed in a constant
temperature water bath. The inner cylinder was also water filled. This
water was heated by means of electrical immersion heaters to simulate the
decay heat of the flask contents and was circulated by means of a pump to
minimise temperature stratification.

The experimental arrangement allowed for:

(i)   a variation in annular gap between the cylinders
(ii)   a variation and measurement of the input power delivered to the
        inner cylinder
(iii)   measurement of the temperatures of the inner and outer cylinders
(iv)   measurement of the annular fluid temperature
(v)   means to allow thermal equilibrium to be established for (iii) and
        (iv)

     Three annular gaps were investigated during the experiments and for
each gap size twenty seven runs were undertaken ie three input powers at
three water (ullage) levels performed at three water bath temperatures.
In addition a separate series of runs was undertaken to investigate the
following phenomena: water stratification, inner cylinder ullage effects
and fire simulation (ie reverse heat flow). As over 100 temperatures
were required to be measured for each set of operating conditions the rig
was extensively instrumented; temperatures being measured by thermistors
linked to a sophisticated data logging system.

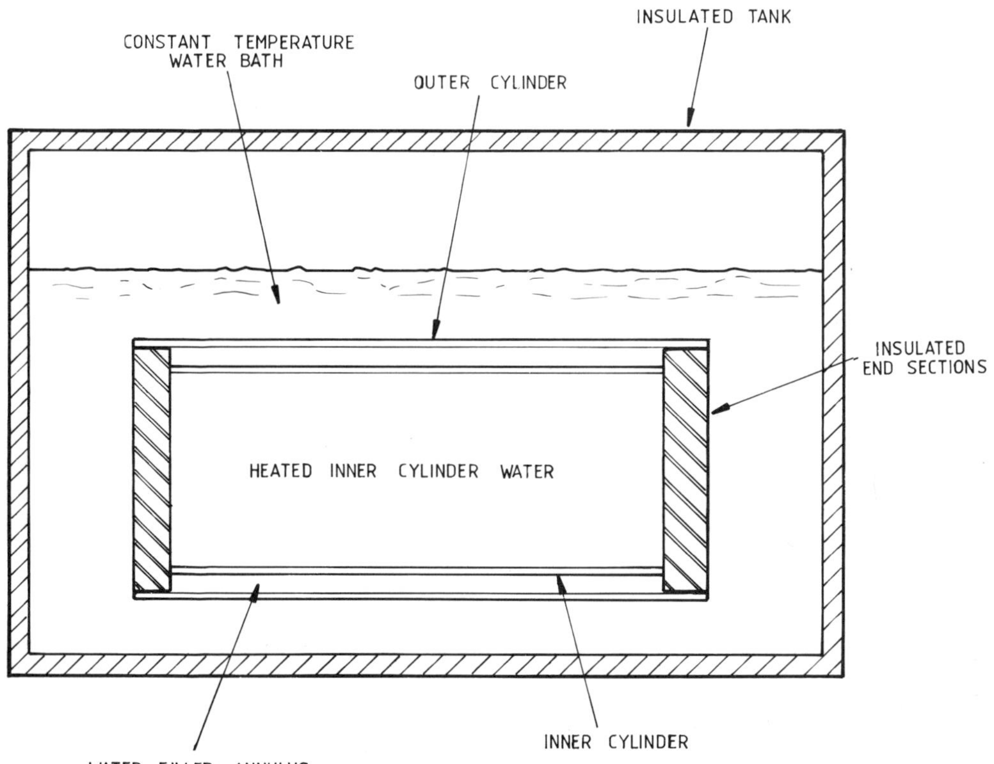

Fig. 4. Experimental Assembly: Annular gap heat transfer experiments.

A simple assumption of conduction results in high temperature drop across the annulus. Consequently the influence of convection is of prime importance. A measure of the relative magnitudes of these two components is given by the Nusselt Number. The experimental data was therefore, following common practice, analysed in terms of Nusselt/Rayleigh number correlation. The Rayleigh number contains terms which involve the annulus fluid properties, temperature differences across the annulus and characteristic gap dimensions. As a result of the experiment, the appropriate heat transfer correlations were developed and their accuracy established for flask operational/design conditions.

<u>Fin Heat Transfer</u> For the flasks described in TRANSPORT FLASKS, heat transfer from the flask surface is primarily via the finned portion of the flask body. It is therefore essential that quantitative assessment of the heat- transfer processes can be made to optimise flask designs in relation to such parameters as, for example, fin geometry.

With this aim in mind a series of heat transfer experiments was conducted on a simulated finned surface of a transport flask. Heat was applied to the inner surface of a number of different test sections with different fin geometries and the steady state temperatures on the test sections measured.

A typical test section is illustrated in Figure 5. It consists of a half cylinder of aluminium alloy of approximately 1.5m OD. The test section was heated electrically by pads attached to the rear surface and the heat was dissipated to the air at the front face by natural convection and radiation.

The section was heated uniformly over an axial length of approximately 0.5m and guard heating applied at each end to minimise axial heat flow. The power to the guard heaters was adjusted to give uniform axial temperatures along the test section. Temperatures were recorded when thermal equilibrium was achieved using thermocouples positioned on the body of the test section and along the fins.

Five finned sections and one unfinned were used in the tests with variations in fin geometry. Measurements were made with each test section at several different power levels. In addition, air temperature and velocity data were recorded (using a hot wire amemometer) during each test. In total 72 thermocouples were used in the experiment: 52 for shell and fin temperature measurement and 20 for air temperature recording. As heat is lost from the finned surface but natural convection and radiation, supplementary tests were carried out to determine emmissivity and thermal conductivities.

The data were analysed to provide convection coefficients for the various fin geometries, allowing an interpolation formula to be derived giving quantitative dependency on fin height, fin gap, fin root temperature and ambient temperature. Currently further analysis of the experimental data is being undertaken to assess the variation of heat transfer coefficient around the flask.

<u>Emissivity Data Assessments</u> For a dry flask (or wet flask after a LOCA) operating with an internal pressure at around atmospheric, radiation is an important heat transfer mechanism. In particular for internal heat transfer, because of the normally high operating temperatures of the contents, it is often the dominant mode.

Analytical techniques and numerical models of radiation exchange systems are readily available and temperature predictions can usually be

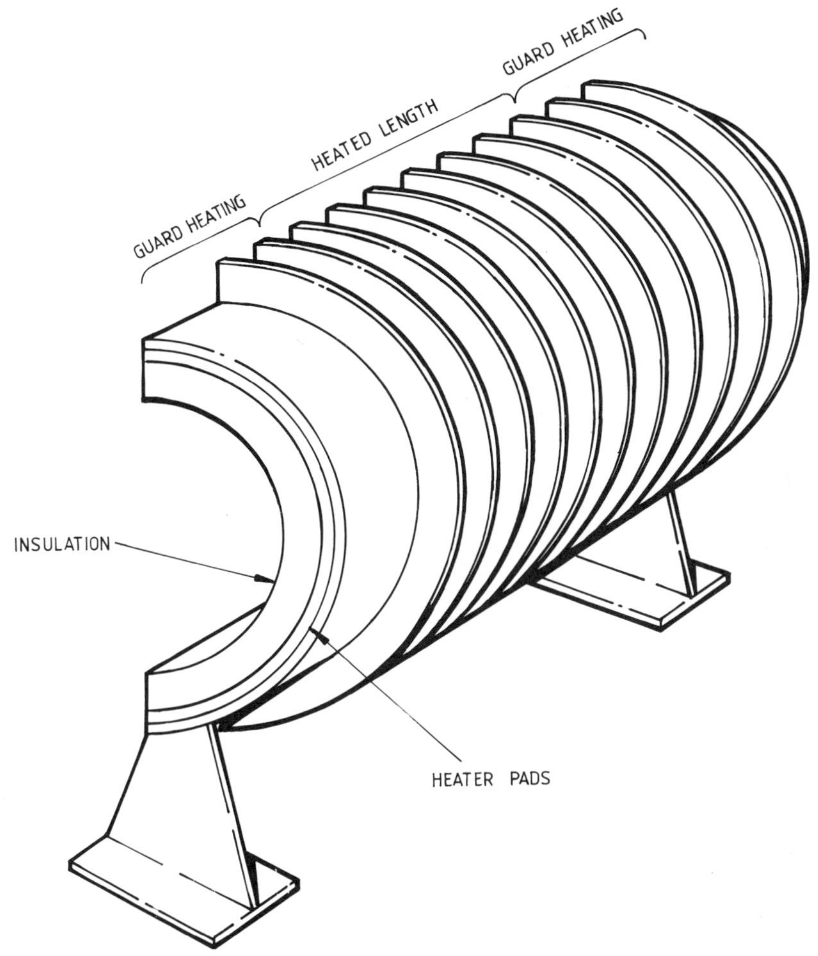

Fig. 5. Experimental assembly: fin heat transfer experiments.

made with relative ease and accuracy when compared with convection dominated systems. However, the most important area of uncertainty lies in the choice of component emissivities, particularly when absolute values are low and hence small errors significant. Unfortunately the choice of materials for flask internal components often results in low emissivity surfaces (eg stainless steels). This not only introduces a possible source of error but can also produce unacceptably high component temperatures. There is, therefore, considerable incentive to have both reliable emissivity data and have acceptable means of increasing the value of low emissivity surfaces, leading to lower temperatures or higher flask payloads.

The programme of work undertaken consisted of an extensive review of available emissivity data coupled with a determination of actual flask material emissivities and their enhancement by mechanical, chemical and coating treatments. The overall objective of the work was to provide a list of recommended emissivity data to be used in future transport flask thermal assessments.

Materials of interest in flask designs can be broadly divided into metals and opaque non-metals such as paint. Pure, smooth metals are often chaaracterised by low emissivities and non-metals by large emissivity values at moderate temperatures. The difficulty in specifying accurate property values arises because these properties depend on many variables such as surface roughness and degree of polish, material purity, thickness of coating, temperature, radiation wavelength and angle at which the radiation leaves the surface.

The experiments involved the measurement of thermal radiation for a heated specimen using a calibrated total radiation pyrometer, having a very uniform response to thermal radiation in the spectral range 0.6 to 40 micron. Measurement of the surface temperature of the sample enabled the corresponding black body radiance to be calculated and hence the specimen emissivity. Measurements were carried out at a range of temperatures between 150°C and 400°C and directions between 0° and 80° to the surface normal.

From the results of an extensive series of test on stainless steel and aluminium with surfaces modified by mechanical, chemical and heat treatments, together with an evaluation of surface coatings (metallic and non-metallic) it was possible to produce a design guide for future selection and assessment of flask construction materials.

<u>Flask Thermal Tests</u> While the other experiments described in this section have been directed towards understanding and quantifying specific aspects of the heat-transfer behaviour of LWR transport flasks, this particular series addressed overall flask thermal performance in an integrated manner (Burgess et al, 1983).

An EXL 3A flask, typical of the current design of wet fuel transport (Figure 1) was loaded with 14 electrically heated dummy BWR fuel assemblies and tested at several power levels in different environments. A supplementary series of measurements was made with the flask orientated vertically. In addition, a loss of coolant accident was simulated.

The flask was extensivly instrumented for temperature measurement both internally and externally. Thermocouples within the central MEB were located on the centre line while those outside were distributed along the "fuel" length, covering the central high heat flux region; each end of the located region and near the lid of the MEB. Radial distributions included the MEB outer surface, MEB and liner water gap, liner inner and outer surfaces, liner/flask water gap, flask body midway between the fin and fin-tips. Air temperatures were measured below and above the flask and at mid-fin depth. In addition, MEB and flask pressures were also recorded.

Within the experimental programme the following aspects of flask behaviour were addressed.

(i)   the influence of test environment – to simulate the regulatory ambient as required by IAEA regulations.
(ii)   the infuence of thermosiphon water flow through the ports in the lead-liner on heat transfer between the MEB and the flask body.
(iii)   external heat transfer from the outer finned surface of the flask body.

An important outcome of the experiments described above was that a very good physical understanding of the important heat transfer processes was developed. In addition to this benefit the large amount of data generated enabled an integrated numerical model of the heat transfer in cylindrical wet transports flasks to be developed and quantitatively validated.

A simple two dimensional finite difference model was developed and can be simply described as follows:

The MEB surface is assumed to have a uniform heat flux. Resistance to water flow is dominated by the lead liner port resistance and heat transfer from the water to the flask body is by natural convection. In addition to incorporating this convection process, conduction through the flask walls is represented and for the finned surface, convection and radiation used as external boundary conditions.

The model described above allowed temperatures within flasks of similar designs to be predicted to about $5^{\circ}C$ accuracy and the effect of changing heat input predicted with confidence.

Experiments Supporting Accident Assessments

Fire Test   The assessment of the thermal response of a flask to an enveloping fire is usually carried out by calculational modelling. The specification of the fire conditions to be applied is given in the IAEA transport regulations and in summary are, an $800^{\circ}C$ surrounding fire lasting for 30 minutes. An area of uncertainty in the modelling is the heat transfer from the flames to the finned surfaces of the flask. To resolve this uncertainty a series of fire tests has been carried out on a cylindrical finned surface. The tests were carried out at RAE Cardington and the test rig is shown in Figure 6.

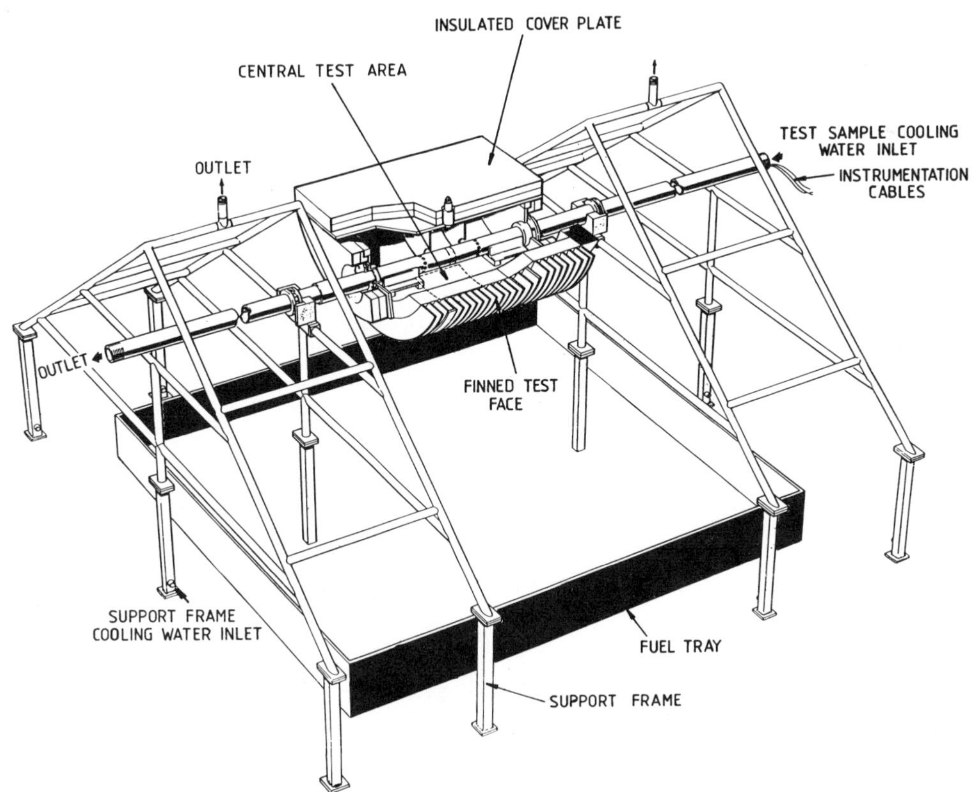

Fig. 6. Experimental assembly: open fire test rig.

The test section represents a 120°C segment of the finned surface of an Excellox 3B flask. It was 1m long, with 21 fins on a 50mm pitch,each fin being 140mm high. The back and sides of the test section were closed with insulated cover plates. Water was passed through the test section to model the large thermal capacity of the contents of a flask and to cool the instrumentation leads.

The test section was supported from a water cooled frame, above a tray partly filled with water. Fuel was floated on the water and replenished during the fire from a header tank. The test rig was placed inside a large hanger at RAE Cardington to isolate it from the distrubing effects of weather.

The instrumentation consisted of 78 thermocouples and 6 radiometers. Most of the thermocouples were concentrated around the central test area. They were positioned at varying depths within the flask body and fins and also within the inter fin cavities and in the area beyond the fins. The radiometers were intended to measure both radiation only and total heat fluxes. However, these proved unreliable and the thermocouple data were mainly used in the subsequent analysis of the tests. Readings from instruments were recorded every 15 seconds. Each test was recorded by video cinematography.

Eight fire tests were carried out at three different orientations of the test section; vertical, inclined at 45° and horizontal with the fins facing down in the fire. Kerosene was used as the fuel for most of the tests but Heptane was also used for comparison. During the test fires the flame cover was variable, producing a random variation of the heat flux into the test section. It was found that any variation of the heat flux with orientation or fuel type was small compared with this random variation between tests.

An analysis of the test data was carried out and the time dependent heat flux determined for each test and an envelope curve drawn, representing a fire test with constant flame cover. This was used as the target for comparison with theoretical fire heat transfer models in direct support of the analysis of th response of a flask to an enveloping fire.

<u>LOCA Thermal Tests</u> To determine the thermal performance of a water filled flask after a hypothetical total loss of cooling water, BNFL have conducted a series of experiments on a wet flask operated dry (Middleton et al, 1986).

The flask used for the tests was an Excellox 3A which had previously been used for transport of irradiated LWR fuel from Japan to the UK. It was decontaminated so the internals could be accessed for instrument- ation. An MEB designed for carrying 5 PWR elements was fitted into the flask. The cross section of the arrangement was similar to that shown in Figure 1 but with the MEB and flask cavity completely dry. Each fuel element was simulated by a 17 x 17 array of heater rods and dummy control rod guide tubes.

These were assembled into dummy grid plates to give an accurate representation of the PWR element geometry. The total rating of the five heater bundles was 52.8 KW, giving a significant reserve over the 30 KW flask maximum heat load.

Thermocouples were located on all the major flask components, including the heater bundles, the MEB, the lead liner and the flask body. The majority of the thermocouples were located at the centre plane of the

flask. The top heater bundle carried the most thermocouples since this
region was expected to reach the highest temperature.

Thermocouple signals, power supply and pressure measurement were
recorded on a floppy disc through a programmable data logger and
micro-computer. The flask was mounted horizontally and the lid and base
lagged to simulate the thermal insulation normally provided by the shock
absorbers.

To preserve the integrity of the rig, temperature limits were placed
on critical components and the rig was fitted with automatic trip devices
to prevent these from being exceeded.

An extensive series of transient tests was conducted to demonstrate
the thermal performance of the flask under various conditions. These
included; a range of heater powers from 5 to 40 KW; a range of air
pressures in the MEB, from 0.15 to 4.44 bar; air replaced by helium in
the MEB; the flask cavity filled with water.

For the ambient pressure air-filled tests the MEB was vented to
atmosphere. The reduced pressure and helium tests were carried out to
determine the significance of convection and gas conduction heat transfer
in the MEB. The water-filled annuli tests were to show the effect of
partial refill of the flask with water, (ie the MEB remaining
air-filled). The pressurised air tests were to investigate the reduction
in equilibrium temperatures which might be obtained by resealing and
pressurising the flasks.

From the test data it was concluded that:

(i)   The maximum equilibrium fuel temperature, for an air filled flask
       at normal ambient pressure and a heat load of 30 KW, is 665°C.
(ii)   Temperature changes produced by reducing the air pressure or
       changing the gas within the MEB were relatively small, indicating
       the dominance of radiation heat transfer.
(iii) Introducing water into the flask cavity results in a reduction in
       the maximum fuel temperature of about 110°C. This reduction is
       insensitive to power.
(iv)   Pressurising the MEB produces a significant reduction in fuel
       temperature due to the increased convective heat transfer at higher
       air densities.

CONCLUSIONS

In this paper a programme of experimental work in support of the LWR
irradiated fuel transport flask design and operation business undertaken
by BNFL/PNTL has been described. This programme addresses heat transfer
behaviour in a range of flask designs for both operational and accident
conditions and is aimed at both understanding and quantifying the various
heat transfer modes and underpinning and validating the numerical models
used in flask design, operation and safety assessment. Where possible
full scale experimental testing has been undertaken to eliminate the
difficulties associated with scaling, particularly in association with
such complex phenomena as fire heat transfer.

The experimental programme itself represents an investment of
approximately £1M and, in conjuction with the numerical modelling
expertise referred to, indicates the extent of BNFL/PTNL involvement in
transport design and assessment.

As a final point to note, the overall programme of experimental work is continuously being extended as flask designs evolve or as novel analysis methods are developed and require support and validation.

REFERENCES

Burgess, M.H., Spiller, G.T., Livesey, E. (1983). Thermal Trials on a Water Cooled LWR Flask. PATRAM 83 paper 19.
Drake, S., Sibson, P., Livesey, E. (1988). Experimental Investigation of Natural Convection Heat Transfer in Narrow Water Filled Horizontal Annuli. Second UK National Heat Transfer Conference paper C361/221.
Fry, C.J., Livesey, E., Spiller, G.T., MacGregor, B.R. (1983). Heat Transfer in a Dry Horizontal Spent LWR Fuel Assembly. PATRAM 83 paper 24.
Middleton, J.E., Livesey, E. (1986). Full Scale Experiment to Determine the Thermal Response Following Loss of Coolant Water from a Flask Containing Irradiated LWR Fuel. PATRAM 86, IAEA-SM-286/86P.

THE DEVELOPMENT STATUS OF CASTOR DUCTILE CAST IRON
TRANSPORT STORAGE AND FINAL DISPOSAL CASKS IN THE F.R.G.

K. Janberg, R. Huggenberg, D. Rittscher

Gesellschaft für Nuklear-Service (GNS)
Goethestrasse 88
4300 Essen 1
Federal Republic of Germany

ABSTRACT

If the name "CASTOR" is used, then it generally denotes the production process for a special development line of casks, as the word contains the initials of "CASTING, TRANSPORT and STORAGE".

The origin lies in the year 1975, when GNS designed the first transport casks based on ductile cast iron.

The reasons for the choice of this material were:

- suitability for serial production of significant numbers of casks
- homogeneous body without welds
- good shock resistance combined with high fracture toughness at low temperatures
- easy machining
- high thermal conductivity
- acceptable costs.

In the following presentation, the experimental background and the current development status are explained.

EXPERIMENTAL BACKGROUND AND BASIC QA PHILOSOPHY

The choice of ductile cast iron (DCI) for transport casks met with the scepticism of the specialists in the transport business, as the iron casting process was not known to permit the fabrication of a material with acceptable ductility.

In order to silence eventual doubts, from the beginning, GNS started in 1978 a test programme based on full scale casks. These were subjected not only to the IAEA drop tests at normal temperatures but also at -40°C. In order to simulate even the highest possible stress concentrations, some tests at low temperatures were even executed without shock absorbers.

Extensive test runs allowed bench-marking of simplified analytical tools, this now serves for extrapolation to other designs, in combination with the increasing application of finite element analysis.

The safety of a transport cask body is not only determined by its mechanical resistance, but also by the size of the maximum undetected defect or flaw, as this can lead to stress peaks under accidental loadings.

In extensive ultrasonic tests the sensitivity of this quality control step could be proven.

The detection limit for 400 mm thick castings is now in the range of a 6 mm equivalent diameter flat bottom hole. A test showed the considerable margin provided by the QA procedures, when an artificial laser-sharpened 3in. deep flaw, 24 in.long was placed at the highest tensile stress location of a cask which was subjected to a 9 m drop test at -40°C on an unyielding target, since the post-test examination showed no flaw growth. This fact is explained by the high fracture toughness of ductile cast iron which for a high-quality casting lies in the range of 2200 - 2700 $Nmm^{-3/2}$ at -40°C. These values could be determined from RCT specimens made from "core bars" extracted from the centre wall location. (RCT = Round Compact Tension).

The method of determining K_{IC} (static fracture toughness) via RCT specimens follows the ASTM standard E 813. The procedure is indirect, and verification of K_{IC} has therefore also been made on larger specimens in accordance with the ASTM standard E 399.

Valid K_{IC}-values could only be obtained with this method on reasonable specimen sizes of 4in. thickness at -95°C or less. So the comparison with the RCT measurements had to be made in this temperature range.

The comparison showed that RCT specimens of about 1in. diameter give only slightly different results from the large E 399 specimens. Thus it can be concluded that the ductile cast iron cask can resist very high stresses even in the presence of large flaws without failure.

<u>Industrial Applications for Transport and/or Storage</u>

Based on the broad experimental background already available in 1980, the first type B(U) licence was issued by the German authorities for CASTOR Ia in 1980.

The next development goal was obtaining the licence for long-term interim dry storage in DCI casks.

For this, additional questions had to be answered which were not raised by the transport regulations, i.e. resistance to permeation of tritium through cask walls, resistance to aircraft impact, long term behaviour of all components of the confinement.

The satisfactory evidence supplied to all these questions led to the Gorleben storage licence issued in September 1983.

Meanwhile several cask designs have seen industrial applications:

- CASTOR Ic   for 16 BWR fuel elements
- CASTOR Ib   for 4 PWR fuel elements

-    CASTOR Ic Diorit   for test reactor fuel elements at
                               EIR, Wurenlingen, Switzerland
-    CASTOR IIa/IIb   for 9 PWR fuel elements
-    CASTOR SPX   for 7 Super-Phenix fuel elements
-    CASTOR HAW 21   for 21 canisters of vitrified
                               high-level waste
-    CASTOR V/21   for the storage of 21 PWR fuel
                               elements, licensed in USA for
                               storage
-    CASTOR WWER 1000   for 12 Woronesch-type fuel elements
                               used for storage in the USSR
-    CASTOR IIa HAWC   for the transport of liquid high
                               level residues
-    CASTOR TVO   for the wet transport of 41 BWR
                               fuel elements for TVO, Finland
-    CASTOR BNFL 1   for the transport of spent fuel
                               from Germany to Sellafield
-    CASTOR KRB MOX   used for Mixed Oxide Fuel
                               transports from KRB power station (FRG)
                               to Sweden.

In total more than 50 transport and storage casks have been built so far, which constitutes a broad basis of experience for further applications.

<u>Alternate applications of DCI casks for final disposal.</u>
<u>Recycling of contaminated scrap metal</u>

The experience gained with the iron casting process led to the conviction that it presented a perfect means to produce high quality, low cost industrial packages (IP) and/or type B(U) casks for transport, interim storage, and final disposal of low, medium, and highly active reactor wastes.

This solution could become economically competitive to the standard waste conditioning technique of waste cementation and use of low-cost concrete over-packs only, if the safety criteria linked to the conditioning matrix (i.e. cement) could be transferred to the package.

This means that liquid wastes should possibly only be dried within the cask, thus assuring maximum volume reduction when compared with the standard technology of cementation. In order to obtain the authorities' approval for this novel approach, intensive development was done in order to assure performance in excess of the IAEA requirements:

-    excellent mechanical resistance by 800 m drop tests onto a concrete
     pad to simulate a drop in the final disposal mine
-    thermal resistance to $800^{\circ}C$ for 60 min.
-    long-term corrosion resistance to the internal effects of the waste
     material as well as to the brines in the geological repository.

The test results showed that DCI casks met all the requirements.

Another novelty - recently introduced in cooperation between the Siempelkamp Foundry and GNS - was the introduction of scrap metal recycling. For this purpose contaminated scrap metal from decommissioning was cut down to oven-size within specialized facilities and the directly melted inside an induction oven equipped with a high-efficiency filter system.

The scrap radioactivity content is either homogenized within the melt (Co, Fe, Mn) or transferred to the slag and the filter system (Cs, Sr, etc.).

The melted product is diluted with metallurgically "clean" material and used to fabricate DCI-waste casks and shielding material.

This procedure assures low dose-rates to the personnel in charge of waste handling and conditioning for final disposal.

It further reduces the amount of radioactivity released to the environment, as the recycled contaminated scrap returns to the power station and ends up within the final disposal mine.

The actual fabrication experience for these small casks is now of the order of 2000 units.

MOST RECENT DEVELOPMENTS

In 1993 the first batches of vitrified HAW in canister-form will be returned by COGEMA to the FRG.

In order to simplify the loading procedures, COGEMA has issued acceptance criteria for casks of foreign origin.

Here, GNS had the great advantage of its previous design, fabrication, and testing of the CASTOR HAW 21.

While this cask has to cope within a thermodynamically more demanding geometrical configuration, with 2,5 KW/HAW canister, the CASTOR HAW 20/CG for COGEMA only has to cope with a theoretical heat load of 2KW/canister. So the realization of such a cask and the assurance of a maximum glass temperature of $450^{\circ}C$ under long term storage conditions is a rather easy challenge.

For even lower heat loads it is possible to realize transport/ storage casks for 28 canisters.

The final design of these casks has been coompleted and the licence application is on its way.

It was parallel in design, but preceded in fabrication by CASTOR BNFL 1 which was built in 1987/88 for wet transport of spent fuel to the Sellafield reprocessing facility. The transports will begin in autumn 1988. In this cask design the flexibility of the casting technology becomes evident: due to the water loading there is a gas-filled top during horizontal transports which requires more shielding. So we have made the top part thicker than the bottom part.

In conclusion we wish to state that the technological choice for nodular cast iron may have been a daring one in 1976/1977, but now it is a well-established technology which has spread well beyond the limits of cask fabrication.

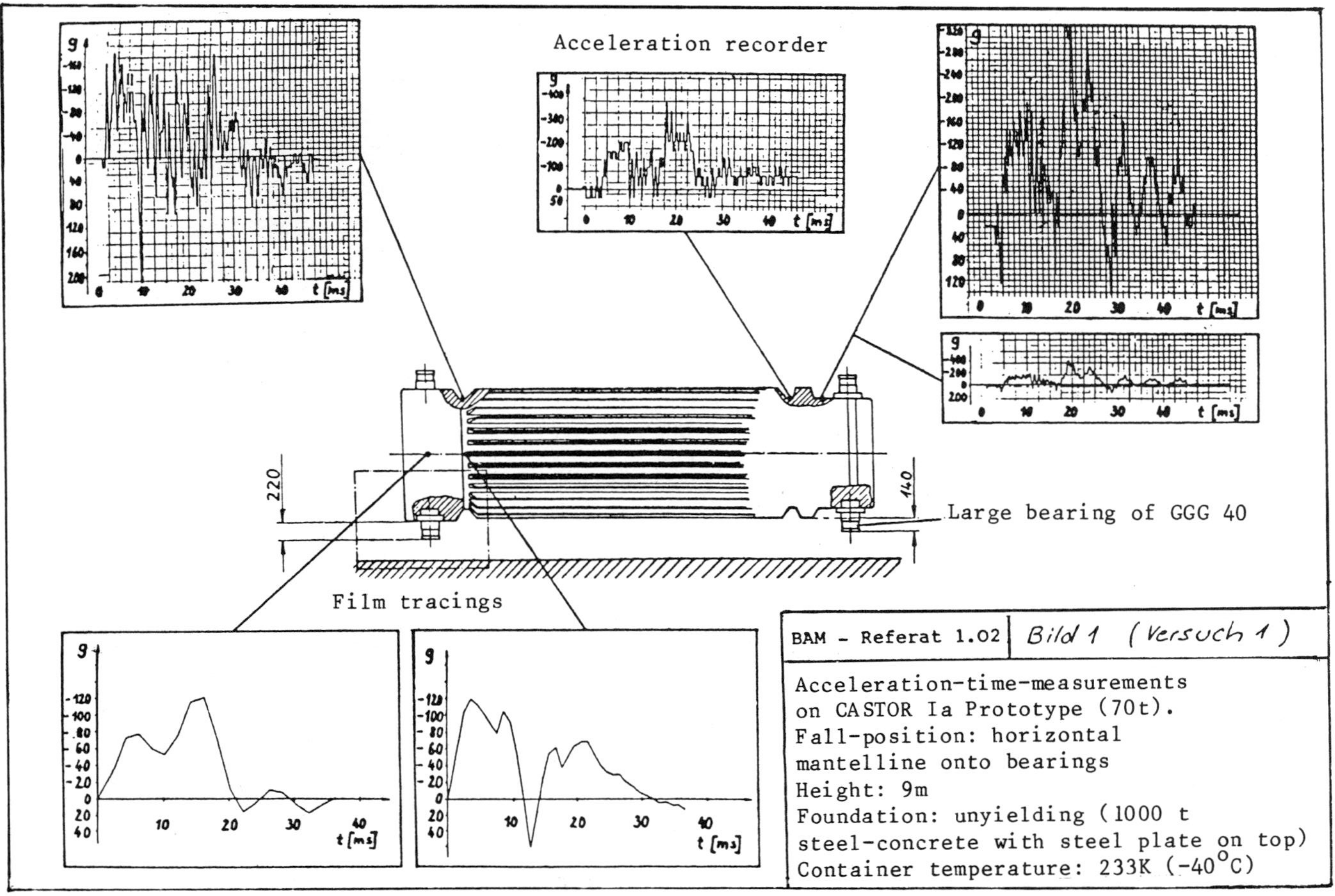

Fig. 1. Full scale drop-test of CASTOR Ia in 1978 at $-40°$C without shock absorbers.

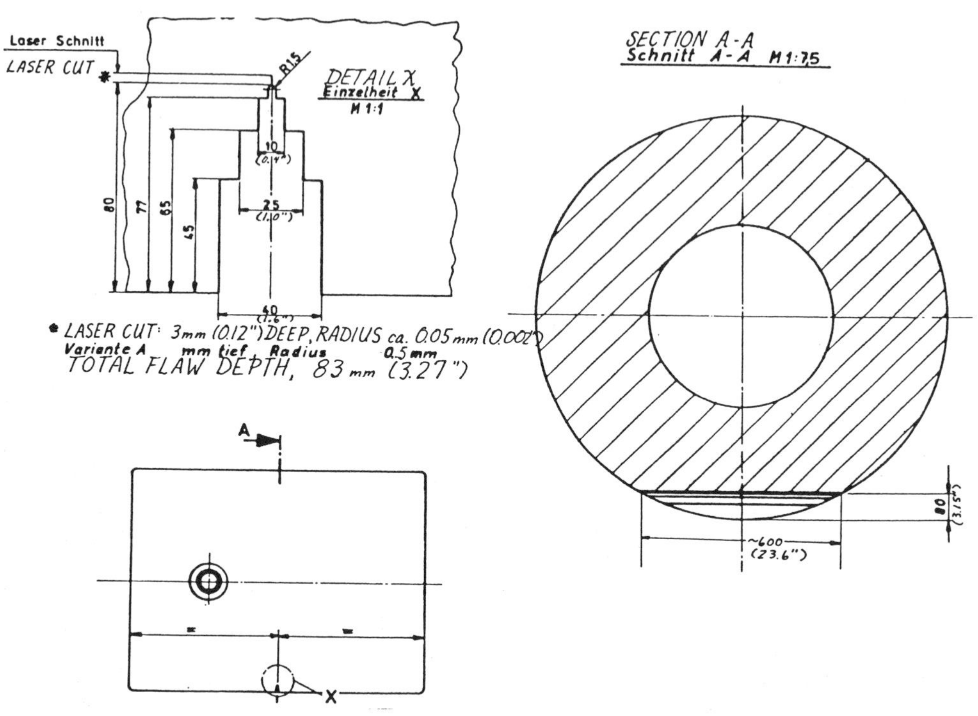

Fig. 2. CASTOR - MTR prepared for a drop-test
with artificial flaw 1/1986

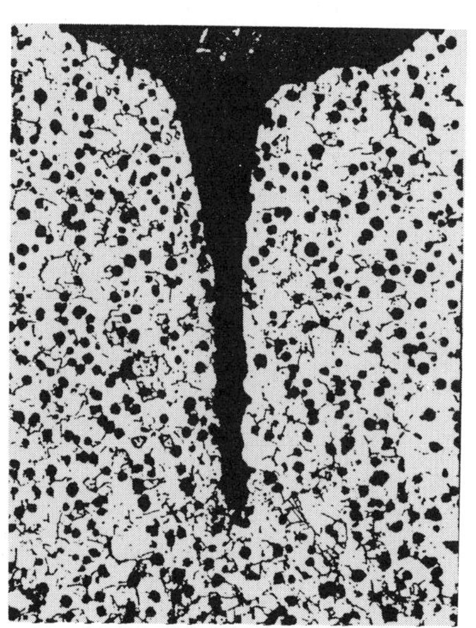

Magnification 25:1

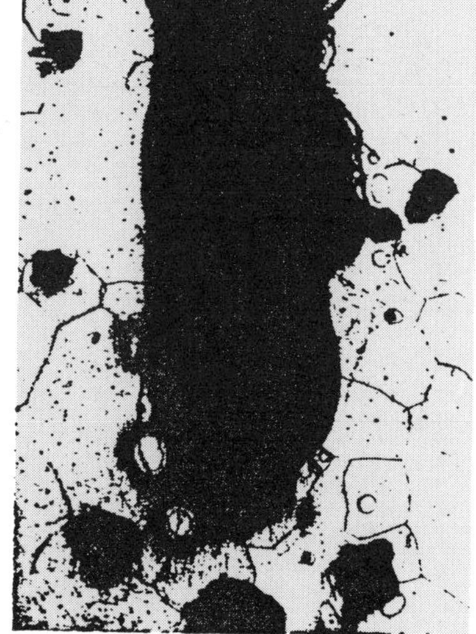

Magnification 200:1

Depth of cut 3,3mm: Root radius 0,06mm
Fig. 3. Cross-section of the flaw groove after the drop-test.
Result: No flaw growth.

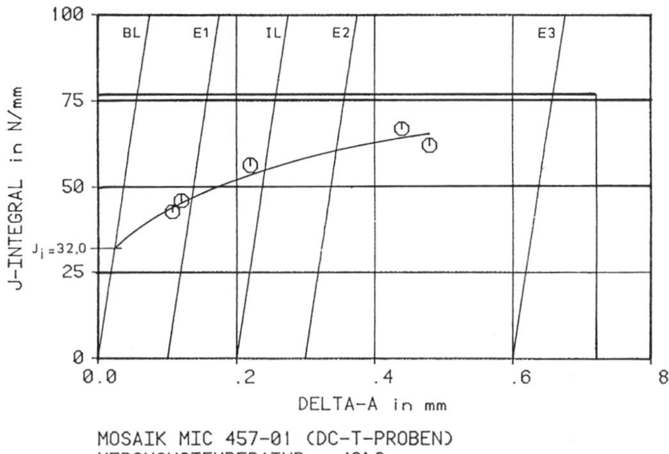

Fig. 4. Typical I-integral measured on RCT specimens

Fig. 5. CASTOR V/21 at Surrey

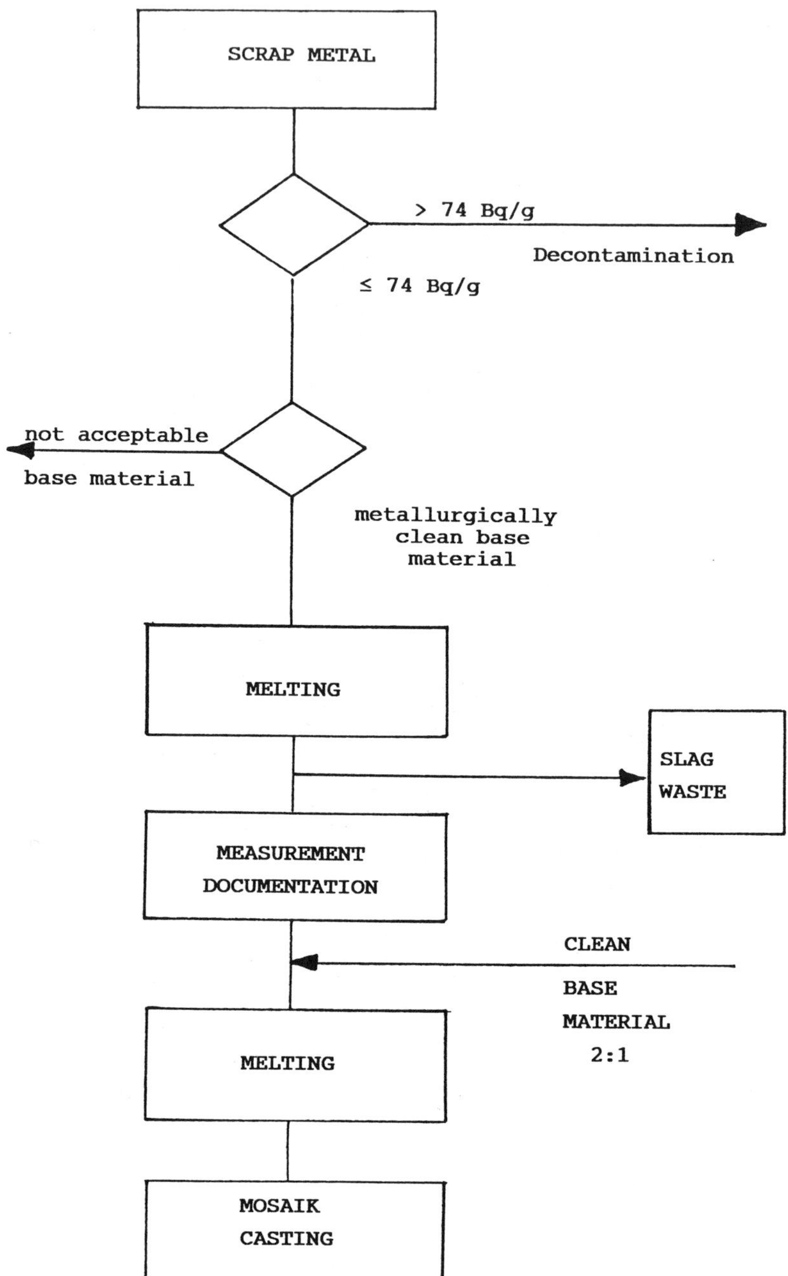

Fig. 6. Scrap metal recycling.

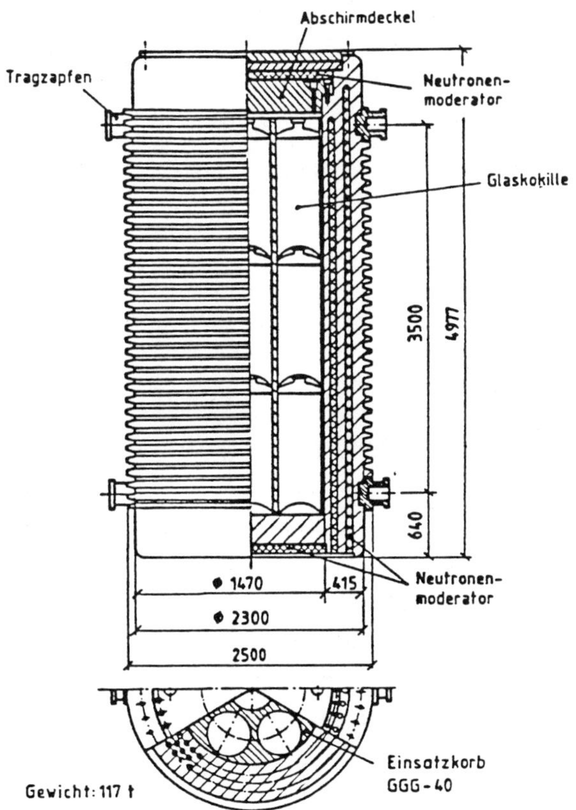

Fig. 7. CASTOR HAW 21.

THE IMPACT OF WASTE GAS GENERATION ON THE DESIGN OF A
SEALED PACKAGING FOR TRANSURANIC WASTE TRANSPORT

W.Bracey, M.Mason, D.Nolan

Transnuclear, Inc.
Hawthorne
New York
USA

ABSTRACT

Radioactive waste generates gas by several mechanisms, which can
result in hydrogen and pressure build-up in a sealed packaging. The
rectangular TRUPACT-II is used as a case study in the design and load
management options for type B transport packagings for radioactive waste
with significant gas generation potential. Data for existing transuranic
waste at Department of Energy sites are reviewed, and recommendations are
made for further research. Catalytic recombiners are recommended for
prevention of flammable gas buildup due to hydrogen or methane
generation. Waste containers that generate large quantities of gas may
need to be mixed in with those that generate little gas in order to be
shipped. A simple calculational method to aid this load management is
described.

INTRODUCTION

Gases may be generated and consumed in organic or water-containing
radioactive waste materials by a number of mechanisms. Radiolysis is the
absorption of ionizing radiation energy in the waste material, resulting
in the breaking of molecular bonds, e.g., the breakdown of water into
hydrogen and oxygen. Elevated temperatures can also cause molecular
breakdown (thermolysis) as well as vaporization and oxidation. Biological
organisms can consume cellulosic waste materials such as paper, cloth or
wood, releasing carbon dioxide, methane, or other gases, depending on the
environment and the organism. Corrosion of iron in an anaerobic
atmosphere with the presence of water will release hydrogen.

Gas generation has long been a concern associated with transuranic
(plutonium-bearing) wastes, currently stored at various national
laboratories and weapons production facilities in the U.S., and scheduled
to be transported to the Waste Isolation Pilot Plant (WIPP) beginning in
1988. Studies have been performed (Molecke, 1979) to determine the rate
at which various mechanisms generate gases, with a particular concern
being the build-up of hydrogen within the waste containers and the
formation of a flammable gas mixture. Strategies to limit the build-up of
hydrogen in storage, such as filtered venting of containers, have been
developed, and hydrogen transport through plastic bagging (Kudera, 1986)
has been analysed.

The first legal weight truck transport packaging designed specifically for contact handled (less than 200 mrem/hr at the container surface) transuranic wastes, the TRUPACT-I, consisted of a rectangar single containment with filtered vents. Because of the venting, the designers were able to demonstrate, using very conservative gas generation data, that neither unacceptable pressure nor flammable gas build-up occurred. When Transnuclear began work on a second generation design, TRUPACT-II, which included double containment and eliminated venting, the problem of gas generation became a major consideration.

THE NATURE OF THE PROBLEM

Two aspects of the gas generation problem are governed by separate regulations. Flammable gas build-up is governed by NRC IE Notice 84-72 (USNRC, 1984), which requires shippers to demonstrate that a mixture of gas which includes <u>both</u> oxygen and hydrogen in excess of 4% will neither be formed in the package, nor in the individual waste containers, within twice the expected period from closure through shipment and reopening of a package. Pressure build-up is governed by the definition of maximum normal operating pressure (MNOP) in 10CFR71 (USNRC, 1986): the pressure at the end of one year, without venting, under ambient conditions of constant 38°C still air and a 12-hours-on, 12-hours-off cycle of solar load. The MNOP must be considered in combination with other regulatory tests, including reduction of the external atmospheric pressure to 241 mbr (3.5 psi) abs. The packaging must be also be tested to 1.5 times the MNOP.

In analyses performed for the Department of Energy (DOE) (Sanchez, 1986, and Smith, 1986) when the idea of a non-vented TRUPACT was first considered, it was determined that a flammable gas mixture would be generated before the design pressure of the TRUPACT-I, 345 mbar (5 psi) gage was exceeded. On this basis it was believed that attention needed to be focused on preventing the flammable gas build-up and that pressure build-up would not be a major design consideration. However, these analyses did not take into accouont the one year period nor the ambient thermal conditions outlined above. Transnuclear's initial calculations of the MNOP found that, using the 10140 Ci payload of the TRUPACT-I, and varying such parameters as the void volume in the waste and the radiolysis constant, the pressure could range from 3.6 to 13.6 atm gage, indicating that pressure build-up was indeed a serious problem.

THE SEARCH FOR A SOLUTION

Various solutions to the problems of gas generation had previously been considered in connection with the transport of waste associated with the Three Mile Island cleanup (Henrie, 1986). The TRUPACT problem was unique however, in several respects, especially in the inherently low pressure capacity of a rectangular packaging, the lightweight design, (possible because of negligible shielding requirements) and the widely varying nature of the waste materials. The options investigated in the course of the TRUPACT-II design included reducing the design radioactivity capacity, increasing the design pressure, and using recombiners or getters to remove generated gases.

Reduction of the design radioactivity results in a reduction of both radiolytically generated gas and internal temperature. Discussions with the waste generating sites indicated that a design capacity of 2000 alpha curies was the minimum acceptable. Efforts were also made to provide a low absorptivity/high emissivity surface coating on the packaging in order to reduce the temperature of the contents. Although 2000 alpha curies at 5.5 MeV/alpha is equivalent only to about 100 watts, analysis indicated that under the regulatory thermal environment, the cavity wall of the TRUPACT-II would reach an equilibrium temperature of 60-70°C after 2 weeks, principally because the TRUPACT-II was surrounded by a thick layer of shock-absorbing foam which also acted as an insulator.

Within the legal truck weight and size limits, the introduction of double containment had already created a formidable problem of maintaining the TRUPACT-I capacity of thirty six 210 liter drums and 6985 kg. This problem was compounded by the need to increase the design pressure. To achieve this increase while minimizing the reduction of capacity, the containment walls were redesigned using honeycomb panels. The final design achieved a capacity of 36 drums and 5895 kg with an MNOP of 896 mbar (13 psi) gage.

RECOMBINERS AND GETTERS

Catalyst beds using platinum or palladium on ceramic pellets have been used to recombine hydrogen with oxygen both in transport packages and in vented storage drums (Henrie, 1986). These applications had demonstrated the effectiveness of the recombiners in preventing the creation of a flammable gas mixture, but prior to TRUPACT-II they had not been used specifically for pressure reduction.

The most obvious limitation on the effectiveness of recombiners is the availability of oxygen. Although radiolysis of water generates stochiometric quantities of hydrogen and oxygen, the radiolysis of organic materials generates free radicals which can scavenge oxygen. Thermal and biological effects can also scavenge oxygen. The concern about the build-up of hydrogen after oxygen depletion has been clearly addressed by the NRC: hydrogen without oxygen does not constitute a flammable mixture and is acceptable under IE Notice 84-72. However, it does present an operational concern, and may require measurement of hydrogen concentration when the containments are vented prior to opening. If the concentration presents a hazard, the containments can be purged.

For the calculation of pressure credit for recombination, the scavenging of oxygen by the waste must be accounted for and quantified. The leakage characteristics of the waste containers and the oxygen available in the containers are additional variables to be considered.

Carbon monoxide and water vapor can reduce the effectiveness of the recombiners. Sampling of waste drums has not revealed significant quantities of carbon monoxide, and at slow rates of carbon monoxide generation, the recombiners will catalyse the formation of carbon dioxide from the monoxide. In addition, Atomic Energy of Canada has developed recombiners whose effectiveness is not reduced by carbon monoxide or water vapor.

A concern has been raised that if all the oxygen is scavenged or recombined, and hydrogen continues to be generated, pyrophoric platinum and palladium hydrides will form on the recombiners (Courtney, 1977). In addition to the fact that this phenomenon has not been seen in the experience with recombiners in transuranic waste storage, there are several reasons why it is not in principal a serious problem. Platinum hydride is very unstable, and will not be formed under normal or accidental transport circumstances. Palladium hydride will form and is stable at a minimum hydrogen partial pressure, which increases with temperature. The necessary hydrogen partial pressure could develop in an

oxygen-depleted transport package, especially at low temperatures.
However, palladium hydride is not pyrophoric in the sense that although
it will release the absorbed hydrogen and generate heat when exposed to
air, it will not spontaneously burn in bulk as some metal hydrides do.

For pressure control, getters which would scavenge carbon dioxide
and hydrogen in the absence of oxygen were investigated. The only carbon
dioxide getter found was caustic soda, which is not desirable for use in
a transport packaging. Organic hydrogen getters have been developed
(Courtney, 1977), but the fact that they become saturated requires that
they be replaced periodically, perhaps after each shipment because of the
uncertainty of the degree of saturation. The quantity required to absorb
hyrogen for one year (recall the definition of MNOP) also appears to be
excessive.

The final TRUPACT-II design incorporated about 2 kg of recombiners
in the cavity to achieve three goals:

a)   To provide pressure control;
b)   To prevent the build-up of a flammable gas mixture in the TRUPACT-II
     cavity;
c)   To provide a gradient in hydrogen and oxygen concentration between
     the TRUPACT-II cavity and those waste containers with filtered
     venting, to prevent the build-up of a flammable gas mixture within
     the containers themselves.

WASTE CHARACTERIZATION

Simultaneously with Transnuclear's design efforts, the DOE
commissioned a review of existing gas generation data for transuranic
waste to determine gas generation rates which could serve as an envelope
for all such wastes. The resulting report (JIO,1986) recommended the
values shown in Table 1. Using these values resulted in an MNOP in excess
of the 896 mbar TRUPACT-II design pressure, as shown in Table 2. Although
there is a consensus among waste generators that the recommended gas
generation data are highly conservative for an "average" waste container,
there is no guarantee that any given shipment will not consist entirely
of above-average containers. As it turns out, the gas generation envelope
is controlled by "combustible" wastes, especially cellulosics.
Non-organic wastes, even those containing water, have far lower rates of
radiolysis, and are not subject to thermal or biological breakdown. It
was clear, therefore, that some method of load management would be
necessary in order to mix low and high gas generating waste containers in
a given shipment so that the design MNOP for the TRUPACT-II would not be
exceeded.

To facilitate the load management, Transnuclear developed a simple
calculation of MNOP, to be used by shippers for qualification of each
shipment. Incorporating the results of thermal analysis under the
regulatory thermal environment in a formula for waste temperature versus
alpha curie content, the calculation determines the following: waste gas
generation for one year from radiolysis, thermal, and biological effects;
gas depletion by recombination and oxygen scavenging; pressure based on
the ideal gas law; and the partial pressure of saturated water vapor. The
recombiners and waste are assumed to have equal access to available
oxygen, i.e., no barrier is assumed between the waste and the TRUPACT-II
cavity. Calculations performed independently using an explicit model for
diffusion across drum vents verified that this simplification did not
significantly affect the final pressure. The required input includes the
alpha radioactivity of the wastes, the mass of wastes, the waste volume,

Table 1. Envelope Gas Generation Data for Transuranic Waste.

Radiolytic:	G(Carbon Dioxide) = 1.0
	G(Hydrogen) = 1.9
	G(Oxygen) = 1.0

Radiolytic: G(Carbon Dioxide) = 1.0
               G(Hydrogen)   = 1.9
               G(Oxygen)   = 1.0

Thermal: Carbon Dioxide
               60°C   : 2.6 moles/drum/year
               70°C   : 5.2 moles/drum/year
               100°C : 44   moles/drum/year

Biological: Carbon Dioxide; below 70°C only
               aerobic : 4.2 moles/drum/year
               anaerobic: 2.6 moles/drum/year

Corrosion: negligible

Average void volume
in waste containers: 50%

Water available in waste: Sufficient to saturate volume of TRUPACT-II
                            at normal and accident temperatures.

Notes:

1. The radiolytic gas generation factor "G" is in units of molecules of gas generated per 100 eV of energy absorbed in the waste. For transuranic waste, alpha energy is between 5 and 5.5 MeV, all of which is absorbed in the waste.

2. Thermal and biological activity are presumed proportional to the mass of waste. The rates given are for a drum containing 51.4kg of waste.

Table 2. Components of Pressure in TRUPACT-II Using Envelope Data.

Contributor	Pressure	
	mbar	(psi)
Ideal gas heating	139	(2.01)
Water vapur (saturated)	199	(2.89)
Radiolysis	793	(11.50)
Thermal	259	(3.75)
Biological	422	(6.12)
Total	1812	(26.27)
Recombination	−430	(−6.23)
Oxygen Scavenging	−104	(−1.51)
Net	1278	(18.5)

Basis: 36 waste drums, 2000 alpha Ci, 20° at loading
       60°C equilibrium waste and gas temperature

Note: Only two-thirds of the oxygen in the drums was assumed available for scavenging by the waste. The remainder, and the oxygen in the TRUPACT-II cavity, was assumed available for recombination. This assumption is not necessarily conservative.

and the waste material. Waste generators are already required to provide this data, except for waste volume, for WIPP emplacement. Waste volume could be estimated from weight and density, or from radiographic viewing of the waste.

To test the load management idea, Transnuclear assumed that organic wastes could be characterized by the envelope gas generation data. Non-organic wastes whcih include water (sludges, concretes), and dry non-organic (glass, ceramic, metal), which usually include some plastic bagging, were assumed to generate hydrogen by radiolysis at about 0.6 molecules per 100 eV (G(hydrogen) = 0.6). The calculated pressures shown in Table 3 indicate that it would be feasible to mix cellulosic wastes with other wastes and remain below the TRUPACT-II design pressure with a reasonably high alpha radioactivity capacity.

IMPROVED GAS GENERATION DATA

In order to implement the load management program, it was necessary to develop a minimum number of waste categories, and corresponding gas generation data, to encompass all wastes. To facilitate the DOE's development of these data, Transnuclear recommended further research to resolve a number of questions that had been raised in the course of reviewing the existing literature.

Probably the most significant question which relates to all three gas generation mechanisms is the relation of oxygen consumption to carbon dioxide and monoxide formation. Reported rates of radiolytic carbon oxide formation are close to the reported rates of oxygen depletion for combustible wastes, and field sampling (Clements, 1985) often finds drums with significant carbon dioxide to be highly depleted in oxygen. Similarly, aerobic bacterial activity and thermal breakdown (oxidation) would be expected to couple oxygen consumption with carbon oxide formation. The implication is that much of the generation of carbon oxides does not contribute to a net pressure increase, and to assume so may be unnecessarily conservative. In addition, a quantification of oxygen consumption is necessary if recombiners are to be given credit for pressure reduction.

Molecke reports a temperature dependence of radiolysis. It needs to be determined if this is true, or whether it is merely a superposition of thermal and radiolytic effects. Additionally, thermolysis may be enhanced by radiolysis, and the two processes may be inseparable.

Table 3. Load Management Pressure Calculations for TRUPACT-II.

TRUPACT-II Load	Pressure	
	mbar	(psi)
1. 36 drums organic waste, 2100 kg @ 700 kg/m^3, 25 Ci	896	(13)
2. 36 drums non-organic waste, 5000 kg @ 2000 kg/m^3, 4200 Ci	896	(13)
3. 12 drums organic, 500 kg, 500 Ci 24 drums non-organics, 5000 kg, 1500 Ci 758		(11)

It may be possible to take credit for the integrated dose dependence of radiolysis due to depletion of the matrix surrounding the radiation sources; some data indicate that radiolysis is reduced significantly in only a period of months for drums containing as little as one curie.

Molecke concludes that biological breakdown offers the greatest potential for gas generation. Waste generators have insisted that in real wastes, as opposed to simulated wastes in laboratory conditions, biological activity is negligible; the source of carbon dioxide in field-examined waste is, however, a subject of dispute (Clements, 1985). It also seems anomalous that the data reported by Molecke attribute higher rates of carbon dioxide formation to anaerobic than to aerobic bacteria, without the expected methane formation.

The data on thermal breakdown are sparse and conflicting. For example, the product of polyethylene breakdown is reported as 93% oxygen. The relative importance and time dependence of vaporization, oxidation, and thermolysis need to be investigated at temperatures up to $200^{o}C$ in order to include regulatory thermal accident conditions.

Because the thermal and bacterial rates are so uncertain, because of the large impact they have on the ability to transport combustible wastes, and because they are easily investigated experimentally (relative to radiolysis), Transnuclear recommend further experimental work in these two areas.

EPILOGUE

In May of 1987, DOE decided that it would seek an NRC licence for its contact handled transuranic waste transport packaging, and based on this decision it abandoned the rectangular design in favour of a cylindrical one. Subsequently, the NRC indicated that it was willing to accept a time period shorter than one year for the calculation of MNOP.

Together, these changes substantially expand the proportion of existing waste which can be transported under an envelope of gas generation values, without load management, as initially attempted with the rectangular TRUPACT-II. High radioactivity combustible waste remains the one item difficult to ship using the current data. One might reasonably argue that it is unnecessary to design to questionable thermal and bacterial data, given the fact that packagings for shipment of non-transuranic, i.e., commercial, radwaste have been approved without considering these effects. In any case, research on these effects should continue, and should it turn out that some wastes can indeed not be carried under bounding values for gas generation, the calculational method developed by Transnuclear will remain a useful tool in the load management necessary to carry such wastes.

CONCLUSION

Pressure build-up due to the generation of gases by radioactive waste is a significant consideration for the design of sealed waste transport packagings, especially those that do not require a heavy structure for shielding purposes. The use of catalytic recombiners is an economical, passive, and reliable method to prevent the build-up of flammable gas mixtures due to the production of hydrogen or methane. Oxygen consumption by the waste must be accounted for if credit is to be taken for pressure reduction due to recombination. Thermal and biological gas generation data for cellulosic and other organic wastes are sparse and questionable, and need to be further investigated. If further research verifies that certain categories of waste, especially high alpha radioactivity organic waste, generate gas in far greater quantities than other categories, load management will be necessary to ship these wastes.

ACKNOWLEDGEMENTS

This work was performed by Transnuclear under contract number 59-WDR-9113-SD to Westinghouse Electric Corp. The authors acknowledge the assistance of Larry Sanchez of Sandia and Craig Smith of SAIC in providing supplementary calculations and data during the course of this work.

REFERENCES

Clements, T.L., Jr., and Kudera, D.E. (Sept. 1985). TRU Waste Sampling Program: Vol I - Waste Characterization. EG&G Idaho, EGG-WM-6503.

Courtney, R.L. and Harrah, L.A. (1977). Organic Hydrogen Getters, Part 1. Journal of Materials Science, 12, 175-186.

Henrie, J.O., Quinn, G.J., Greenborg, J. (August 1986). Hydrogen Control in the Handling, Shipping and Storage of Wet Radioactive Waste. Rockwell International (Hanford Op.) RHO-WM-EV-9 REV 1 P.

Joint Integration Office (Nov. 1986). TRUPACT-II Criteria for Gas Generation. DOE-JIO-016.

Kudera, D.E., et al (Sept. 1986). Evaluation of the Aspiration Rate of Hydrogen from a Waste Drum. EG&G Idaho, EGG-WM-7228.

Molecke, M.A. (1979). Gas Generation from Transuranic Waste Degradation: Data Summary Interpretation. Sandia National Laboratory, SAND79-1245.

Sanchez, L.C. and Sandoval, R.P. (25 July 1986). Unpublished Study. Sandia National Laboratories.

Smith, C.F. and Miller, D.E. (15 July 1986). Preliminary Evaluations for an Unvented TRUPACT. Scientific Applications International Corp, SAIC-86/1768, draft.

U.S. Nuclear Regulatory Commission Office of Inspection and Enforcement (10 Sept 1984). Clarification of Conditions for Waste Shipments Subject to Hydrogen Gas Generation. I.E. Information Notice No. 84-72.

U.S. Nuclear Regulatory Commission (Jan 1986). Packaging and Transportation of Radioactive Material. Code of Federal Regulations, Title 10, Part 71.

THE DEVELOPMENT AND TESTING OF A CONTAINER
FOR THE TRANSPORT OF DECOMMISSIONING WASTES

J.R. Wakefield

United Kingdom Atomic Energy Authority
Windscale Laboratory
Seascale
Cumbria CA20 1PF

ABSTRACT

The Windscale Advanced Gas-cooled Reactor was shut down in 1981 after 18 years operation at high load factor. The UK has a commitment to dismantle all its nuclear power plants as they cease operation and some of the early stations are approaching that stage. The decision was therefore taken to decommission the WAGR to the ultimate Stage 3 (complete clearance of the site) as a demonstration project and as a development tool for the techniques involved.

One of the key elements in the process is the packaging, transport and disposal of several hundred tonnes of irradiated steel, graphite and concrete. This paper describes the design, development and testing of a suitable container for this purpose.

A reinforced concrete box of some 2.2 metres side weighing up to 50 tonnes has been produced. The strength and radiation shielding properties of concrete have been combined in the structure of the container. The wall thickness of 240 mm in normal concrete is sufficient to shield against most of the radioactive burden. For the more highly irradiated pieces the same design in high density concrete is equally suitable.

The containers have been rigorously tested and developed to meet the IAEA Transport Regulations with a considerable margin in hand. They are suitable for transport by road, rail, and sea, within the public domain.

INTRODUCTION

The Windscale Advanced Gas-cooled Reactor (WAGR) was constructed in 1961 as a 33 MW prototype for the current designs of UK commercial nuclear power plants. After 18 years operation at high load factor, the reactor was shut down and the decision was made to decommission the plant to Stage 3. This involved the discharging of the fuel, the removal of the turbine, generator and other non-nuclear plant, and ultimately the dismantling and complete disposal of the reactor and peripheral equipment.

The main objectives of this work are to develop equipment and processes and to demonstrate their role in a nuclear decommissioning task. Much emphasis is placed on the recording of information and its application to the dismantling of present and future generations of nuclear plant.

This has involved the UKAEA in an extensive programme of development work, one of the key items of which is the production of an acceptable transport container for Intermediate Level Waste arisings. When the project was initiated the sea disposal route was available, and this method was, and still is, the preferred technical solution. However, following Government reappraisal of sea disposal, it has proved necessary to determine alternative routes involving disposal on land.

The iterative design and development programmes have produced a reinforced concrete container weighing up to 50 tonnes in which a payload of 18 tonnes of steel or graphite is held within a cement matrix.

THE WAGR WASTE INVENTORY

When the reactor was shut down in 1981, the total induced activity was estimated to be:

$$0.4 \times 10^{15} Bq \ (2 \times 10^{5} \ Ci)$$

This has now (1988) been reduced to:

$$2.2 \times 10^{15} Bq \ (6 \times 10^{4} Ci) \ \text{by natural decay.}$$

The total weight of the plant is approximately 16000 te of which only 1,900 te is active. This comprises 1,100 te of Low Level Waste (LLW) for disposal to the Drigg Site (Drigg is a Low Level disposal site for solid waste operated by British Nuclear Fuels plc on behalf of the UK Government) and 800 te of Intermediate Level Waste (ILW) for which the transport containers (ibid) are required. (The current definition of ILW in the UK (RWMAC,1984) is solid wastes whose activity level is greater than:

$$12 \times 10^{9} \ Bq/te \ (0.32 \ Ci/te) \ beta/gamma \ \& \ 4 \times 10^{9} \ Bq/te \ (0.1 \ Ci/te) \ alpha$$

but is sufficiently low not to generate significant heat.) After a further 43 years, ie 50 years after shut-down, these proportions will have changed to 1600 te LLW and 300 te ILW. At 100 years, the values will be 1700 te and 200 te respectively, demonstrating that a wait of 50 years before packaging would save some 500 te of ILW. However, this saving would be off-set by additional maintenance and supervisory costs associated with the plant.

It was known from operational experience that contamination levels were low and would be swamped by activation levels. An exception is the four heat exchangers which were not irradiated but contain relatively high levels of contamination. Contrary to the procedures adopted for the rest of the plant, these components will be decontaminated and cut up by hand. It is expected that the resulting scrap material will fall into the LLW category.

An inventory of the principal activated components of the WAGR system is given in Table 1. The quoted activities refer to levels at 7 years following the shut-down.

The complete inventory has been placed in material classifications in Table 2. Here it will be seen that 64% of the total activity is attributable to stainless steel which constitutes only 5% of the active mass.

Table 1

Component	Material	Mass (te)	Activity Bq (Ci)	Sp.Activity Bq/te (Ci/te)
Fuel stringer components	Stainless steel	30	6.6×10^{13} (1,800)	2.2×10^{12} (60)
Test loop tubes	Stainless steel	4	6.6×10^{13} (1,800)	1.7×10^{14} (4600)
Core Restraint system	Mild/stainless steel	17	9.1×10^{14} (24,700)	5.4×10^{13} (1460)
Neutron shield supports	Stainless steel	5	2×10^{14} (5,500)	4.1×10^{13} (1100)
Thermal shields plates	Mild steel	186	2.2×10^{14} (6,000)	1.2×10^{12} (32)
Pressure vessel	Mild steel	210	1.8×10^{13} (480)	8.5×10^{10} (2)
Core support plates	Mild steel	18	1.4×10^{14} (3,800)	7.8×10^{14} (210)
Core support bearings	Stainless steel	2	2.4×10^{14} (6,600)	1.2×10^{14} (3240)
Moderator	Graphite	110	3.7×10^{13} (1,000)	2.5×10^{11} (7)
Reflector	Graphite	100	2.8×10^{13} (750)	2.8×10^{11} (7)
Bioshield	Concrete	750	2.8×10^{12} (75)	3.7×10^{9} (0.1)

Table 2

Material	Mass (Te)	Activity Bq (Ci)	Sp.activity Bq/te (Ci/te)	Mass %	Activity %
Mild steel	732	7.0×10^{14} (19,000)	9.66×10^{11} (26)	39	32.8
Stainless steel	89	1.38×10^{15} (37,250)	1.55×10^{13} (418)	5	64.1
Graphite	283	6.5×10^{13} (1,750)	2.3×10^{11} (7)	15	3.0
Concrete	750	2.8×10^{12} (100)	3.7×10^{9} (0.1)	41	0.1

Further analysis shows that there will be a marked decrease in activity levels over the next 50 years due to the decay of iron 55 and cobalt 60. The residual activity will arise mainly from nickel 63 in the stainless steels.

REQUIREMENTS OF THE TRANSPORT REGULATIONS

The waste producer is responsible for convincing the regulatory bodies that his proposals for packaging, transport and disposal meet the published criteria. In the UK, the organisations involved are Government Departments of Transport and the Environment and the Ministry of Agriculture, Fisheries and Food. In additon, the Nuclear Industries Radioactive Waste Executive (NIREX), established in 1982, is responsible for the final disposal of nuclear waste. It follows that NIREX (UK) Ltd must also approve the design and performance of the waste container.

When the WAGR Decommissioning Study commenced in 1975, culminating in a proposal submitted in 1979, dumping Intermediate Level radioactive material at sea was the only available disposal option. This technically correct method was governed by the London Convention on Sea Dumping (Report, 1983-85) and the principal disposal requirements were:

(1)   the curie content of the package must not exceed 3.7×10^{12} Bq/te (100 Ci/te) averaged over 1000 te;

(2)   the package must not release any radioactive material before it reaches the sea bed at a nominal depth of 4000 m.

The transport requirements, covering overland movement from the originator's site to the disposal site or docks were defined in the IAEA Regulations for the Transport of Radioactive Material (Safety Series 6, 1973) (IAEA, 1985). This document has since been revised in the 1985 Edition with further explanatory notes in Safety Series 37, 1987, 3rd Edition.

The design concept of a monolithic concrete block with integral external shielding is based upon the premise that the materials to be transported conform to the Low Specific Activity Category III (LSA III) This is defined as solids in which:

(3)   the radioactive material is essentially uniformly distributed in a solid compact binding agent;

(4)   the radioactive material is relatively insoluble or is intrinsically contained in a relatively insoluble matrix;

(5)   the estimated average specific activity of the solid, including the matrix medium but excluding any shielding material, does not exceed a specified value which is determined by the radionuclide content of the solid.

The IAEA Safety Series 6 Regulations introduce the following additional requirements:

(6)   the quantity of LSA material in a single package shall be restricted such that the external radiation level at 3 m from the unshielded material does not exceed 10 mSv/hr;

(7)   the maximum radiation level at any point on any external surface of the vehicle shall not exceed 2 mSv/hr;

96

(8)   the maximum radiation level at any point 2 m from the vertical
        planes projected from the outer edges of the carrying vehicle shall
        not exceed 0.1 mSv/hr;

(9)   the package must be capable of withstanding an impact on its
        weakest feature following a free fall of 0.3 m without impairment
        of containment or shielding;

(10)   the package shall withstand water spray for 1 hour duration;

(11)   the fully laden package shall be capable of stacking five high;

(12)   package shall not be penetrated by a 6 kg rod dropped through 1m.

It was readily concluded by inspection that a design strong enough
to withstand more than 400 atmospheres pressure as required in (2) above
would also meet requirements (10), (11) and (12). Thus the development
programme was required to demonstrate the drop test requirement (9).
Also since the radioactive content of the entire WAGR inventory is:

$2x10^{15}$ Bq (58,000 Ci) at 7 years leading to $1x10^{12}$ Bq/te (30 Ci/te)

when averaged over 2,000 te, it follows that requirements (6), (7) and
(8) present the key radiological factors. These can be met by systematic
packing and higher density walls where necessary, since the shield
thickness was derived from average specific activity values. Some
exceptionally high level components such as the control rods and
stainless steel loop tubes will be accommodated in high density boxes
within an additional high density matrix.

A further requirement is a statement of the contents of each box.
Individual scrap components will be weighed and gamma scanned as part of
the packaging operation. With knowledge of the geometry and isotopic
content of each piece, these measurements will be converted to
radioactivity content in Bq.

INITIAL DESIGN OF THE CONTAINER (see Appendix and Figure 1)

Material of Construction

Consideration was given to a range of materials for the manufacture
of the box walls. This material is required to provide the functions of
strength, containment and shielding. In the case of sea dumping it is
also necessary to retain these features during the descent of the
container the sea bed. Similarly, for land burial, the containment should
remain effective preferably for several centuries.

In both cases, the transport requirements described earlier must be
met. When these considerations are coupled to the economics of supply,
the optimum choice of material becomes reinforced concrete. This is
cheap, plentiful, strong, and is available in a range of densities. Its
only disadvantage is the considerable bulk required for shielding
purposes.

Shape and Size

The majority of the contents of the WAGR disposal boxes will be of
rectangular form; for example, graphite blocks, thermal shield plates and
pressure vessel sections. For efficient packing, the internal shape of
the box should conform to this pattern, and, in order to provide a

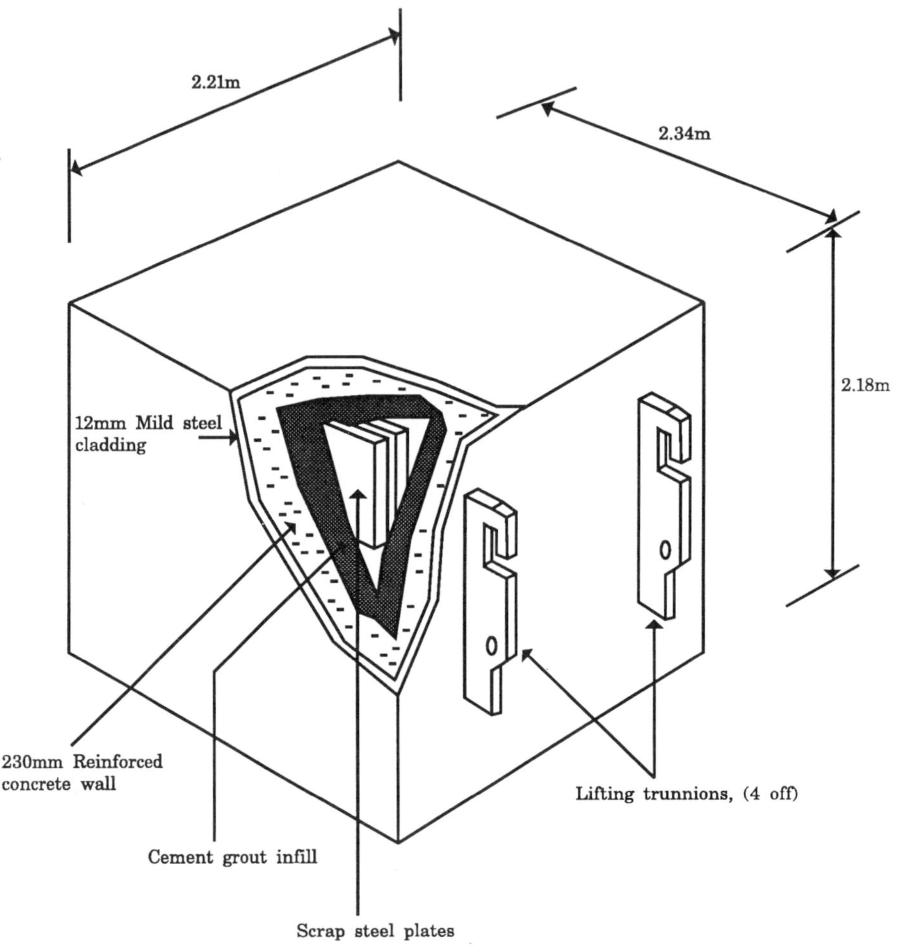

Fig. 1. Steel clad container loaded with pressure
vessel plates, weight 50Te

regular shield thickness, so should the external profile. Calculations,
based upon a payload of the average activity levels and Ordinary Portland
Cement (OPC) concrete, indicated a shield wall thickness of 230 mm in
order to meet the transport regulations. For given external dimensions
this represents a compromise between the high cost of high density
concrete, the reduced payload resulting from additional standard density
concrete and the expected frequency with which material will exceed the
assumed average activity levels. For uniformity in handling, it was
decided to keep the box dimensions constant and improve the shielding
values by using supershot or high density concrete for the walls. It has
been predicted that out of a total of 140 boxes needed for the complete
WAGR inventory, 40 will require to be of the supershot variety. These
will each cost approximately £2000 more than the standard boxes.

On the assumption that these containers will at some time travel by
rail, the external dimensions were determined by the British Railways
loading gauge W5 in conjunction with a currently available eight axle
truck. Steel lifting trunnions were cast into the concrete walls and
terminated in reaction plates which were flush with the inner surface.
Gaps were left in the internal webs for the reinforcing bars to pass
through. The four trunnions per box required a rectangular frame to
locate the lifting links, any pair of which must be capable of supporting

the full weight of the loaded box (50 te). This frame was costly and heavy and was added to the required lifting capacity of the hoists.

Outer Cladding

The initial design of container was fitted with an all-welded outer mild steel skin of 12 mm thickness. This was used as a former to cast the wall/shield concrete but was not invoked in the strength or shielding design calculations. It offered a smooth steel surface which could be repainted as required and was considered to be desirable for presentation to the general public. Following impact test on the boxes it was shown to be effective in retaining shattered concrete, thus preserving the shielding, and also lent some strength to the whole structure.

INTERMEDIATE DESIGN OF THE CONTAINER (see Fig 6)

This was introduced in order to reduce the cost of the container by eliminating the outer steel sheath. This saved the prime cost of the cladding and the maintenance that would be required during long-term surface storage. The lifting trunnions were retained and a half-thickness wall joint for the cast lid was introduced.

PRESENT DESIGN OF THE CONTAINER (see Appendix and Fig 2)

The present design is identical in external dImensions to the earlier designs. For the reasons given above the outer steel cladding has been replaced by an equal thickness of concrete. In addition the massive steel lifting trunnions have been eliminated. These were costly (£1200 per box), heavy (600 kg) and presented a low resistance path for water into and out of the box at the interface between the steel and concrete. It has been decided to use polypropylene straps external to the box for general lifting and a fork-lift truck for on-site lifting. The straps will be taylored to the box dimensions and will be removed and used again for a number of boxes. Furthermore, they can be stored for re-use over a long period. Minor features will be cast into the containers to guide and retain the straps and means will be provided (eg base pads) to enable the straps to be threaded under each box prior to the lift.

In order to improve the wall/lid joint, a steel collar 600 mm high and 8 mm thick has been incorporated at the neck of the container. This provides a full width joint, is formed to provide a rounded upper edge and also acts as shuttering for the lid concrete. It is hoped that this will reduce the spillage of remotely poured concrete in the Waste Packaging Building.

Infill Material

The regulations require that the radioactive material be encapsulated within a solid matrix. This has several functions: it fixes any surface contamination that may be present; it fills voids and spaces within the container and thus resists implosion; it secures the loose components and reduces internal damage of the container walls on impact; it contributes to the monolithic structure and thus adds considerably to the crushing and impact strength; it provides a measure of shielding which is addditive to that of the external walls; it reduces the movement of water towards and away from the radioactive contents. Some of these features are present by design but some may be regarded as enhancements to the concept. A further example of the latter categtory is the beneficial influence of the infill material on the chemistry of the steel waste under the repository conditions.

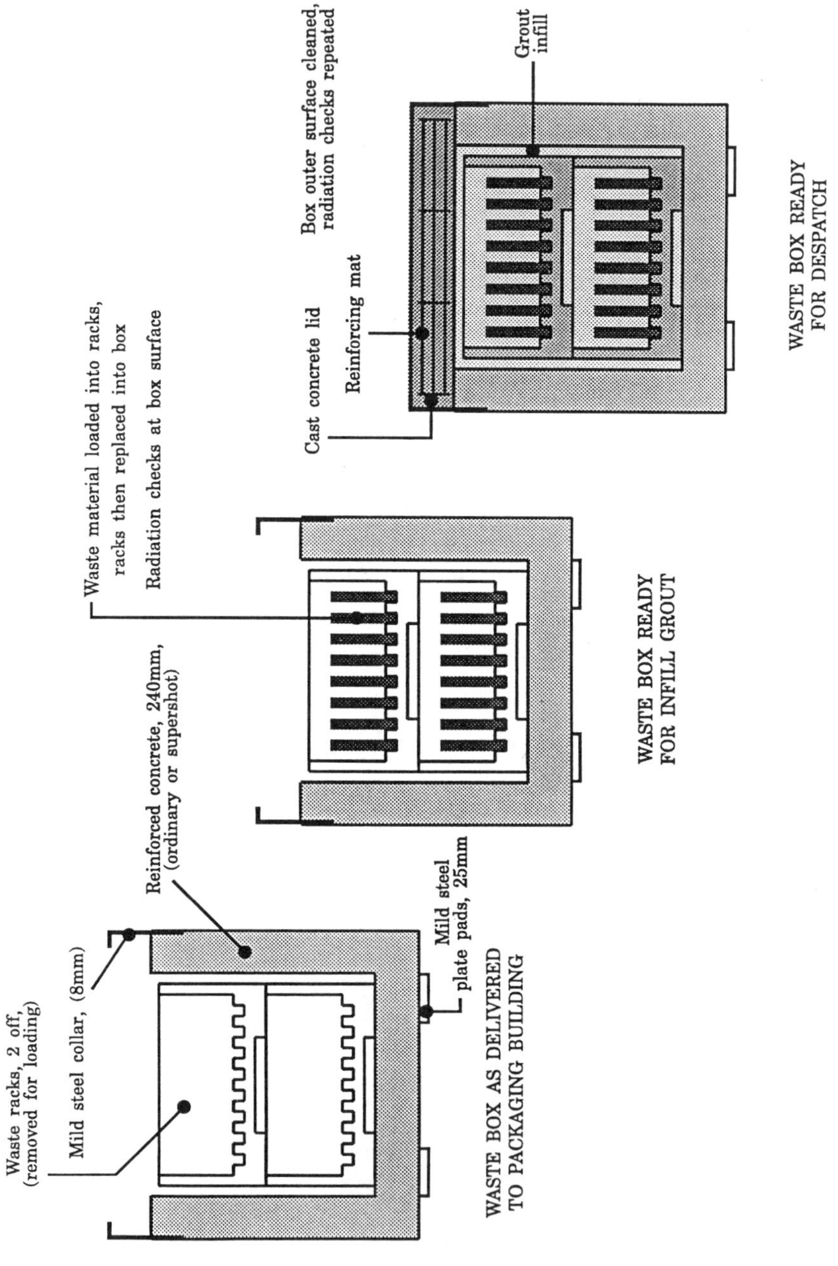

Waste racks, 2 off,
(removed for loading)
Mild steel collar, (8mm)
Reinforced concrete, 240mm,
(ordinary or supershot)
Mild steel
plate pads, 25mm
WASTE BOX AS DELIVERED
TO PACKAGING BUILDING
Waste material loaded into racks,
racks then replaced into box
Radiation checks at box surface
WASTE BOX READY
FOR INFILL GROUT
Cast concrete lid
Reinforcing mat
Box outer surface cleaned,
radiation checks repeated
Grout
infill
WASTE BOX READY
FOR DESPATCH

Fig. 2

A cement grout formulation giving low porosity and high strength was chosen. In order to reduce heat generation some cementitious material was replaced by a proprietary blast furnace slag (Cemsave). It was originally intended to vibrate the complete box and contents whilst pouring the grout but it was found during trials that the available energy input was insufficient to achieve this. A decision was taken to abandon any attempt at vibration and to rely upon the free flowing properties of the grout for complete filling and consolidation. To aid this, a proprietary superplasticer (Sikament) was added.

DEVELOPMENT PROGRAMME

This commenced in 1983 and has comprised a series of experimental tests and mathematical analyses. The former investigated the flowability of infill grout and continued with drop tests on model and full scale loaded containers. The analyses attempted to model the performance of the containers in relation to damage arising from impact and crushing forces. Additionally an assessment of the risks associated with transporting the containers from Windscale to a nominal disposal site some 300 miles away was made. A further study estimated the behaviour of the boxes in an underground repository in relation to the migration of the active species to the site boundary.

Flow and Drop Test on 1/3 Scale Models

An array of tubes and components was fixed within a framework contained in wooden shuttering. Cement grout to two different specifications both with and without vibratory assistance was added. After curing, the blocks were sectioned and the penetration of the grout measured. Based upon the results of this examination a further four blocks were manufactured and drop tested at 1m free fall. No significant damage was produced.

Test on Steelclad Full Sized Boxes

Four boxes were manufactured to the design of Figure 1 and filled with steel plates loaded into racks. Cement grout to the stated specification was added followed by the lid reinforcing mat and lid concrete. Finally a 12 mm steel plate was welded into place to complete the outer steel cladding. Instrumentation, consisting of accelerometers mounted on the internal plates and the box walls, was provided. In addition four vertical steel tubes were cast into each corner of the box for insertion of a gamma source. This was used to check damage to the shielding of the box by taking measurements on the box walls before and after each test.

Each box weighed 40 te and was dropped onto a 600 te concrete block capped by a 150 mm steel plate. The regulatory drop height of 0.3 m was increased to 0.65 m to allow for scaling up of the box weight to 50 te and to include an ample contingency to cover statistical variations in materials and construction. Four different dropping attitudes were used in order to explore and determine that which was most damaging. These were base, corner, edge and lifting trunnion. In each case, the damage was slight and no detectable change in the shielding properties took place. A photograph of the damage to the box following a drop onto a corner (steel cladding removed) is shown in Figure 3.

All four boxes were then retested from 5m free fall onto similar undamaged features. The box struck the target at 10 ms^{-1} and the resulting deceleration for the base drop was 555 g. This naturally

Fig. 3. Damage to corner of container following 0.65m drop, (steel
cladding cut away to reveal concrete).
Peak deceleration 20g., impact velocity 3.6ms^{-1}

caused most damage but the only visible sign was splitting of the
vertical welds and the loss of a cupful of concrete dust at each corner.
In all cases the shielding was unimpaired and there was no loss of
contents. Sections of the steel clad were removed and cores were taken
form the wall concrete to confirm the absence of internal damage. The
appearance of the box following the edge drop is shown in Fig 4. Had
this unlikely incident occurred whilst the box was in transit, the damage
could easily have been repaired and the box and its contents continued on
the journey.

Fig. 4. Damage to edge of container following 5m drop.
Peak deceleration 92g., impact velocity 9.9ms^{-1}

<u>Test on Moulded Concrete Boxes</u>

Two boxes were manufactured to the intermediate design described above (Fig 6). These were loaded with steel plate and filled with grout without vibratory assistance. The instrumentation and experimental programme was similar to that above.

The result of the 0.65 m drop onto an edge is shown in Figure 5 where it will be seen that superficial damage resulted. The shielding checks demonstrated that no changes had taken place. This same box was then dropped through 5 m onto its base. In this case, the outer steel casing was not present to contain the broken concrete. The deceleration experienced was 714 g and the damage pattern is shown in Figure 6. The increase in shock loading over the steel clad box is thought to be due to the absence of the 25 mm steel pads welded to the base which in the former case acted as rudimentary shock absorbers. Cores were taken through the wall and infill and revealed that the cracks did not penetrate the matrix. It was concluded that this monolithic structure which had been overtested by a factor of 12 was very strong and would provide a satisfactory ILW container. Once again, the damage could have been repaired and the box which is shown lifted intact from the test pad could have continued its journey.

A further conclusion is that a test in any other attitude would have produced a much lower deceleration value, shock loading and hence damage. For example it has been calculated that a drop from <u>15m</u> onto a corner, which is a more probable attitude, would produce a peak value of only $(120^{+}_{-}60)$. (Jowett & Roberts, 1984).

MATHEMATICAL ANALYSES

A number of studies have been carried out by contractors in support of the test programmes. These are listed below with brief details and conclusions.

Fig. 5. Damage to concrete container following 0.65m drop onto edge.
Peak deceleration 46g. Impact velocity 3.6ms^{-1}

Fig. 6.   Damage to concrete container following 5m drop onto base.
View on underside, (dark patches are rainwater stains).
Peak deceleration 714g. impact velocity 9.9ms^{-1}

Impact Behaviour of Edge and Corner Contact (Steel Clad Box) (Jowett & Roberts, 1984).

This study modelled a concrete cube with steel cladding for both edge and corner drop attitudes. The two drop heights selected for the experimental trials were used and close agreement was obtained in terms of both peak deceleration and displacement. The results are summarised in Tables 3 and 4.

Table 3. Comparison of Calculated and Experimental Peak Decelerations: Steel Clad Box

Drop Height (m)	Corner Impact Deceleration(g)		Edge impact deceleration(g)	
	Calculation	Experiment	Calculation	Experiment
0.65	16.5	20	49	46
5.0	58	56	138	92

Table 4. Comparison of Calculated and Experimental Displacements on 'Knock-back': Steel Clad Box

Drop Height (m)	Corner impact displacement (m)	
	Calculation	Experiment
0.65	0.09	0.07
5.0	0.22	0.19

Table 5. Calculated and Experimental Values of Peak Deceleration:
Unclad Box. Values in Brackets refer to Clad Box.

Drop Height (m)	Corner Impact Deceleration(g)		Edge Impact Deceleration(g)	
	Calculation	Experiment	Calculation	Experiment
0.3	4		18	
0.65		(20)		36 (46)
2.0	16		44	
5.0	30	(56)	66	(92)
9.0	46		86	

<u>Impact Behaviour of Edge and Corner Contact: Unclad Box</u>
(Ove Arup, 1985)

This analysis modelled the unclad box and the study went on to
suggest ways of improving the strength of the wall-to-lid joint. A range
of predicted values to 9m were obtained. Results are given in Table 5.

Direct comparisons will not be possible until the remaining unclad
box is drop tested. However, inspection of the Table indicates reasonable
agreement when account is taken of the added stiffness of the steel clad
construction.

<u>Hydrostatic Loading on Immersed Box</u> (Taywood Engineering Ltd., 1987)

The thermal shield plates within the reactor are bolted into sets of
three plates with 1/16" gaps between internal faces. It is expected that
they will be loaded into the disposal boxes without unbolting and it was
considered unlikely that the inter-plate gap would be filled completely
with cement grout. This study calculated the effect that this voidage
would produce on the box structure when subjected to a hydrostatic
compression force of 40 MNm^{-2} (5800 psi). It was found that the
stiffness of the shield plates and separators prevented significant
closure of the included voids; stress levels would not cause yield of the
plates or separators; the infill grout would be overstressed in some
locations and stress redistribution can be expected to relieve maximum
stresses; levels of stress in the box wall nowhere approached a general
failure state. Some cracking on the inside faces may occur but no
through-thickness cracks were were predicted.

<u>Transport Risk Analysis</u> (Appleton & Poulter, 1986)

Road and rail transport to a nominal land disposal site (Elstow) and
rail transport to Barrow Docks for steel clad boxes were considered. The
damage to the containers was extrapolated from the drop test results and
the accident scenarios covered collission, falls from a bridge or crane
and fire. The probabilities of all incidents involving structural damage
to the container were shown to be very low (less than 2.5×10^{-5} for
the total WAGR inventory over 3 years) and in most cases the exposure of
the contents was shown to be impossible. The probability of an accident
leading to radiological consequences was even lower and included the case
of a graphite fire (less than 1.2×10^{-8}). Thus the overall transport
risks are very small and a similar result is expected for the moulded
concrete design.

QUALITY ASSURANCE

The manufacture of the empty containers will be subjected to vigorous Quality Assurance procedures and these will also apply to the filling and final closure operations. In addition an inventory of the contents of each container in terms of composition, weight and radioactivity levels will be prepared. This will involve monitoring and on-line sampling of the materials together with archiving specimens of the more active materials for future analysis.

CONCLUSION

The WAGR project is the first large-scale reactor decommissioning task in the UK. It has therefore been necessary to initiate designs and procedures which are acceptable to the regulatory authorities. This paper describes the complex development route needed to establish the concept of an economical self-shielded strong container which is suitable for the transport of decommissioning waste within the public domain.

ACKNOWLEDGEMENT

The work reported here was undertaken as part of the WAGR Decommissioning Project for which the UKAEA is pleased to acknowledge the support and financial contribution of the CEGB acting on behalf of the two Generating Boards.

REFERENCES

Appleton, P.R., Poulter, D.R., UKAEA, SRD, Preliminary Risk Assessment for the Transport of Decomissioning Waste in the UK. IAEA Symposium (PATRAM 86) Davos. June 1986.

International Atomic Energy Agency, Regulations for the Safe Transport of Radioactive materials. IAEA Safety Series No. 6, 1973 Edition. Superseded by current 1985 Edition.

Jowett, J., Roberts,T.M., UKAEA, SRD Internal Document 1984.

Over Arup and Partners Ltd., London, UKAEA Contract 1985.

Radioactive Waste Management Advisory Committee, 5th Annual Report (June 1984)

Reports of London Convention on Sea Dumping of Radioactive Waste (1983-1985.

Taywood Engineering Ltd., Southall, UKAEA Contract 1987.

SPECIFICATION OF STEEL CLAD RC WASTE CONTAINER

Outside dimensions   2.34 m x 2.21 m x 2.18 m high
Inside dimensions   1.82 m x 1.69 m x 1.66 m high
Wall thickness (concrete)   230 mm
Steel clad thickness   12 mm
Overall weight   50 te (max.)

Concrete for base, walls and lid (standard shielding)

30 Grade 30 N mm^{-2}
OPC 360 Kg m^{-3} (min.)
Max. free water/cement ratio   0.5
Min. density 2350 Kg m^{-3}
Min. cover to reinforcement   30 mm

Concrete for base, walls and lid (extra shielding)

30 Grade 30 N nn^{-2}
OPC 360 Kg m^{-3}
plus supershot in ratio 7: 1 to cement by weight
plus superplasticiser (Sikament) 1.2% by weight
Max. free water/cement ratio   0.4
Min. density 4000 Kg m^{-3}
Min. cover to reinforcement   30 mm

Steel work

Cladding   - mild steel to BS 4360 Grade 43A
Lifting trunnions - mild steel to BS 4360 Grade 50D
Reinforcement   - h.t. steel to BS 449 and BS 4461
Welding to BS5135 with electrodes to BS 639

Infill Grout

30 Grade 30 N mm^{-2}
OPC   200 parts by weight
Cemsave (BFS)   200 "   "
Sand   650 "   "
8 mm gravel   1100 "   "
superplasticisers (Sikament) 1.25% by weight
Min. free water/cement ratio   0.5
Workability flowing 55 cm min. spread, ie self levelling

SPECIFICATION OF INTERMEDIATE MOULDED RC WASTE CONTAINER

Outside dimensions   2.34 m x 2.21 m x 2.18 m high
Inside dimensions   1.82 m x 1 73 m x 1.70 m high
Wall thickness   240 mm
Overall weight   50 te (max.)

Concrete for base, walls and lid (standard shielding)

As for steel clad container

Concrete for base, walls and lid (extra shielding)

As for steel clad container

Steel work

No external cladding, but wire mesh added to retain concrete
Reinforcing scheduled revised - spacing reduced
Minimum cover increased to 40 mm
Lifting trunnions - as above but reduced in weight

Infill grout

As for steel clad container

SPECIFICATION OF CURRENT MOULDED RC WASTE CONTAINER

Outside dimensions	2.34 m x 2.21 m x 2.18 m high
Inside dimensions	1.82 m x 1.73 m x 1.70 m high
Wall thickness	240 mm
Steel collar thickness	8 mm
Overall weight	50 te

Concrete for base, walls and lid (standard shielding)

As for steel clad container

Concrete for base, walls and lid (extra shielding)

As for steel clad container
Lid composition may differ if trials indicate that supershot concrete
cannot be placed and levelled without vibratory assistance

Steel work

No external cladding, no wire mesh for retention
No lifting trunnions
Reinforcing as intermediate design of reinforced concrete container
Mild steel collar, 600 mm high keyed into concrete, surrounds neck of box

Infill grout

As for steel clad container

DISCUSSION FOLLOWING SESSION 1: Papers 5 - 8:

MR.J.R.HOGSTON, SCIENCE AND ENGINEERING RESEARCH COUNCIL, RUTHERFORD
LABORATORIES for Paper 1:5
Have any experiments been carried out on flasks without fins, subjected
to the fire test?

MR.E.LIVESEY, BRITISH NUCLEAR FUELS PLC, RISLEY:
I am the joint author with Dr.Swindlehurst. We have done some fire tests
on packages without fins, fairly recently. The actual data from them
hasn't been published yet and are subject to some discussion within BNFL.
Therefore, I cannot give technical data on heat fluxes. The tests were
carried out on thin skinned packages with a wood infill.

MR.G.WILCOCK, ROLLS-ROYCE & ASSOCIATES LIMITED, DERBY for Paper 1:6
(1) Can you provide any indication of the UK competent authority's view
on the use of ductile cast iron for transport flasks?
(2) Can you give some idea of overall timescales for this project from
start to finish in September 1988?

DR.H.GEISER, GNS, Essen:
(1) This is a difficult question to answer, I don't know the regulatory
position of the UK. The history of ductile iron development is a long
one, and we have carried out more than 100 drop tests and also a good
number of fire tests. The ductile iron quality has improved over the last
ten years and the last product which I showed, the CASTOR 10, will have
even better qualities than the ones produced three or four years ago.
This material is no longer cast iron material. It has similar properties
to steel and, talking about facture toughness for example, especially at
low temperatures (-40°), it has even better fracture toughness than
reactor vessel steel. This has to be borne in mind if you are going to
test to -40°. We went the experimental way instead of the analytical
way and we did tremendous over tests, including shock absorbers, and the
most difficult, aircraft crash impact tests, where we had to shoot a one
metric tonne missile with 300 kilojoule energy to the lid area and the
cask really survived. It seems to me that it is a long discussion process
with the licensing authorities to show step-by-step how such casks can
work.
(2) In the current project, the CASTOR BNFL which we are just now
commissioning, one cask which will be assembled and we are going to have
shipments at the end of this year. Thus with the recent development
CASTOR BNFL, where we are going to ship fuel from Germany to Sellafield,
this will be done during this year. For the CASTOR 10, designated for 10
year old fuel, we have made a prototype, with no fins, the manufacturing
time for such a cask is in the range of 6 to 8 months and - for example -
the cask we delivered to Virginia Power, 10 units, we had a delivery each
month. So with the CASTOR 10 we expect our first delivery in the
beginning or middle of 1989.

MR.D.BLACKMAN, UK DEPARTMENT OF TRANSPORT:
I feel that the question about the UK competent authority attitude to
CASTOR flasks was probably directed at me, rather than Dr.Geiser. I think
I must say at this stage that we have not yet received an application for
a licence for a CASTOR flask at the competent authority, but
nevertheless, my quality assurance staff have been to GNS and have
watched the casting of these flasks, they have also been to a seminar in
Germany and we are monitoring the situation very closely. We know that
the application will be coming and we know that it is a very revolution-
ary concept as far as the UK is concerned, but my assessment staff are
gathering all the possible information that they can so that when the
application comes we will be ready for it.

DR.G.C.McCREESH, BRITISH NUCLEAR FUELS PLC, RISLEY for Paper 1:6
How was the area of highest stress determined? What stress level was
obtained?

DR.H.GEISER:
There were several drop tests with an artificial floor, and this was an
analytical pre-drop test analysis, done by finite element analysis to
find out the worst case for drop orientation and at the location of the
highest stress, calculated in a pre drop-test analysis. You then
introduce your floor and even if you don't know exactly where it is you
can introduce more than one artificial floor.

As a design basis from a regulatory point of view in Germany, just to
give you an idea, the stresses during a drop should not be higher than
half of the allowable during accident conditions. That means the stresses
are in the range of 100 newtons per square mm. The minimum tensile stress
of ductile iron is 240.

MR.M.S.T.PRICE, AEE WINFRITH for Paper 1:8
I have some comments to make which are relevant to Dr.Wakefield's paper.
These comments derive from an ongoing study of large transport containers
which is partly funded by the Commission of the European Communities.
This study is being carried out by AEE Winfrith, Windscale Laboratory and
Ove Arup and Partners with SRD Culcheth as consultant.

For large reinforced self-shielded concrete boxes there has been a long
debate about whether to clad the boxes with carbon steel. We have looked
at the environmental impact of decommissioning waste in disposal and find
it to be extremely small. Thus cladding is not needed on technical
grounds.

We have also looked at whether the shielding around such boxes should be
integral with the boxes (and disposed with them) or whether it should be
returnable. This has been studied in relation to the volumes of waste
disposal. We find that these volumes are not markedly different and so
the incentive to move to the technically more difficult returnable
shielded option is small.

The weight of a reinforced concrete box, which we would prefer, would be
slightly greater than the WAGR box so as to take maximum advantage of the
carrying capacity of railway wagons. We can note that British Rail has
recently slightly increased the loading gauge in the W6 freight gauge,
details of which were issued in March 1988.

An interim progress report on this work has been issued as AEEW-M 2507,
Issue 2, and it will be reported at the Institution of Mechanical
Engineers Decommissioning Conference to be held in London on 11 and 12
October 1988.

DR.J.R.WAKEFIELD, UKAEA:
I am entirely in agreement with what Mr.McCreesh has said. I would like
to ask the audience a question if I may. Following on from what was just
said, we have in fact dispensed with the enormous lifting trunnions which
penetrated the wall of the box, and represented a way in for water, and a
way out for other material. We now have a completely smooth box. We
intend to lift them using polypropylene straps. Around the site we shall
use a 60 tonne fork lift truck. But because we have dispensed with the
trunnions - they cost something like £1200 per set, we now have no
features for tying it down on to the railway waggon. I have a letter on
file from British Railways (I must admit it is a very old letter) in
which they stated quite categorically that they prefer the box not to be
tied down. It must not slide, but shouldn't be tied down because if there
is a serious derailment they would prefer that the box tumbled off and
didn't drag the rest of the train with it. I do know that there is an AEA
code of practice which specifies tie-down and explains how to do it, so
at the moment we are quite uncertain whether we need to provide any tie
down features. I would welcome a comment on that.

DR.R.VAUGHAN, CROFT ASSOCIATES LIMITED:
As a member of the UKAEA Transport Container Standardization Committee
(TCSC) which wrote and revises the Code of Practice AECP 1006 I can say
that I do not see any conflict as the AECP allows tie-down to be effected
by chocking as well as by positive mechanical tie down. The code of
practice indicates methods of providing chocks and how high they have to
be designed to avoid a package coming off under normal conditions of
transport - but not under accident conditions. So I think the code of
practice copes with the problem.

MR.D.J.RIBBANS, ONTARIO HYDRO for Paper 1:8
I see that your testing to date has been done at full scale, but you
mentioned that you were going to test on half scale models, do you have
any information on the scaling of concrete?

DR.H.R.WAKEFIELD:
I would like to pass that question to others who are more expert in this
than I. We do have information on the scaling of concrete, in fact it is
a well documented science and we have certainly made use of it, but we
have tended to contract this work out, both to our own SRD organization
and also to Ove Arup, to name but one. So if any of those gentlemen would
care to answer this I would be grateful.

DR.J.MILES, OVE ARUP AND PARTNERS:
You mentioned that you had spoken to us about that subject. I believe our
response was that one has to be really circumspect about it. We have done
a certain amount of work in that direction which is not conclusive. I
would agree it is a fairly well documented science but not particularly
in this area, so I would have some reservations though it is very much
dependent on the information you want to get from the series of tests
that you are carrying out. Because, of course a test can be constructed
to reach any one of a number of different ends so, not being quite sure
what your intended programme is I wouldn't like to say more than that.

MR.A.NIELSEN, UKAEA WINFRITH:
Although not directly involved in the work on these concrete boxes, we
have over the past decade been doing quite a lot of scaling experiments
at Winfrith for impact response of concrete structures, and the general
comments I would make are that whilst scaling is perfectly acceptable for
gross damage indicating what the extent of the cracking is on the box and
such like, the details of the damage and widths of cracks etc. are very
much more difficult to scale. In fact you do not replicate, without a

great deal of trouble, the detail damage to a concrete structure under dynamic loadings.

DR.H.GEISER, GNS for Paper 1:8
How unyielding was your unyielding foundation?

DR.J.R.WAKEFIELD:
It was _very_ unyielding. It was in fact a 600 tonne concrete block, let into solid rock, capped by, I think, a four inch steel plate, which in turn was grouted in and bolted down. So I think as far as unyielding targets go this was a pretty good one. This is situated at Winfrith – Mr.Price may have something to add to that – but that is certainly a description of the target.

MR.G.OLDROYD, OVE ARUP AND PARTNERS for Paper 1:6
Could give us some values, if you have them available, for dynamic fracture toughness data on your ductile iron – I mean K1D values as opposed to K1C?

MR.H.GEISER:
Yes there are papers dealing with the properties of ductile iron, though not my paper. Usually it is a monolithic casting. We refer to c.100 tonnes of caste material and we take core bars from the middle of the wall, which is the area of the lowest fracture toughness to be expected, and the fracture toughness data we took was for static and dynamic, and the licensing limit for the production series of CASTOR 5 are 45 ksi. Several papers were given at at PATRAM, which might be of help.

MR.C.R.H.STRANG, STRACHAN AND HENSHAW, BRISTOL for Paper 1:8
Could you give me some idea of the cost of the concrete box, and also how do you intend to prove that it doesn't leak in 4,000m of water?

DR.J.R.WAKEFIELD:
The current estimate of the cost of a box is about £4,000. When you describe water, do you mean dumping at sea? Well, of course this method of disposal has long since been abandoned, although we still retain the option to dispose at sea. What we have done is to initiate calculations on the implosion strength of our box. We have demonstrated by calculation that it will descend to that depth, quite safely, without any damage. What it does after that the estimate of release of radioactive products assumes that the box will in fact open up – of course it won't – but the main criterion is that it should be able to descend that depth without damage and this has been demonstrated. We have done no practical tests, though it would be possible. I think there is a facility available that British Telecom have that they use for testing under water repeaters for their cabling which can introduce some of these enormous pressures but of course this would have to be a modelling exercise again. We haven't made use of this, we have used calculation.

MR.P.DONELAN, OVE ARUP AND PARTNERS for Paper 1:7
What is the status of the Trupack 1 container and why are you now developing the Trupack 2?

MR.W.BRACEY, TRANSNUCLEAR INC. USA:
The Trupack 1 container was abandoned. They went through several stages in their theory about how they were going to license it. The first was that they would make a package that was stronger than existing packages and closer to NRC regulations but it was not going to be double contained and it was going to be vented, that was the Trupack 1. I cannot say technically that the idea was not right, and that it wasn't a good package, but for political reasons, there are problems in the USA with

the perception of DoE operations relative to NRC regulations. Especially
since Chernobyl. Questions have arisen regarding DoE reactors and they
have been very sensitive to that issue. So for political reasons they
decided to go to a package that looked like Trupack 1 and develop the
initially rectangular Trupack 2. It looked very much like the first model
but had double containment and did not incorporate venting and it was in
theory meeting NRC regulations but was to be licensed by DoE. Now they
have banned the idea of licensing themselves and although the NRC is
forbidden by law from licensing DoE operations they are going to provide
a Certificate of Compliance for any packaging the DoE uses. As far as the
status of Trupack 1 is concerned there is currently no more work being
done on it, there are a couple of models in the desert collecting
tarantulas I guess. Some of them were used for testing of recombiner
effectiveness etc. but they have abandoned any licensing efforts.

DESIGN OF SHIPS FOR THE TRANSPORT OF SPENT NUCLEAR FUELS

H.E.Spink

Mechanical Engineering Design Office
British Nuclear Fuels plc
Risley
Warrington WA3 6AS

ABSTRACT

Pacific Nuclear Transport Ltd. (PNTL) operates a small fleet of ships for the carriage of spent nuclear fuels.

Some details are presented - including aspects of the hull design, power plant, operation, safety and reliabiltiy - to provide an over-view of the vessel design.

The reference design cited is the 'Pacific Pintail' - the latest addition to the fleet.

INTRODUCTION

Pacific Nuclear Transport Ltd. operates a fleet of five ships for the purposes of transporting spent nuclear fuel between Japan and the Northern European ports of Cherbourg (France) and Barrow (UK). The latest vessel 'Pacific Pintail' joined the fleet in late 1987; it is the reference ship in this paper - see Appendix A.

The fleet (Table 1) has been built in the period from 1976 to the present. The ships have evolved from the first ship in the series, the 'Pacific Swan', by incorporating changes found necessary from operating experience, safety audits, new flask designs, and changing regulations. The changes have not been extensive thus confirming the quality of the initial design. Essential changes have been retrofitted into the earlier vessels.

Table 1

Ship	First Voyage	Ship Yard
Pacific Swan	Apr 1979	Swan Hunter, Hebburn, UK
Pacific Crane	Oct 1980	" " " "
Pacific Teal	Dec 1982	" " " "
Pacific Sandpiper	Aug 1985	Appledore Shipbuilders, Devon, UK
Pacific Pintail	Dec 1987	Mitsubishi Heavy Industries, Kobe, Japan

The ships are normally routed between Europe and Japan via the Panama canal, but they are capable of completing the voyage around either the Cape of Good Hope or Cape Horn without refuelling.

All of these ships are British registered and come under the regulations of the Department of Transport (Marine Division) for cargo ships on long international voyages.

This paper is concerned with aspects of ship design related to the transport of spent nuclear fuel. Spent nuclear fuels are transported in very substantial flasks designed to protect people from damaging ionising radiation and to prevent the dispersal of radio-active material in the event of a maximum credible accident : consequently the cargo does not present any major difficulties. However, the design of the transporting vessel is dictated, to a large extent, by local requirements viz. the need to avoid entering ports for unscheduled repairs or supplies; national (local) controls governing the movements of nuclear materials; cargo handling capabilities in the terminal ports and harbour conditions such as the depth of water, turning basins and exposure to ocean swells.

THE CARGO : TRANSPORT FLASKS

Spent nuclear fuels are transported in flasks designed and approved in accordance with the regulations of the International Atomic Energy Agency (IAEA, 1985). A number of different flasks exist for carrying fuel elements from the various types of reactors; consequently they differ in physical size, weight and heat output. The types and dimensions of flask, inclusive of transport frames, as carried by 'Pacific Pintail', are listed in Table 2.

Table 2

Flask Designation	Dimensions (metres)	Loaded Weight (Tonnes)	Heat Output (max) kW
Excellox 3B	5.994 x 2.286 x 2.301	78.6	30
Excellox 4	6.258 x 2.286 x 2.540	110.0	40
TN12 MK1	5.898 x 2.590 x 2.830	106.6	75
TN12 MK2	6.150 x 2.590 x 2.830	111.4	75
TN17 MK2	6.150 x 2.546 x 2.405	84.0	42

Flasks which comply with the IAEA regulations are designed within limits set out in the codes. Those relevant to the design of ships specifically for the transport of nuclear materials are:-

(a) Radiation level (para 469)

not exceeding 10^{-4} Sv/hr (ie 10^{-2} rem/hr) at 1 metre from the surface of the flask when it is carrying the maximum radio-active content.

(b) Ambient temperature range (para 556)

-40C to 38C

(c) Max flask surface temperature (para 555)

85C (no solar heating)

(d) Maximum normal flask operating pressure (gauge) (para 554)

700 k Pa (ie 7 atmospheres)

The IAEA regulations also require the segregation of radio-active materials from transport workers and the public. Thus, the maximum dose within normally occupied areas must not exceed 5.10^{-3} Sv/yr (ie 0.5 rem/yr). This regulation, allowing for a time occupancy of about 30%, is satisfied by a maximum permitted dose rate of $1.8.10^{-6}$ Sv/hr (i.e. $1.8.10^{-4}$ rem/hr) at the interface between the cargo space and the working/living areas of the vessel.

THE HULL

A prime requirement for a nuclear materials transport vessel is to 'stay afloat' after sustaining damage from collision or grounding. The most stringent definition of this requirement is that of the Japanese Ministry of Transport (JMoT, 1979) for ships carrying spent nuclear fuels and sailing in Japanese waters.

To meet JMoT requirements the ships are constructed with a double hull. The inner shell embracing the cargo space is formed by watertight longitudinal and transverse bulkheads as shown in Figure 1.

The structure and sub-division of the hull is designed so that the vessel will stay afloat after it has sustained damage defined in reference JMoT, 1979, which is in excess of the extent specified for Class I chemical tankers. The wing tanks formed by this construction are used for normal ballast and trimming requirements except for the tanks abreast of No. 5 Hold (Figure 1) which are allocated for holding bilge water. This bilge water could be contaminated, in remote circumstances, with radio-active materials; consequently it is not normally discharged directly to the sea.

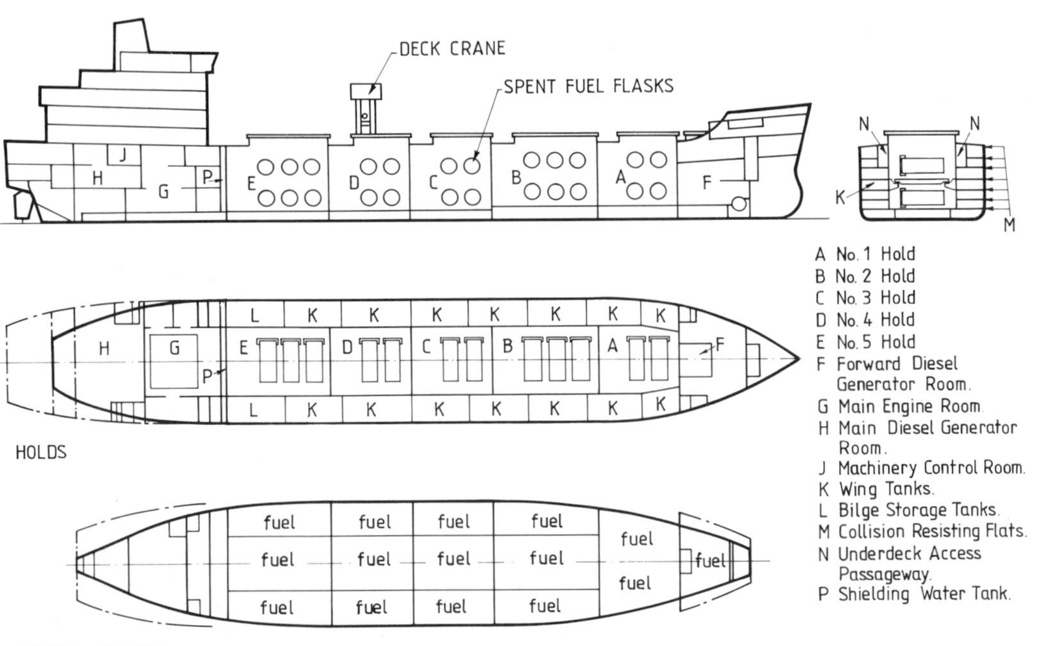

Fig. 1. 'Pacific Pintail' general arrangement.

The wing tank space is also structurally stiffened in order to prevent impact damage being sustained by flasks within the holds, in the event of a collision with another vessel. For design purposes the colliding vessel is assumed to be about 24,000 tonne displacement travelling at 15 knots.

The wing space is also used to provide all weather passage-ways on both sides of the ship, immediately below deck level for access to the holds and forward plant rooms. These access ways are necessary for checking the flask securing arrangements in heavy weather and for the routine checking of the hold cooling systems, flask pressures and radiation levels. The passages also provide convenient routes for segregating electrical power cables, hold cooling system control panels etc. The sub-division of the hull is preserved throughout the passage-ways by the use of water-tight doors.

CARGO HANDLING

The transport flasks are loaded into the five available holds on two levels as shown in Figure 1. The stowage plan devolves from the shipping schedule for reach voyage - taking account of the need to maintain adequate stability and satisfactory trim at all stages of the voyage, and also ensuring that any utility (ie the electrical power producing authority operating a reactor site) does not have to handle the flasks of another utility in loading the ship. The flasks are handled using either shore based or floating cranes of about 150 tonnes capacity.

Flasks are invariably mounted in frames for sea transport - the combined flask and frame is considered as an integral unit for shipping purposes. The lower tier units are fastened to seats at the bottom of the hold and the upper tier units are bolted individually to removable flask support beams which span the hold (transversely) and fasten directly to the ship's structure.

The transport frames are bolted down to permanent seatings which are drilled to a standard arrangement shown in Figure 2 which ensures the necessary flexibility in planning the cargo layout. The frames are located on the seats by fixed corner guides and they are fastened down with bolts which are tensioned by tightening them to a pre-determined torque. For convenience, the nuts for the fastening bolts are retained in cages and unused bolts are stored carefully in adjacent racks.

The hold hatch covers consist of two panels per hold - they are lifted and handled by the on-board crane. Generally, only one or two holds are opened at any time, therefore it is possible to stow the removed covers on top of the remaining ones. This simple and satisfactory arrangement is self contained and allows the ship to be made sea-worthy without the need for shore based support.

RADIATION SHIELDING

The segregation between the cargo space and the normally occupied space is provided by radiation shielding and energy absorbing barriers in the form of a water tank extending the full width and depth of the cargo hold at the aft end of No. 5 Hold. The tank is formed by two transverse bulkheads (each 40 mm thick) separated by 750 mm of water space. The radiation shielding is extended forward from the bridge by concrete overlaid on the deck and beneath the hatch covers as shown in Figure 3.

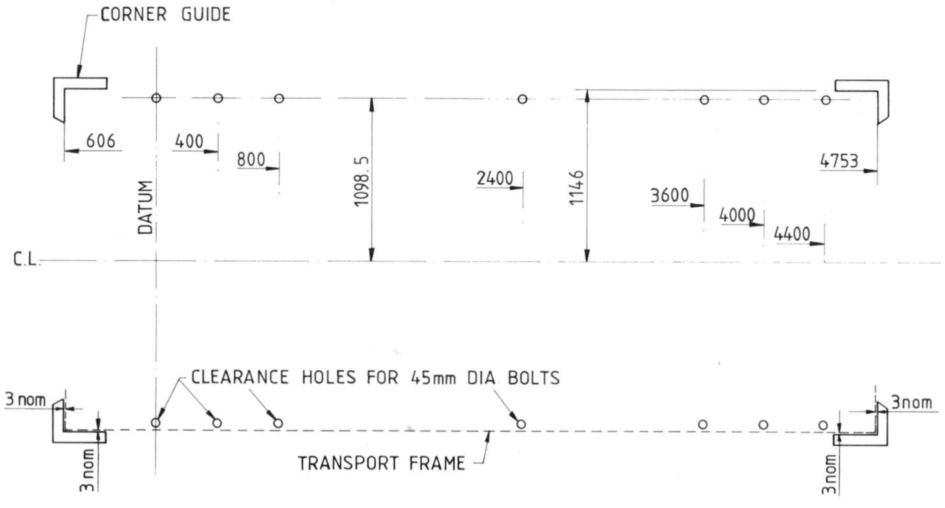

Fig. 2. Flask seating arrangements.

CARGO COOLING PLANT

The control of the ambient temperature in the cargo holds within the limits – 40C to +38C presents some problems in the design of the ships. The lower limit may be ignored because the vessels are unlikely to be sailing in such severe conditions. In order not to exceed the upper limit it is necessary to provide forced cooling within the holds.

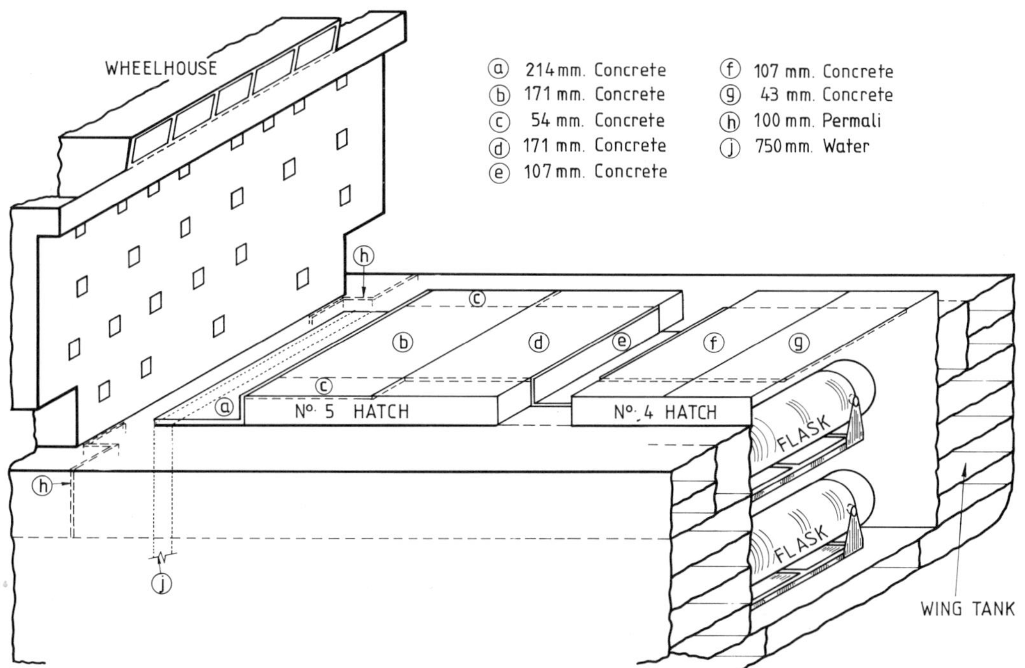

Fig. 3. Radiation shielding arrangement

The quantity of heat to be removed varies over a considerable range depending upon the flask loading and the prevailing weather conditions. For the purposes of ensuring the essential cooling service it is necessary to duplicate the equipment.

The cooling requirements have been met by providing two forced circulation air chillers in each hold whch reject the heat directly to sea. The chilled air is ducted to distributors low down in the hold and extracted at high level using axial flow fans - see Figure 4. This choice of equipment avoids the problems of centralised plant - ie it avoids the extensive use of air ducting which would destroy the water-tight subdivision of the ship, it minimises the electrical power demand, and it maximises the availability of plant.

Each air chiller consists of two independent refrigerator units (50% capacity; Freon 22) sharing a common air circulating duct. Thus part loading conditions can be met, usually by running only one refrigerator.

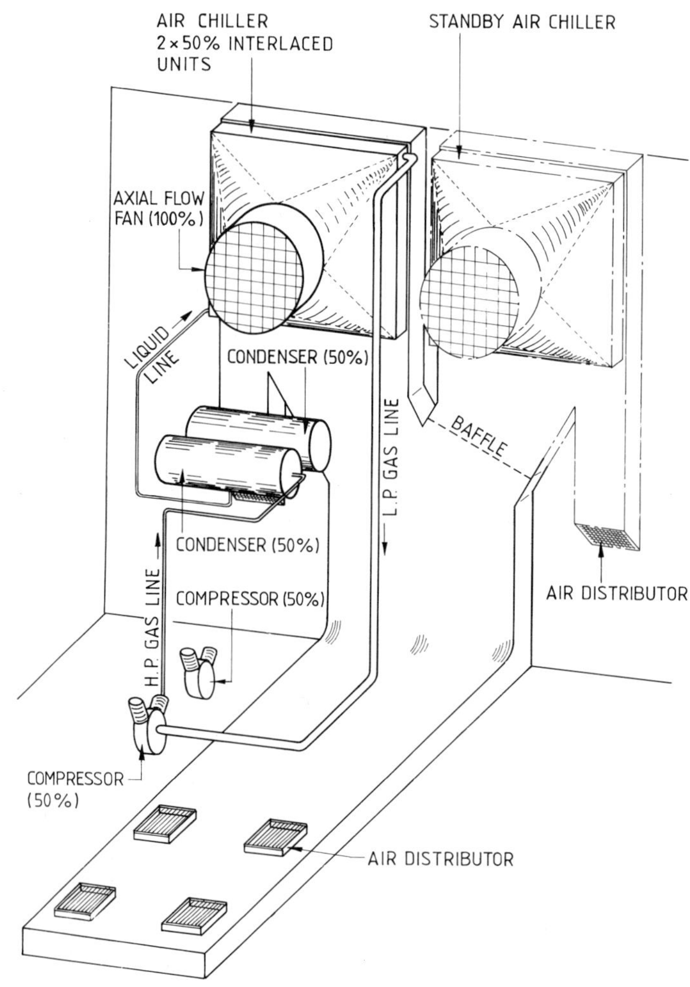

Fig. 4. Cargo cooling system

The heat is rejected to sea directly. The sea water circulating pumps for each of the 4 refrigerators in a hold share common suction and discharge pipes (mains). The suction main is fitted with two low level and one high level inlets which include coarse filter grids and compressed air clearing connections. The typical duty of the air chillers is:-

Flask heat	:	300 kW
Other heat	:	90 kW [motors,lights, external environment]
Capacity margin	:	40 kW
Total Load	:	430 kW
Air circulation rate	:	9000 m^3 /hr
Maximum air temperature	:	25C
Sea water circulation rate	:	100 m^3 /hr

Short term outages of the cargo cooling plant lasting several hours are not a serious problem because of the high thermal mass of the flasks and their contents. The upper limit of time is set by either the ambient temperature at high level in the hold exceeding 38C or the surface temperature of the flask exceeding 85C. In the period of outage, habitable conditions allowing maintenance/repair work to proceed, are prolonged by the conventional ventilating systems fitted to each hold.

In addition to the chilling and ventilation systems above, spray nozzles and pipework fitted to the hatch covers can be connected to the ship's fire main. these sprays in conjuction with the bilge pumps can be used to provide an emergency flask cooling system.

FLASK PRESSURE RELIEF SYSTEM

The Excellox flasks (Table 2) carried by the PNTL fleet, are filled with water for the purpose of transferring heat from the fuel elements to the outer surface of the flask where it is dissipated by natural air convection over fins. An ullage is allowed for the differential expansions of the flask and its contents following conventional practice. The flask is not provided with pressure relief devices in the normal way; consequently it is necessary to check flask pressures routinely and to provide means for relieving any excess pressures. A valved connection to the ullage space is incorporated in the flask for this purpose. In practice excess pressures have not been experienced in Excellox flasks.

A pressure monitoring and relief system is provided in each hold. When the flasks are loaded they are connected by flexible hose to the installed equipment. The system is extensively valved and gauged to prevent the inadvertent interconnection of flasks. By the sequential operation of valves in accordance with a strict procedure the condition of each flask can be checked regularly and gases transferred to the system can be diluted and exhausted via high efficiency particulate filters (which retain any radioactive material) to atmosphere at the head of the foremast.

POWER PLANT

The ship's power plant has been designed to provide a high degree of reliability. To this end, all the main plant is duplicated and installed in ways which avoid common mode failures as for as it is economically practicable.

The main propulsion system consists of 2 x 6 cylinder diesel engines each driving its own propeller through a reverse/reduction gearbox and clutch. Each engine has its own auxiliary service systems as shown in Figure 5. The engines however share a common compartment, sea water suction and discharge mains, electrically driven standby pumps, and main fuel oil storage tanks (but not the daily service tanks).

The propulsion system has been very reliable in service. In practice, one engine can be stopped and declutched while the ship maintains progress at about 10 knots on the other engine.

The electrical power distribution system is shown in Figure 6. The installed plant consists of 2 independent generators, each of 100% capacity, for supplying all of the ships services and a separate emergency generator for essential services such as navigation communications and standby lighting systems. For these fleet ships, two further pairs of generators are provided for cargo related duties; each pair is capable of taking the full cargo cooling load, thereby satisfying the JMoT requirements for separate and duplicated equipment for these areas.

Each of the generators is driven by a diesel engine with its own auxiliary services systems (similar to the main engines). The installation is split between two generator rooms (Figures 1 and 6) with the emergency generator separately housed in the forecastle.

OPERATION

The manning of the ship is outlined in Appendix A.

It is PNTL practice to maintain a bridge watch of one officer, a lookout, and a helmsman if the auto pilot is not in operation. This requires the carrying of 3 watchkeeping officers and 6 seamen. The

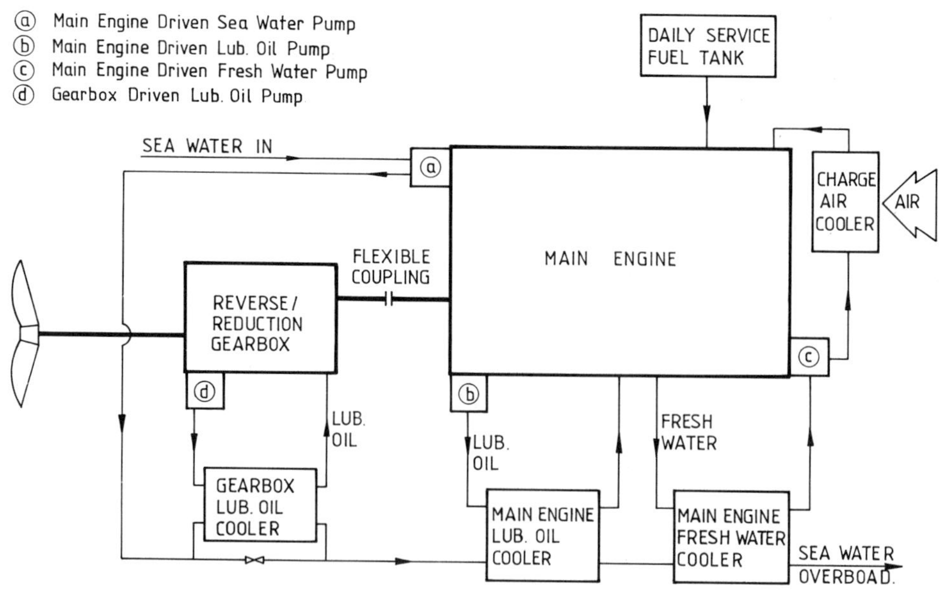

Fig. 5. Main engine services

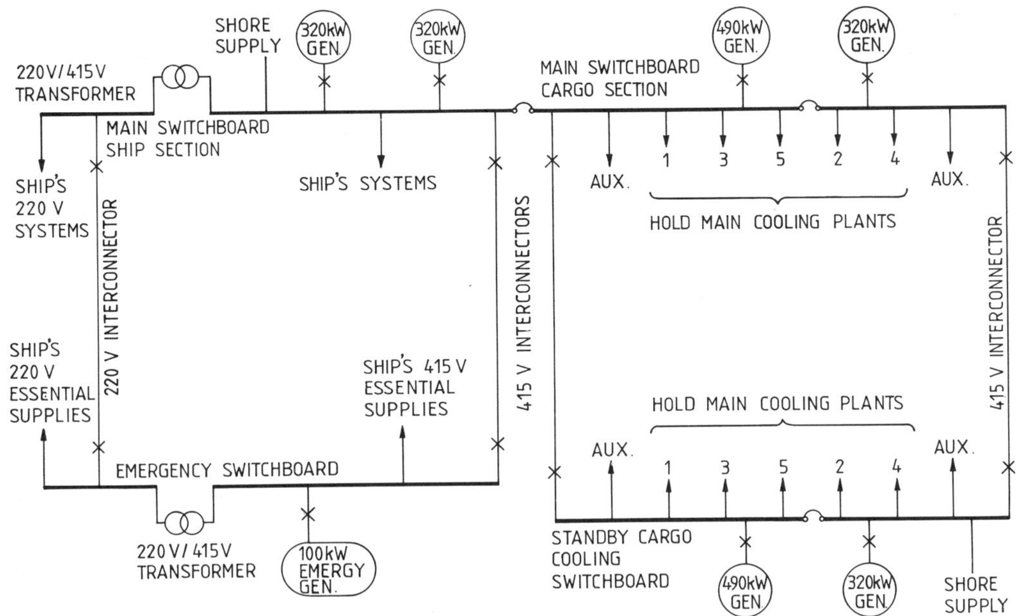

Fig. 6. Electrical distribution system.

Master is not expected to keep watch and the Radio Officer has his own statutory listening periods.

A similar system is operated for the power plant – requiring 3 Engineering Officers and 3 Petty Officers to man the watches. The Chief Engineer is not a designated watchkeeper. The ship's Electrical and Refrigeration Officers are both day workers with duties ranging throughout the ship. There is potential for stand-in engineering staff in the event of injury or illness viz:-
- the Chief Engineer and Electrical Engineer are available for watchkeeping
- the Chief Engineer and Second Engineer are qualified to carry out the duties of the Refrigeration and the Electrical Officers
- the Radio Officer is available for the maintenance and repair of electronic equipment.

Access to the cargo holds is via passage-ways in the wing spaces. In operation, when the ship is carrying loaded flasks, the holds are designated radiation areas with controlled personnel access via a change-room. The change-room is situated at the aft end of the passage-ways. Its purpose is to prevent the spread of radio-active material into the normally manned areas of the ship by enforcing a change of protective clothing (especially footwear) at a designated barrier. The change-room incorporates personnel radiation monitors, a wash basin, a shower, and separate lockers for normal and radiation area clothing.

The installation of air chillers and pressure monitoring equipment in the holds is dictated by a number of practical reasons but it increases the radiation exposure of the Refrigeration Engineer – especially during maintenance/repair work. In practice the increased exposure has not been significant but it is possible for him to exchange duties with other officers if necessary.

COMMENTS ON SHIP DESIGN

For practical and economic reasons the vessels in the PNTL fleet have been designed and built to normal shipbuilding standards of engineering and manufacture : however they incorporate features which enhance their safety and reliability in service.

At the outset of design it was considered impractical to impose higher than normal engineering standards of inspection and quality control on shipbuilders but it was recognised that normal merchant ships were not reliable enough for the nuclear industry.

A review of the longitudinal strength of the ship, taking account of both the 'still water' and the 'wave' stresses shows that the maximum total bending stress is about 0.8 kgf/cm^2 which is about 40% of the allowable level (classification rules). It is obviously uneconomic to design ship structures with these margins. The additional stiffening of the wing space, to cater for collision damage, is incorporated in the hull where it contributes significantly to the longitudinal strength of the vessel - hence the low value of the bending stress - but it adds about 400 tonnes of additional steel. In mitigation, the strength of the design provides a good margin to cater for increased bending moments arising from any loss of buoyancy following damage.

The International Maritime Organisation (IMO) is currently considering a new standard for cargo ships built specifically for transporting spent nuclear fuel flasks. At this early stage, it appears that PNTL ships will meet these new standards and may better them in some aspects such as the water-tight subdivision requirements.

CONCLUSION

The existing fleet of ships has completed 50 round trips between Japan and Europe carrying irradiated nuclear fuel without any radiological incident. There have been failures in various items of plant, but none has gone beyond the first line of defence of bringing immediately available spare plant into operation.

ACKNOWLEDGEMENTS

PNTL is a subsidiary of British Nuclear Fuels plc (BNFL). The author is grateful for permission from PNTL for the publication of this paper. The design of the spent fuel transport ships described above was undertaken by the Mechanical Engineering Design Office (MEDO) of BNFL.

The success of this enterprise to date is due to the combined efforts of shipbuilders, crews and shore management; together with the designers and builders of the flasks and the extensive organisation of people and companies worldwide on which the enterprise depends.

REFERENCES

International Atomic Energy Agency (IAEA), 1985 Edition, "Regulations for the Safe Transport of Radioactive Materials"
Japan Ministry of Transport document, submitted to The International Maritime Organization, Maritime Safety Committee 14th Session, 1979, MSC XL/25/1, "Safety Requirements for Sea-going Ships Carrying Spent Nuclear Fuel Shipping Casks".

APPENDIX A

Particulars of "Pacific Pintail" are:-

Length overall   103.9 m
Breadth   16.5 m
Depth   10.0 m
Draught   (maximum)   6.047 m
Deadweight (maximum)   3,865 tonnes
Draught   (Service)   5.647 m
Speed   (Service)   13 knots
Endurance at service speed   77 days
Range at service speed   24,000 nautical miles
Lightweight   3,870 tonnes
Cargo   24 flasks

Complement: Master, Chief Engineer, 3 Deck Officers, 3 Engineer Officers,
Radio Operator, Electrician, Refrigeration Engineer, Catering Officer,
Bosun, Cook, 3 Engineer Petty Officers, 10 Seamen, 4 Cadets

----

31 Normal Crew, 6 Supernumeries (max), 37 Maximum Number on Board

Propulsive power   3,712 B.H.P
Number of propellers   2
Number of rudders   2
Electrical power   2,260 kW
Emergency generator   100 kW
Cargo-cooling capability (flask heat output)   1,280 kW
Classification   Lloyds + 100A1 + LMC = RMC
Registration   British
Class   Cargo ship - Long international voyages

CEMENT SOLIDIFICATION USING LARGE CONTAINERS :
THE PREDICTION OF PRODUCT QUALITY

D. Saul

NEI International Research & Development Co.Ltd.
Fossway
Newcastle Upon Tyne
NE6 2YD

ABSTRACT

Cement solidification is an increasingly important activity in the
field of radioactive waste management.

Solidification processes which are based upon 'in-drum' mixing
methods offer a number of advantages, and if large containers are
employed further benefits can be gained in terms of transportation,
operator dose and cost. The large container enables a given waste volume
to be processed in fewer operations, therefore subsequent handling and
transport is minimised. Further benefits can be gained by performing the
solidification operation with the container pre-installed within the
transport container.

A programme of testwork has revealed that cement based laboratory
formulations, derived under controlled conditions, can be scaled up to
produce large 2000 litre size products. The stability of such products
appears to be comparable or superior to small scale laboratory products.

INTRODUCTION

The operation of nuclear plant or the use of radioactive material
inevitably produce by-products which are themselves radioactive. Waste
materials therefore arise in many different fields of work e.g. power
stations, researsh laboratories, hospitals, fuel reprocessing facilities
etc.

The level and type of waste material generated is obviously varied.
Consequently, the regulations or guidelines governing storage, transport
and disposal also vary as these are dictated by the nature of the waste.

The largest proportion of waste materials can be classified as Low
Level Wastes (LLW):

i.e. less than 4×10^9 bequerels per tonne of alpha, and

1.2×10^{10} bequerels per tonne of beta/gamma.

These have traditionally been disposed of at near surface disposal sites or via sea discharge. Future considerations to improve the methods of disposal are based upon reducing the waste volume by means of compaction, incineration and changes in waste management procedure (UK Nirex Ltd., 1987).

The second largest category of radioactive waste materials is the Intermediate Level Wastes (ILW). The levels of radiation associated with these types of waste exceed those of the low level types. However the amount of heat generated is not sufficient for them to be classified as High Level Waste materials.

Due to the higher levels of penetrating radiation, the ILW types are shielded to prevent exposure during storage, transport and disposal.

ILW consists of irradiated fuel cladding, reactor components, chemical process residues, ion exchange resins and filters. These have accumulated at the waste producers site as a disposal policy has not yet been implemented in the UK.

Recent proposals for ILW disposal involve retrieving the waste from storage, immobilising in a cement based matrix, and packaging within a steel or concrete container. It is therefore apparent that cement solidification, i.e. fixing in cement/concrete, will become an important activity.

The use of solidification will eliminate the need to transport liquid type wastes. These wastes present additional hazards due to their ability to spread readily if not contained.

Solidification obviously increases the overall cost of the waste management cycle. Although this is relatively small compared to the total cost, it is important that the technique selected is cost effective as well as efficient. In addition, the solidification operation will contribute to operator dose and it is therefore essential that the technique minimises exposure.

Various solidification systems exist, although in general terms these can be classified as mobile or fixed. The type chosen will be dependent upon the volume of waste arising at the producer's site. For large volumes a fixed facility will be beneficial and this will justify the construction and maintenance costs incurred. For smaller volumes of waste the use of a mobile plant is the more cost effective option, as this can be employed by a number of waste producers all contributing to the cost of the system.

In the USA, private specialist companies offer a solidification service to the nuclear industries. The techniques employed by these companies have been developed and optimised over the years in order to maintain a competitive service. These developments have included the following:-

-   Increasing waste loading, i.e. increased volume of waste in the solid product. This is exceedingly important when high cost disposal methods are used (over £1,000 per m^3).

-   Use of mobile, in-drum mixing, solidification systems, i.e. one container for waste reception, dewatering and solidification, in which the container is disposable, see Figure 1. Such a system has few components in contact with the radioactive waste/waste product, therefore contamination is minimised. Consequently transportation of

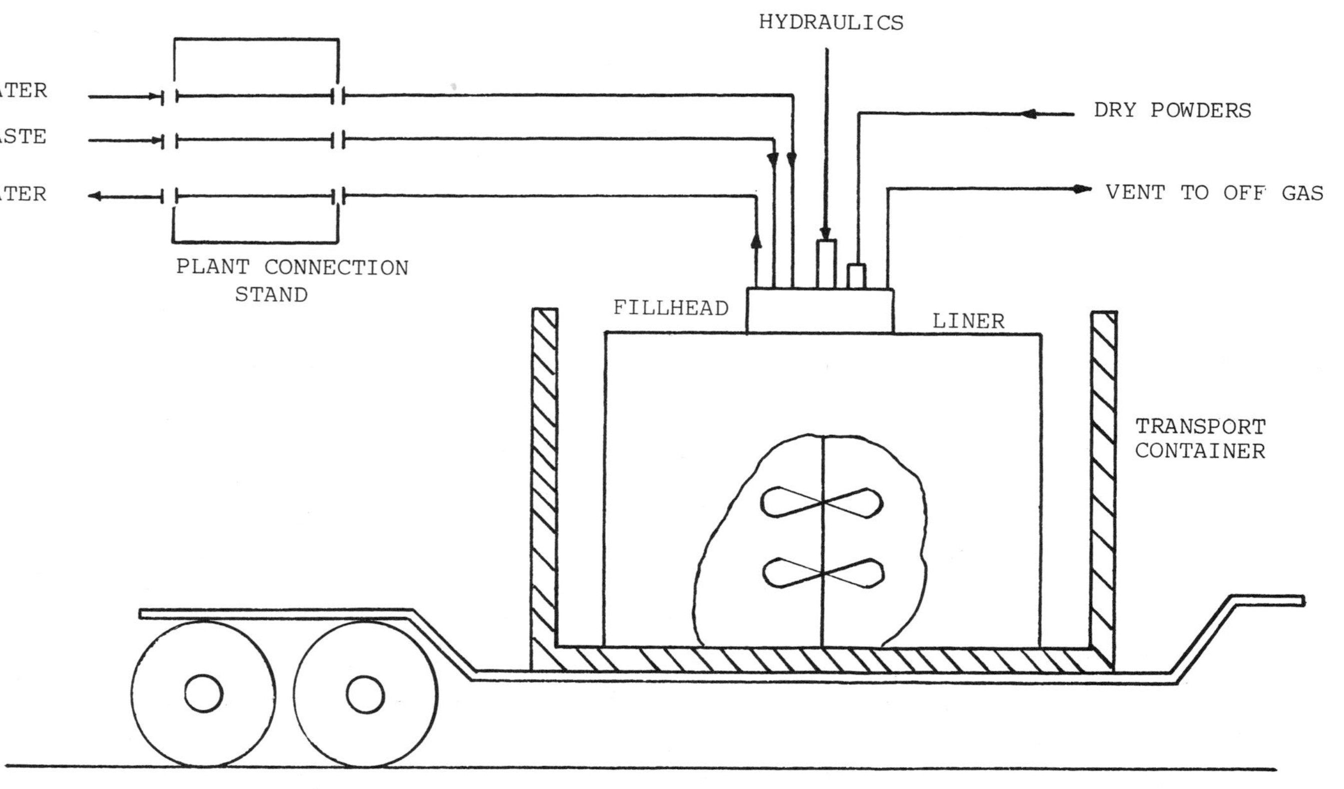

Fig. 1. Large container solidication concept schematic.

active plant is not a problem as any contaminated component can be packaged.

- Use of cement and cementitious powders.

- Use of large containers, (called liners in USA, large drums in UK). These are large cylindrical steel vessels, ranging from 2000 up to 6000 litre volumes.

All of the above developments have been investigated by NEI during a 5 year solidification research programme.

Until recently the use of large drums in the UK had not been considered likely as the trend was towards 200 litre drums. However, current proposals have suggested that 500 litre drums could become standard, indicating that the benefits of large waste containers have been recognised (Grave, 1987). Such benefits include a reduced number of solidification operations per volume of waste; reduced number of drums and transport containers and hence handling operations; reduced processing costs.

It can be seen that increased container size can reduce operator dose. In addition, further dose reduction can be achieved by placing the large drum within the transport container, prior to the solidification operation, see Figure 1. This eliminates the need to load an unshielded large drum.

The transport containers proposed by UK Nirex Ltd are capable of holding four 500 litre drums. Alternatively, one 2000 litre standard liner could be accommodated and therefore it would be possible to employ large drum systems without affecting current proposals for waste transport and disposal.

EXPERIMENTAL WORK

Background

The nature of Portland cement makes it a particularly useful solidification medium (Staehr, 1986). The solidification of cement is a hydration reaction which is exothermic in nature. It can be influenced by external sources, in particular temperature and certain chemicals, therefore these factors must be taken into consideration during formulation and process development.

Solidification development work based upon Portland cement has been carried out within NEI for 5 years, and in 1985 a mobile 200 litre drum solidification system was commissioned using low level 'active' lubricating oil. Continued testwork has been carried out to assess the feasibilty of solidifying a wide range of typical UK waste streams. The work has comprised small scale laboratory and full scale 200 litre solidifications, with products from both being subjected to environmental stability tests. All testwork has employed inactive waste simulants, although leach tests have employed active tracer materials or samples of the active waste material.

Small scale laboratory tests offer the most practical means of assessing the waste/cement formulation, as large numbers can be examined simultaneously. This enables waste and cement chemistries to be matched and modified as necessary. Such tests do not simulate the full scale environment however, as heat generation and loss are different. The small scale samples dissipate exothermic heat readily and any temperature increase is small, therefore any influence on solidification is negligible.

The larger scale samples act in a more adiabatic manner and heat is retained. This increase in temperature causes the cement hydration reaction to accelerate, resulting in further heat generation.

This effect was carefully examined during early solidification work, and a technique to cure laboratory samples at elevated temperature was adopted.

This enabled the conditions of large scale processing to be simulated, and gave formulations which could be scaled up to 200 litre size.

Larger drum sizes result in an even more adiabatic system and the importance of small scale simulation is greater. The objectives of this testwork were therefore as follows:-

-    To examine the feasibility of solidifying typical UK powor station wastes at large scale i.e. 2000 litre drum size.

-    To examine the feasibility of predicting large scale (e.g. 2000 litre) formulations by small scale laboratory tests.

-    To examine and compare the integrity of products manufactured at laboratory and large (2000 litre) scale.

Laboratory

Fifteen inactive waste simulants were identified for the test work. These comprised the following:-

(a) Ion exchange resins (8 types) including bead, granular and mixed bed.

(b) Filter Media (3 types) including coarse sand, diatomaceous earth precoat and molecular sieve material.

(c) Granular dessicant material (3 types).

(d) Pond sludge.

Small laboratory mixes, approximately 1 litre in volume were manufactured and 3 samples produced from each mix.

Sample number 1 was cured at elevated temperature.
Samples 2 and 3 were cured at ambient temperature.

Following a 24 hour period, samples which were not dry and hardened (VICAT penetration test BS12 part 2) were rejected and the formulation considered unsuitable.

Acceptable samples were subjected to a short term (2-4 days) water immersion test to ensure product stability. Degradation of sample number 1, i.e.cured at elevated temperature, resulted in rejection of the formulation.

Stable cement based formulations were identified for all of the simulant materials.

<u>Large Scale Solidification</u>

Four of the simulant materials were selected for large scale (e.g. 2000 litre) solidification. These were as follows:-

1. Mixed bed ion exchange bead resin
2. Coarse granular dessicant material
3. Diatomaceous earth filter precoat
4. Coarse filter sand

The simulants were transported to Barnwell, USA, and a programme of testwork was carried out using a Chem Nuclear System Inc mobile solidification system with 2.2 cubic metre drums. The plant was found to be capable of handling and solidifying all of the waste simulants supplied. Processing was carried out by one operator working remotely from the drum.

Samples of each large scale mix were removed, and laboratory cube samples manufactured. These were cured at elevated temperature for 72 hours. The centreline temperature of each mix was monitored for a period of approximately 24 hours. Temperature profiles are shown in Figures 2 and 3. For simulants 2,3 and 4 start temperatures were higher than those for simulant 1. This was thought to be due to the effects of mechanical energy input. Inspection of the products was carried out after 24 hours and all were found to be dry solids. Sections of the drum and core samples were removed after 2-19 days age, and all products were found to be homogeneous.

<u>Stability Testing</u>

Additional laboratory samples were maufactured for each cement formulation. These were subjected to the following stability tests along with core samples, and the laboratory cube samples manufactured from the large scale mix.

1. Freeze/Thaw: 20 cycles - 20C/+20C

2. Water Immersion: 30 days in demineralised water at ambient temperature.

3. Elevated Temperature: 150C until weight constant.

Under freeze/thaw conditions, the small laboratory samples and core samples of simulants 2, 3 and 4 i.e. dessicant, diatomaceous earth precoat, and filter sand, were found to behave in the same way.

The laboratory and core samples containing simulant 1, i.e. ion exchange resin bead, were found to behave differently under freeze/thaw conditions. All small laboratory scale products were found to suffer degradation whilst core samples were unaffected.

Water immersion testing was, as expected, found to cause weight increase. In all cases the laboratory samples and core samples were found to behave in the same way.

Samples containing simulants 2, 3 and 4 did not suffer degradation, and a typical plot of weight gain vs time is shown in Figure 4a.

Laboratory and core samples containing simulant number 1 were not considered satisfactory during water immersion. Once again large scale samples proved superior to bench samples. Examination of the core sample

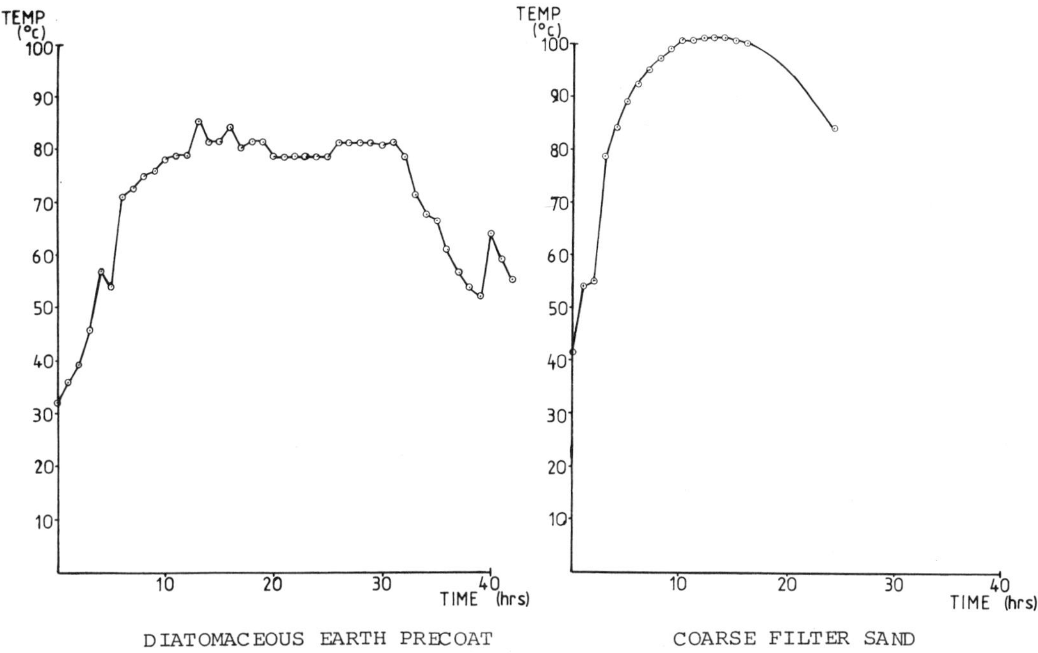

Fig. 2. Temperatures monitored during large scale solidication.

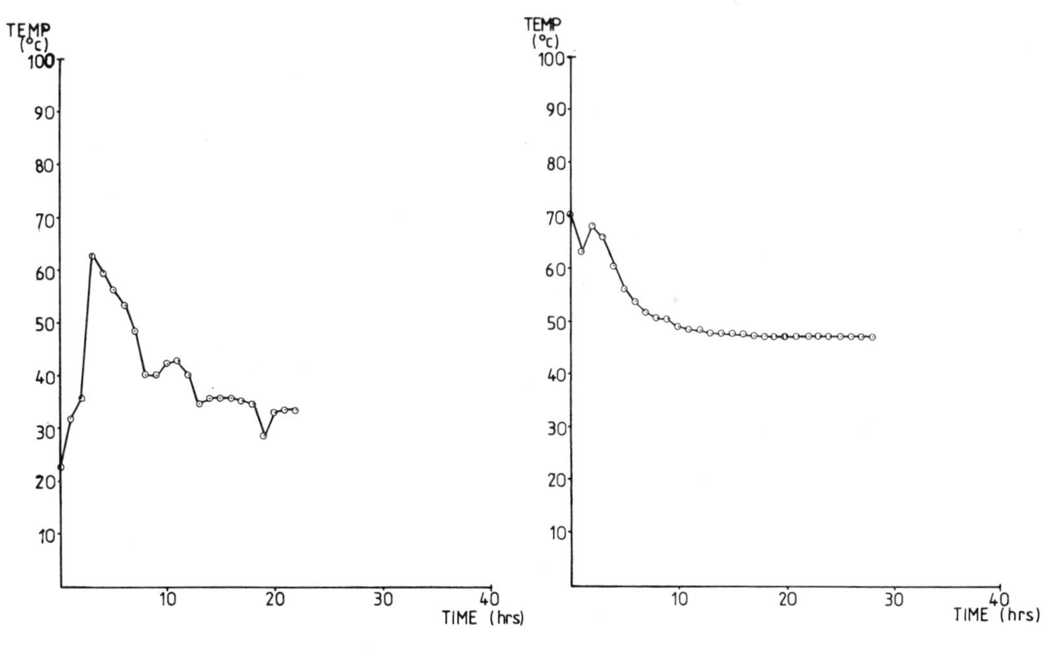

Fig. 3. Temperatures monitored during large scale solidication.

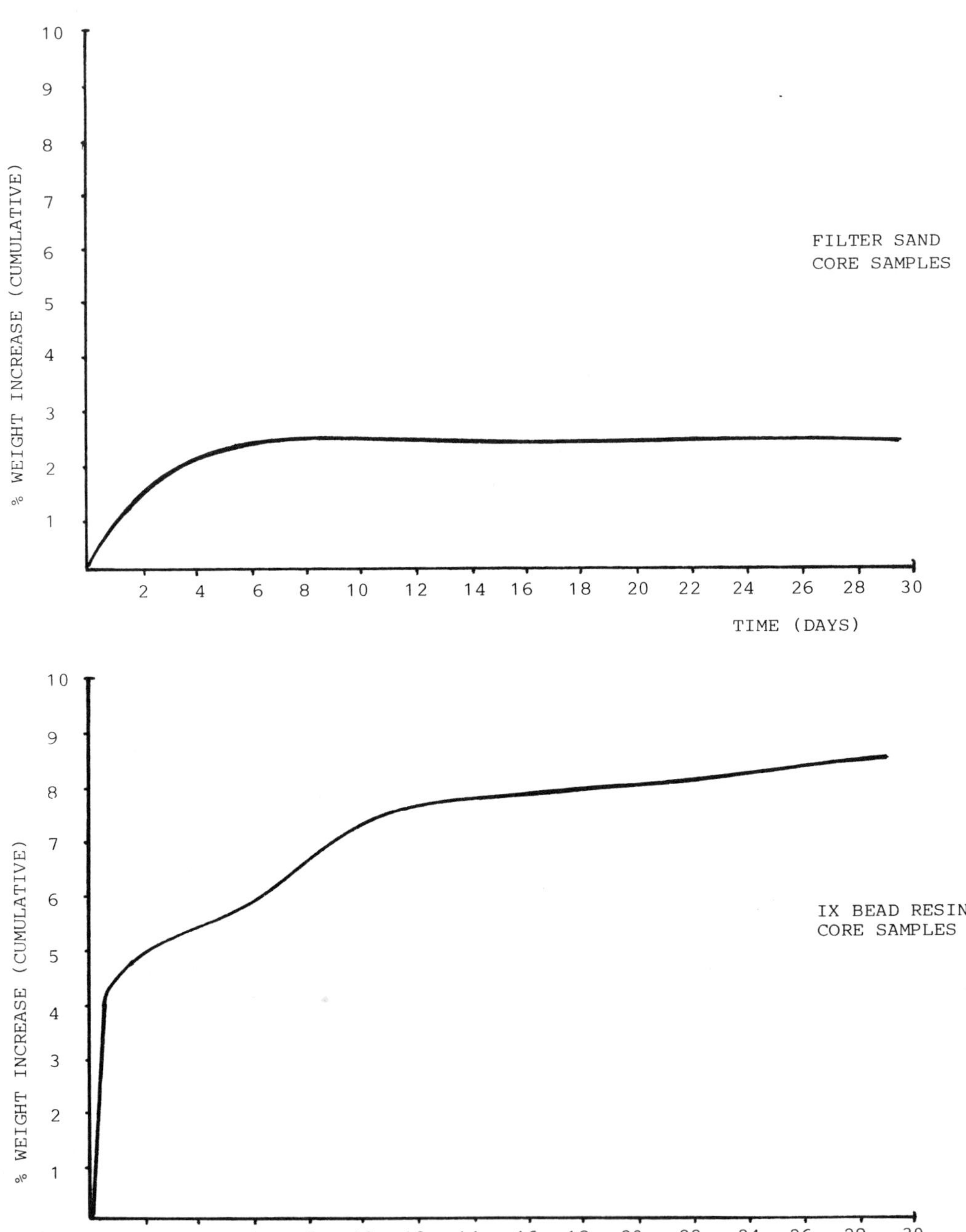

Fig. 4. Comparison of water immersion test results.

weight vs time plot for this simulant in Figure 4b shows different characteristics from other simulant materials, i.e. continuous weight gain. This observation may be of interest in future testwork, as continuous weight gain may be indicative of an unstable formulation.

The laboratory and core samples of simulants 2, 3 and 4 appeared to behave in an identical way during elevated temperature exposure.

For simulant number 1 core samples were found to be superior to the small laboratory samples.

DISCUSSION

The testwork has demonstrated that a mobile, large drum solidification system could be employed to process U.K. waste streams. Such a system has few re-useable items which become contaminated and those which do, i.e. fillhead and plant connection stand, are small enough to be suitably packaged for transport by the mobile plant vehicle itself.

Operation of such a system is remote and can be carried out by one operator, hence operator dose may be made as low as reasonably practicable. Dose reduction is assisted by processing with the solidification drum pre-installed within the transport container.

Because the drum was processed whilst standing in the transport container, the time and consequences of handling following solidification was limited to removal of the fill head following curing (24 hours). This required a lifting device of 1 te capacity e.g. forklift truck. The system design and process combine to make risk of contamination of the transport container extremely unlikely.

It appears from the test results that small scale laboratory formulations can be employed as a means of predicting large scale product integrity. In order to achieve this it is necessary to cure at elevated temperatures in order to simulate the large scale condition. The temperatures employed did not duplicate those recorded at full scale and further testwork to examine such temperatures at laboratory scale is considered necessary. It is considered that the bench scale formulations are pessimistic.

The level of temperature produced during large scale testwork appeared to be affected by the degree of mechanical energy required to rotate the mixing paddle, high start temperatures being associated with high mechanical energy input. It seems apparent therefore that the physical nature of the waste being processed is an important factor to be considered. It is noted that the high ambient temperatures in South Carolina affected the results.

In practice temperatures of up to 101C were observed during the testwork and these were considered to be typical for large container solidifications in the USA.

Previous work (Palmer and Smith, 1986) had suggested that even higher temperatures would be associated with such large scale solidifications and the possibility of thermal cracking was suggested. This, however, was not confirmed during the testwork carried out and all products were found to be solid, homogeneous monoliths. It is thought that the waste loadings achieved (i.e. typically wet settled, 70% by volume) contribute to the limitation of exothermic heating.

From the stability test work it would appear that the large scale product is comparable or superior to the small laboratory scale sample cured at lower temperature. The technique therefore appears to offer a practical means of predicting large scale formulations as it did with 200 litre solidifications. It should be remembered, however, that applicability of the technique will be dependent upon the waste type and it is important that the waste is examined prior to solidification work. An example of this concerns the encapsulation of low boiling point solvent liquids. Such liquids readily separate from cement slurry when subjected to elevated temperature, thereby producing rejectable laboratory products. However the same formulation manufactured at 200 litre scale will produce a stable monolith. This illustrates the importance of categorising waste streams and experience in this field is essential.

CONCLUSIONS

Mobile soldification plants which facilitate both transport of the solidification plant and the resulting conditioned waste package are in commercial use in USA. A fundamental aspect of the system is the use of large drums. Before such systems are utilised in the UK it is necessary to establish product performance as it is relevant to both transport and disposal activities.

Testwork has shown that typical UK power station wastes can be solidified at 2000 litre scale using cement formulations identified in the laboratory. It is therefore apparent that laboratory testing offers a practical means of assessing large scale product integrity.

The stability of large 2000 litre products appears to be comparable or superior to that of the small laboratory samples.

One mobile solidification system is capable of processing waste materials of different physical character.

ACKNOWLEDGEMENTS

The author would like to thank the management and staff of Chem Nuclear Systems Inc, USA, and NEI International Research & Development Co. Ltd. for the assistance and advice given in preparation of this paper, and the directors of NEI International Research and Development for permission to publish this paper.

Special acknowledgement is given to the Central Electricity Generating Board whose financial assistance and technical support contributed greatly to much of the work reported in this paper.

REFERENCES

Grave, M.J., 1987, The Application of American Container Operations to the UK Waste Disposal Programme, in: "Nuclear Containment", D.G.Walton ed., Publisher, Cambridge University Press
Palmer J.D., Smith D.L.G., 1986, The Incorporation of Low and Medium Level Radioactive Wastes (Solids and Liquids) in cement, Nuc.Sci. and Tech., CEC Report, EUR 10561 EN.
Staehr J.P., 1986, Cement Solidification : A Perspective of Changes in the '80's, Technical Paper Chem Nuclear Systems Inc, Columbia, USA.
UK Nirex Ltd, 1987, "The Way Forward - A Discussion Document: The Development of a Repository for the Disposal of Low and Intermediate Level Radioactive Waste".

20 FOOT CONTAINER SYSTEM FOR SHIPMENT OF SPENT MTR-FUEL ASSEMBLIES

G.Armbrust,[1] H.Geiser,[2] D.Rittscher [2]

[1]Kernforschungsanlage Jülich GmbH (KFA)
Nuclear Research Center
Postfach 1913
517 Jülich
Federal Republic of Germany

[2]Gesellschaft für
Nuklear-Service mbH (GNS)
Rosastrasse 15
43 Essen
Federal Republic of Germany

SUMMARY

A new development based on Type B(U)-casks integrated in a 20 foot container has recently been successfully introduced. The reinforced 20 foot container has a tie-down function and holds the shock absorbers. The system has been in use for 2 years now, and during this time about 240 spent fuel assemblies from German material test reactors have been shipped from Germany to the reprocessing plant in the United States.

One of the advantages of the container system is its compatibility with the different transport systems. It can be fully integrated into existing truck, train, or ship requirements used for the transport of usual (non-nuclear) goods. This makes the system very flexible in relation to different transport applications.

DESCRIPTION OF THE 20 FOOT CONTAINER SYSTEM

In addition to the shipment of a single cask, which is always possible, two casks can be shipped in a 20 foot standard container which is reinforced in the bottom region and at the load attachment points. This special-container system with the integrated Type B(U)-shipping casks is shown schematically in Figure 1.

The casks are shipped in an upright position. For tie down and fastening the cask shock absorbers, which are bolted to the reinforced bottom region, are used. A lockable door is located at the front end for entering the container. Two lockable openings through which the casks can be loaded and unloaded are located in the roof of the container. All openings are sealed off in such a way that the casks are protected from spray water and rain water. The container also has ventilation holes in the bottom and top regions so that air exchange and circulation is possible in the interior of the container.

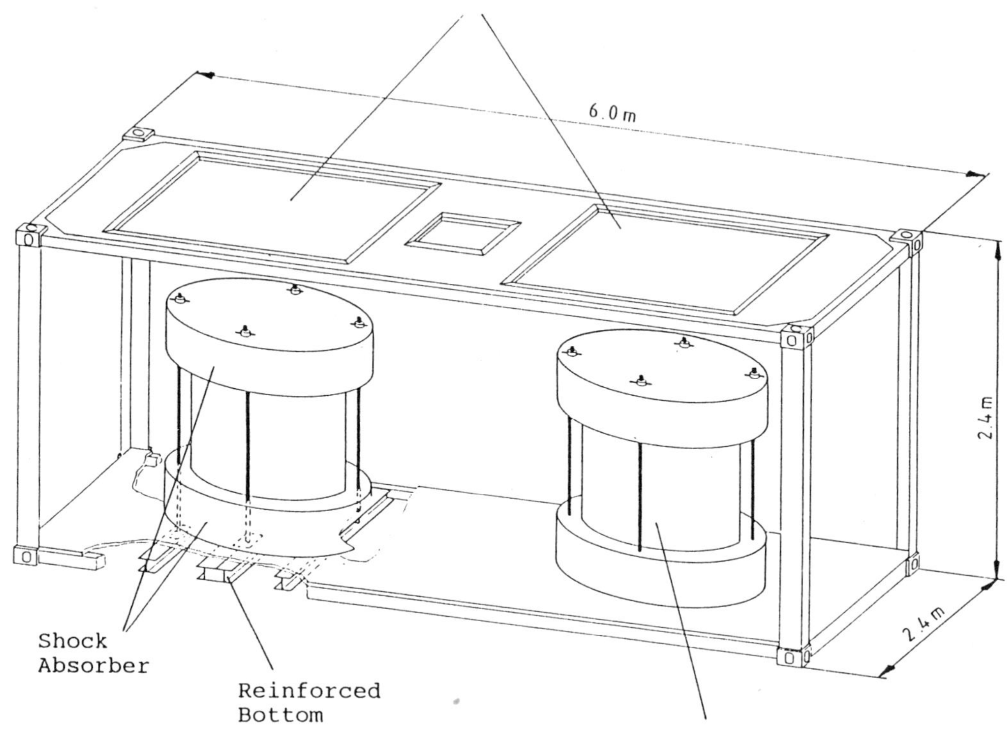

Fig. 1. 20 foot container system for shipment of two type B(U)-Casks

Extensive testing of the whole system was done prior to receiving the authority's approval of the whole system. Figure 2 shows a photograph of testing in progress at the facilities of the German Railways. A loaded container, on a rail wagon, was shunted against another rail wagon to simulate handling accidents during rail transportation.

Fig. 2. Measuring of loads on the tie-down system during simulation of handling accidents.

Fig. 3. View through the open container roof door.

Figure 3 gives a view through the open roof door of the container top. The tie-down system, as well as the load attachment point in the middle of the container roof, is shown in this photograph.

DESCRIPTION OF THE CASKS

GNS has developed a whole fleet consisting of different cask types which can be transported in the 20 foot container. All these different cask types were licensed in compliance with Type B(U) regulations. Figure 4 shows examples of various cask types which can be transported either singly on a transport cradle or in a pair inside the 20 foot container. The casks, developed by GNS, are:

- CASTOR MTR
- CASTOR MTR-F
- GNS 11
- GNS 12

All these casks are licensed and currently in service. The following kinds of shipment can be made in such casks:

- spent material test reactor fuel
- radioactive material in 200 l-drums
- large intensity radiation sources
- vitrified high level waste in canisters

Basically, there are two types of cask body materials

- monolithic ductile cast iron
- sandwich-type steel/lead/steel

All casks are suitable for underwater loading due to their inner/outer corrosion-resistant surfaces. For the monolithic cask bodies of the CASTOR MTR and CASTOR MTR-F, corrosion protection is maintained by a nickel coating and stainless steel lids and bolts. The sandwich type casks GNS 11 and GNS 12 are inherently protected by their surfaces, lids and bolts.

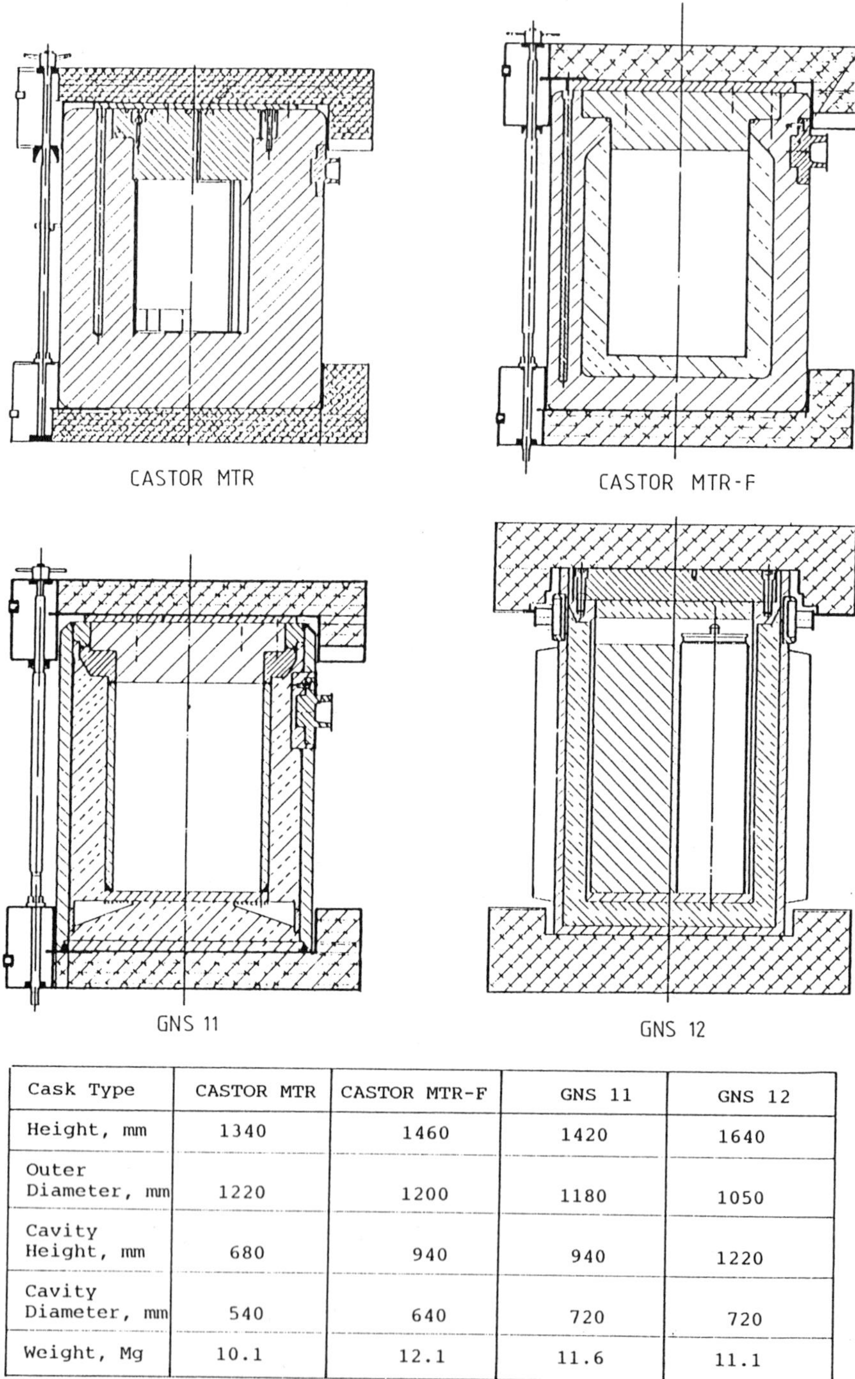

Cask Type	CASTOR MTR	CASTOR MTR-F	GNS 11	GNS 12
Height, mm	1340	1460	1420	1640
Outer Diameter, mm	1220	1200	1180	1050
Cavity Height, mm	680	940	940	1220
Cavity Diameter, mm	540	640	720	720
Weight, Mg	10.1	12.1	11.6	11.1

Fig. 4. Cask types for transport in a 20 foot container.

The main feature of this cask fleet, besides Type-B properties, is that the weight of a single cask is in the range of 10 tons, so that the transport weight of a loaded 20 foot container is less than 30 tons. This weight can be handled in international harbours as well as for rail and truck transportation.

SHIPPING EXPERIENCE

With the 20 foot container system many shipments have been carried out to date, including dry loading/unloading in hot cells as well as underwater loading/unloading in spent fuel pools.

Shipments of spent material test reactor fuel are routinely made from nuclear research centers, e.g. Jülich, to the Savannah River Reprocessing Plant. With these shipments, especially, all modes of transports are used:

- rail from research center to harbour
- ship from Europe to USA
- truck from U.S. harbour to Savannah River

Seven shipments with a total of 240 spent fuel assemblies have been made up to the present time.

Figure 5 shows a photograph of the underwater loading of a GNS 11 with spent fuel at the Nuclear Research Center Jülich.

In addition to transport of spent fuel, several shipments of other radioactive material, i.e. cemented fuel assembly structures in 200 l-drums, cemented reprocessing waste, test specimens and radiation sources, have been made.

The system has proven to be a good method, especially for international shipments with different modes of transportation, due to the use of a standard 20 foot container. The advantages lie, obviously, in the use of standard equipment for handling, which reduces the turnover times.

Fig. 5. Underwater loading of a GNS 11 in the Jülich spent fuel pool.

PACKAGING AND LOADING STRATEGIES FOR THE
TRANSPORT OF DRUMABLE LOW LEVEL WASTES

A. Johnstone[1], J. Fitzpatrick[2]

[1]Electrowatt Engineering Services (UK) Ltd
Grandford House
16 Carfax
Horsham
West Sussex RH12 1UP

[2]UK Nirex Ltd.
Curie Avenue
Didcot
Oxon OX11 0RH

ABSTRACT

An outline study was carried out to investigate the feasibilty of
using 200 litre steel drums as the standard packaging for the transport
of low level solid radioactive waste (LLW) to a disposable site. A survey
was made of the major waste producers in the UK nuclear industry to
estimate both the total quantities of LLW and the fraction of this
suitable for packaging in 200 litre drums. Possible methods of handling
and loading the drums, with special attention to the feasibility of using
ISO type freight containers, were studied; together with possible methods
of transporting loads of drums from the site of arising to the disposal
site. The outcome of the study was that about 70% of the LLW under
consideration was suitable for packaging in 200 litre drums; this
corresponds to a total of about 45,000 drums per year requiring
transportation. The drums were assumed to be handled in stillage of
twelve (2 X 3 X 2) and transported in 20 foot freight containers with
five stillages per container. The transport network considered was based
exclusively on road movements, with a total of about 134,000 loaded
vehicle miles per year. A fleet of forty freight containers and six road
vehicles would be required to service such a network.

INTRODUCTION

This paper describes a study for the Nuclear Industry Radioactive
Waste Executive (Nirex) in 1985. The objectives of the study were to
identify quantities of solid low level radioactive waste (LLW) arisings
in the UK suitable for packaging in nominal 200 litre drums, and to
investigate options for the handling and transportation of these drums.
The findings of the study are still of relevance to the problems of LLW
transportation to a deep disposal facility. The definition of LLW was
taken to be that which was acceptable at the Drigg disposal site, i.e.
the average beta activity (of pure beta emitters) does not exceed

2.2 GBq m^3, the average alpha activity does not exceed 0.74 GBq m^3 and the dose rate at the surface of substantially unshielded waste must not exceed 7.5 m Gy.

The preferred package for handling and transport of LLW is a steel drum of 200 litre nominal capacity. The general form of such a drum is described in BS 2003. Various sizes are permitted. A drum with an internal diameter of 571.5mm and an internal height of 813mm was adopted as standard in this study.

CURRENT AND ANTICIPATED DISPOSAL PRACTICES

At present the principal UK disposal site for LLW is at Drigg. This site is owned and operated by British Nuclear Fuels plc (BNFL), and the majority of the waste currently disposed there originates from the nearby Sellafield site. A much smaller disposal site is located at Dounreay. This site is owned and operated by the UK Atomic Energy Authority (UKAEA), and is used exclusively by the Dounreay establishment. LLW arising at sites other than Sellafield and Dounreay is transported by road for disposal at Drigg.

For the purposes of this study the following future disposal scenario was assumed: the Drigg site would continue in operation, but would only accept LLW from the Sellafield site. The Dounreay site would continue to operate as at present. The remainder of LLW producing sites would be served by a new disposal site developed and operated by Nirex. At the time this study was being prepared the location of the Nirex disposal site was taken to be at Elstow in Bedfordshire.

QUANTIFICATION AND CHARACTERISATION OF LLW ARISINGS

Both the composition and rate of production of LLW are variable, and are thus difficult to predict. An estimate of the quantities of LLW to be transported was made by carrying out a survey of waste producers in the UK (see Table 1). The tabulated quantities of LLW are the volumes which are transported to the Drigg facility for disposal. Only operational LLW arisings were considered, no attempt was made to predict the volumes of LLW arisings from future decommissioning programmes.

The operational LLW was divided into two groups. The first group was wastes from prospective clients of the Nirex operated LLW repository, and this group's sub-total represents the quantities of waste which were considered in the study. The second group was from producers who are considered to have their own LLW disposal routes, namely Sellafield and Dounreay. For each waste producer an estimate was made of the fraction of the LLW suitable for packaging in 200 litre drums; the average was found to be 70%, but there were significant variations between the waste producers.

The levels of radioactivity within the waste were more difficult to quantify. Historic data from the waste producers indicated an overall average of the order of 0.75 GBq/m^3, although with a range possibly an order of magnitude to either side of this figure. Data on specific nuclide inventories for the waste were not available. Surface dose rate data were available, and indicated that, although the majority of the waste had surface dose levels in the range 0 to 0.1 mSv/hour, some packages had levels as high as 7.5 mSv/hour, the maximum permissible for disposal at Drigg. From CEGB data the indications were that the majority of the packages had dose levels less than 0.02 mSv/hour.

144

Table 1. Projected UK Arisings of LLW

Source of Arisings	Quantity (m^3/year)	Fraction (%) Suitable for 200 Litre Drum
Group 1		
Amersham International	2,200	90
BNFL - non Sellafield	3,000	30
CEGB	3,000	80
National Disposal Service (NDS)	2,000	100
SSEB	800	90
UKAEA - Harwell	1,000	50
UKAEA - Winfrith	1,000	60
sub total	13,000	70 (average)
Group 2		
BNFL - Sellafield	23,000	
UKAEA - Dounreay	1,000	
sub total	24,000	
Overall total	37,000	

Another relevant property of the waste is its bulk density when packaged in 200 litre drums. Knowledge of this is required in order to assess weight loadings for handling and transport, and also for the application of the Transport Regulations, which are defined in terms of activity per unit mass rather than per unit volume. The available data gave an average of 400 kg/m^3 within a range 90 to 1600 kg/m^3.

APPLICATION OF TRANSPORT REGULATIONS

During the survey of LLW producers it was found that LLW was generally transported to Drigg under the 1973 IAEA regulations as non-fissile LSA (low specific activity) material. While the study was in progress the 1985 edition of the IAEA Transport Regulations became available in draft form. This version of the regulations was used to evaluate the requirements for transport, since it was anticipated that these would be in force before the operational date of the LLW repository.

No attempt was made to distinguish between wastes classed as LSA-II within the strict wording of the Transport Regulations. The LSA categorisation implies a uniform distribution of a package's radioactive contents, and wastes which are a collection of SCO-II items inside the envelope of a 200 litre drum. All of the wastes were for practical purposes considered to fall within the definition of LSA-II material.

The A_2 value for a drum of LLW was estimated using the Drigg limits, and assuming that the split between beta and alpha activity was in the same proportion as these limits. An A_2 value of 0.6 GBq resulted, implying an LSA-II qualifying limit of 6×10^4 Bq/gram. When combined with the density figures this gives a qualifying range from 5.4 GBq/m^3 to 96 GBq/m^3. The average activity found from the LLW survey was 0.75 GBq/m^3, which is well below these limits.

MULTIPLE HANDLING OF DRUMS

A basic principle of any material handling operation is the unit-load concept, whereby the larger the load the lower the handling cost per unit. This principle is particularly relevant to the handling of bulk materials in drums. It was considered that most of the waste producing sites could probably gain some advantage by assembling their outgoing drums in stillages; however much more significant benefits would be found at the disposal site, at which it was assumed they would be disposed of directly into the trench. The drums were therefore assumed to be transported within an ISO type freight container using stillages. Various multiple load plan configurations were examined to identify the most suitable size of stillage. The best arrangement was found to be either a two by three or three by three plan array with either single or double stacking. The major constraint was the available width on a road or rail vehicle, which favoured three abreast drum layouts. One basic assumption was that a drum would be transported with its axis vertical; any advantages in load utilisation which could be obtained with other orientations were considered to be outweighed by the need for additional equipment to rotate the drums.

The study also examined the most suitable size of freight container for transporting the drums. Custom built freight containers can be manufactured to any length, height, and width which transport regulations will permit. Three standard freight container lengths exist; these are known as '40 foot' (12 192 mm), '30 foot' (9 125 mm) '20 foot' (6 058 mm). Assuming that the drums were placed into a freight container three across and two high, these standard sizes can carry nominally 120 drums, 90 drums and 60 drums respectively. From a consideration of weights, logistics, and a review of current practices for transportation of LLW and other drummed chemicals, the '20 foot' freight container was selected as being most suitable. This freight container would be suitable for transporting five of the two by three double stacked stillages identified above. Assuming a typical bulk density of 400 kg/m^3, and allowing 25 kg as the weight of a drum, a typical freight container load did not, at 6300 kg, present a major weight problem.

A comparable payload of drums may be carried in an enclosed rollon-rolloff demountable skip. This type of conveyance is already used by some waste producers to carry LLW to Drigg. However, these are operated within the context of the tipping practices presently used, and would not be suitable for the deliberate emplacement of wastes which has been assumed for this study.

For the purposes of the IAEA transport regulations either the individual drums or the freight container may be the transport package. There are advantages in either approach. However, this aspect was not considered in detail, but was noted for further consideration. In this study the individual drum was regarded as the package.

Consideration was given to the possibility of not unloading the drums from the stillage at the disposal site for final emplacement, but continuing the use of the stillage to carry a multiple load of drums directly into the repository, and to regulate the stacking of the

emplaced drums. This scheme would require a comparatively low cost design
of stillage to be adopted.

POSSIBLE TRANSPORT NETWORKS

Three possible transport networks were considered, namely:
(1) rail; (2) road/rail; (3) road

The rail only option was considered to be not feasible as many of
the waste producer sites do not have adjacent rail sidings; the necessary
rail extensions could only be provided at great cost and with major
planning difficulties. Assuming the repository site would be provided
with its own rail sidings the road/rail options would involve a short
road journey from the waste producer site to a convenient railhead,
followed by a rail journey to the repository site. The major advantages
of rail transport over road transport are high payload capacity and less
stringent weight considerations. However, transport of LLW does not
really benefit from these. The rate of production of LLW is such that a
large accumulation at the producer site would be necessary in order to
obtain an economic payload for a train. The weight constraints are not
particularly relevant since LLW does not require the massive shielding or
containment required for intermediate level wastes or irradiated fuel.
Therefore road transport was adopted as the standard for this study. Rail
transport may offer some advantages in cases where the distance between
producer site and repository site is greater than could be reasonably
covered by a road vehicle in one day (in this particular study the SSEB
sites would fall into this category). The alternative would be to provide
overnight accommodation for the road vehicle on some licensed site at a
suitable intermediate location.

THE ROAD TRANSPORT NETWORK

The projected rates at which waste will be produced by the
prospective clients of the Nirex LLW repository were used to calculate
the rate at which loads will be transported. The waste production rates
listed in Table 1 are, in most cases, for a particular organisation and
are summations over a number of geographic locations, each being
different distances from the disposal site. The survey of waste producers
considered these locations individually, the breakdowns for the
multi-site waste producers are summarised below:

-   Amersham International has two sites, one at Amersham, and one at
    Cardiff. About 80% of the LLW arisings are dispatched from Amersham.
-   The BNFL "non Sellafield" wastes were dominated by the arisings from
    Springfields, and therefore all of the LLW was considered to
    originate from there.
-   The CEGB waste was considered to be from fifteen nuclear power
    stations spread over ten locations:

Berkeley	(2)
Bradwell	(1)
Dungeness	(2)
Hartlepool	(1)
Heysham	(2)
Hinkley Point	(2)
Oldbury	(1)
Sizewell	(2)
Trawsfynydd	(1)
Wylfa	(1)

Table 2. Magnitude of LLW Transport Network

Source of Arisings	No.of drums per year	Vehicle Loads per year	Distance to Elstow (miles)	Vehicle miles per year
Amersham International				
Amersham	7,920	132	40	5,280
Cardiff	1,980	33	200	6,600
BNFL				
Springfields	4,500	75	200	15,000
CEGB				
Berkeley	1,600	27	180	4,860
Bradwell	800	13	100	1,300
Dungeness	1,600	27	150	4,050
Hartlepool	800	13	250	3,250
Heysham	1,600	27	220	5,940
Hinkley Point	1,600	27	220	5,940
Oldbury	800	13	180	2,340
Sizewell	1,600	27	140	3,780
Trawsfynydd	800	13	230	2,990
Wylfa	800	13	230	2,990
SSEB				
Hunterston	2,700	45	430	19,350
Torness	900	15	430	6,450
UKAEA				
Harwell	2,500	42	70	2,940
Winfrith	3,000	50	150	7,500
Others (NDS)	10,000	167	200 (typ.)	33,400
	45,500			133,960

The list included power stations which were presently under construction or planned. Berkeley Nuclear Laboratories was considered to be equivalent to one power station. There does not appear to be any correlation between the amount of LLW a power station produces and its capacity, reactor type or load factor. For the purposes of this study the quantities of waste were divided equally among the power stations.

- The SSEB has three nuclear power stations; two at Hunterston and one at Torness. These have been assigned 75% and 25% respectively of the LLW arisings.

In Table 2 the magnitude of the road transport network is presented. The LLW suitable for packaging in 200 litre drums (see Table 1) is expressed as a number of drums arising per year. This number of drums provides the number of vehicle loads per year (assuming 60 drums per load), and hence the total vehicle miles from the site of arising to the repository site at Elstow. Thus the magnitude of the road transport network may be expressed as a number of (loaded) vehicle miles per year.

The method of operation assumed was that the transport vehicles would be based at the repository site, and dispatched with an empty freight container to the waste producer site when required. On arrival the empty freight container will be exchanged for one full with drums, and the vehicle returned to the repository site leaving an empty freight container to be filled for next load. The minor waste producers, labelled "Others (NDS)" in Table 2, only produce a small number of drums each (although their combined total is significant). In this case the method of operation assumed was that a vehicle made a tour around several waste producing sites collecting drums until fully loaded, and then proceeded to the repository site.

An analysis of the necessary vehicle movements, and rates at which freight containers would be turned around at the waste producer sites, indicated that a fleet of forty freight containers and six road vehicles would be required to service this transport network. Preliminary cost estimates indicated a capital cost (in 1985) of 0.5 million.

Investigations were also made into the probability of an accident involving a loaded container. An attempt was made to quantify reported incidents involving loads of hazardous chemicals. However, there are no systematic collections of such data at a national level. The Department of Transport's road accident statistics offered some basis for an estimate. The tabulated data for 'fatal' or 'serious' accidents involving heavy goods vehicles suggest a probability of the order of 6×10^{-7} per vehicle mile; only this category of accident was considered, implying that an accident would need to be at least classed as serious to cause significant damage to a package. This would imply that the road transport network described above (about 134 000 loaded vehicle miles per year) would have a probability of 0.08 per year of a serious accident involving a load. The above figure applies to all HGV movements, whether carrying hazardous or non-hazardous loads. There is some evidence to suggest that the bulk movements of chemicals (which was used as a model for this study) has a much smaller accident rate; this is typically 300 incidents per year involving packaged hazardous chemicals, from a total of the order of 10^8 package movements per year. Only about 10% of these incidents result in a loss of containment.

CONCLUSION

The study has shown that the 200 litre drum is a suitable package for LLW, and that these could be handled by a road transport network. Although the overall quantities of LLW may be predicted with reasonable confidence there is still a need to establish more precise data relating to radioactive content and bulk density when packaged. In particular it was concluded that the package surface dose rate bands at the lower end of the scale should be subdivided to give a better picture of the exposures likely to occur. Volume reduction was not considered in detail in this study, but it was felt that this merited some further investigation. One major area of uncertainty is the location of, and rates of arising from, the minor waste producers. Although these may be individually small, together they contribute over 20% to the total number of drums, and require 25% of the capacity of the assumed transport network.

ACKNOWLEDGEMENT

This paper is published by kind permission of UK Nirex Ltd.

DISCUSSION FOLLOWING SESSION 2 Papers 1 to 4:

MR.F.NUHAAN, WESTINGHOUSE INTERNATIONAL, BRUSSELS for Paper 2:4
I haven't heard the criteria for durability of the container addressed,
could you answer this please?

MR.A.JOHNSTONE, ELECTROWATT ENGINEERING:
To a certain extent that was beyond the scope of the study. I think it is
a problem that all the individual waste producers have had to address on
their own sites anyway, in that a steel drum of about 16 gauge steel has
a rather limited life unless it is in a protected environment.
Essentially we were looking at a one-day trip from the producer's site to
the disposal site with the idea that it either goes into a shelter there
or is disposed of almost immediately. I didn't mention this but one of
the concepts we actually looked at was taking the stillages straight off
the container and into the trench for immediate disposal.

MR.F.NUHAAN:
You are speaking about waste which is to be buried and this container
should be able to withstand conditions during a certain period. Is a
factor being calculated into the material specification?

MR.A.JOHNSTONE:
Have you seen the Drigg site? It is not particularly well contained at
present although I believe they are improving the standards now.
Generally I think we considered this to be beyond the scope of our study,
we were looking at moving these things, but not designing the disposal
life.

MR.M.J.S.SMITH, UK NIREX LTD:
As Mr.Johnstone mentioned, the work he described was carried out several
years ago. There have been a number of developments subsequently. We no
longer envisage that the 200 litre drum will be used as the disposal
container at a Nirex repository. Instead most of the drums will be super
compacted to conserve space and reduce voidage in the repository. The
compacts will be placed in Nirex standard $3m^3$ or $12m^3$ boxes and the
voids between them filled with grout.

MR.D.RIBBANS, ONTARIO HYDRO for Paper 2:3
The use of the impact limiters for tie downs is unusual. Can you comment
on the relative strengths of the tie down vs. the strengths of the impact
limiter attachments?

DR.H.GEISER:
This is a key question. You use more than one cask inside a 20'
container. You have to calculate the tie down, the necessity for the tie
down and the necessity for holding the shock absorbers together means the

shock absorbers basically belong to the cask and are secondary to the over pack which is in this case a 20' container.

MR.T.W.STEPHEN, JAMES FISHER AND SONS PLC for Paper 2:3
You refer to a drum which is contained in a 20' container, the shock absorbers are holding the drum in the container, what force - presumably the friction between the shock absorber and the container was holding it in place? Is that correct?

DR.H.GEISER:
The shock absorbers are bolted to the reinforced container bottom. They are bolted and designed to accommodate the usual handling accidents and a little bit beyond, but the bolting is not designed to be in proper operation for 9m drop or similar. The testing of the casks will be done separately without the container as well. The shock absorber should be always present in case of an accident with the cask, not with the container.

DR.H.CODEE, COVRA NV, THE NETHERLANDS for Paper 2:2
Mixing of cement and waste is very important to obtaining a good quality product. Did you compare the quality of your in drum mixed product with the quality of, for instance, screw-mixed products?

MR.D.SAUL, NEI:
No the project itself was purely to do with the in-drum mixing technique.

MR.D.RIBBANS, ONTARIO HYDRO for Paper 2:4
You mentioned that you have chosen road transportation over rail. In surveys in Canada road transportation is the least acceptable politically and you are going to be transporting a large volume, have you considered that side of things, the public acceptability of it?

MR.A.JOHNSTONE:
It isn't explicitly addressed in the paper but we did give it some consideration in that we are looking at a dedicated fleet of containers. These would I think from an acceptability point of view be presented as a very clean-cut fleet of well managed, well looked after freight containers that are not in a gashed or battered condition as a lot of freight containers for commercial shipping are. Even since this work was done the public acceptability has become a bigger barrier than it was at the time. Certainly there is no question about it that rail transport is perceived as safer and we attempted to quantify the relative accident frequencies of road movements and rail movements. The data are particularly good. There are very broad figures for road accidents involving heavy goods vehicles, but there is no systematic collation of data in the UK as to how many of these involved hazardous loads and how many were of sufficient violence to cause the load to lose containment. Even things like transporting acids and chemicals around in bulk tankers on roads is a very large undertaking and incidents with these are very infrequent. I think perhaps the public could be sold something along those lines rather than being forced to the rail option. But perhaps politically in the end the rail transport option might be taken.

CHAIRMAN:
Most of the UK low level waste producers are of course being asked to produce rail transport options for sending their waste to Drigg, or a plan for it, and they have to make this plan by the middle of the year.

MR.I.BRAYBROOK, RAILFREIGHT, BRITISH RAILWAYS BOARD for Paper 2:4
I would like to clear up one misconception which has perhaps crept into

the debate, possibly because of the historical nature of the paper. I
think it is fair to say that Railfreight is now in discussion with UK
Nirex Ltd., BNFL, CEGB and others over the use of rail for individual
containers. There is in fact a network set up for just that sort of
movement. Possibly our own advetising and marketing is at fault – in 1985
it wasn't well enough known. I will in fact be expanding on this tomorrow
in my own paper, but there is an efficient, hopefully cost effective,
network available for moving individual container or waggon loads, though
albeit obviously from certain locations a road journey is involved.

CHAIRMAN:
It's nice to see market forces at work.
MR.I.BRAYBROOK:
Yes, we would far rather win traffic by being the best competitive option
than having to resort to government diktat etc.

MR.M.J.S.SMITH, UK NIREX LIMITED:
This whole subject is to be addressed in a paper by D.Bennett, tomorrow
afternoon, which gives accounts of many subsequent studies on road vs.
rail vs. sea transport.

MR.M.SIMPSON, BRITISH NUCLEAR FUELS PLC, SPRINGFIELDS for Paper 2:4
My question is addressed to Mr.Smith in his supplementary statement. If
super-compaction is envisaged for waste, why should 200 ltr. drums be
used at all for waste movement? It adds to the waste to be handled, the
strength of the material being compacted, and makes maximum concentration
in shipping containers difficult to achieve. Loose loading or bagging
direct to ISO-containers would avoid problems of drum handling.

MR.M.J.S.SMITH:
In fact the super compactors work with drum waste and depend upon the
waste being drummed. I might add that the super compaction might be done
at the site of arising if the waste producer wished to install a
super-compactor as I know several are.

MR.D.BLACKMAN, UK DEPARTMENT OF TRANSPORT for Paper 2:3
Did Dr.Geiser have problems with transport from Germany to the USA? You
mention that your material went from Julich to Savannah River. As these
were B(U) packages, presumably the US Competent Authority recognized the
B(U) concept from the engineering point of view. Was there any reluctance
on their part to validate the fissile aspects of these packages?

DR.H.GEISER:
The seventh shipment is under way right now, and we have had no problems
with validation. We built six of these 20' containers and they are not
only used for shipment of spent MTR fuel from Germany to the USA. This is
one of the benefits of the system because it cuts by half the price of
transportation compared to other packages due to the easier handling
system, but also it is used very frequently for medium and high level
waste transportation in Germany. That means waste packed, cemented in 200
ltr. drums, having surfaced dose rates in the range of 10,000 to 100,000
rads per hour. These casks are then used for dry loading and unloading in
hot cells.

MR.F.J.L.BINDON, INSTITUTION OF NUCLEAR ENGINEERS:
Chairman, if you and the rest of the delegates will forgive me putting a
question to both this morning's authors and those of this afternoon.
Designers of casks and packaging systems have to work to many
regulations, there are regional, national, international regulations and
of course in Europe the EEC directives. Could I ask whether anyone would
like to proffer an opinion as to whether all these regulations involve

conflicts which hinder the nuclear industry and its progress? I know that
is a very big question to answer. Secondly, do those people who addressed
the question of tests on flasks, casks and packaging systems, and who
talked about the tests conducted with fires, involve the outside Fire
Brigades in exercises to see that safety can be paramount when such
accidents occur?

MR.D.RIBBANS:
I cannot comment on the first, but the second point, when we did our fire
tests we involved the Ontario Fire Marshall's office, they were there as
observers and in fact produced the thermograms I showed this morning. We
also have an emergency preparedness programme for all packages in which
we involve the police, fire and ambulance services.

MR.M.S.T.PRICE:
I was away last week, but I believe at Winfrith there was a local fire
and I think 250 firemen came from the SW area.

CHAIRMAN:
There was something about that on BBC radio this morning. I think I can
add that our County Fire Authorities all have emergency procedures laid
down and these are regularly tested in exercises involving BNFL, CEGB and
the major transporters.

MR.C.TAPLEY, AMERSHAM INTERNATIONAL:
In answer to Mr.Bindon's first question. We ship our products to
something like 120 countries around the world and I can say that
complying with the different regulations adds a lot to our costs, but
very little to safety.

MS. B. MORGAN, TECHNICA LIMITED:
I was at a press launch of a report by Large and Associates a few months
ago where I heard the Secretary of the London Fire Brigades' Union claim
that his members had never attended any training exercise for spent fuel
flask accidents in London.

PROFESSOR J.R.A. LAKEY, PRESIDENT, INucE
The Sizewell Inquiry revealed that Fire Service Personnel were subject to
operational radiation dose limits below the ERL for evacuation of the
public. For several years the Institution of Nuclear Engineers has worked
closely with the Institution of Fire Engineers and in 1987 I was able to
discuss the problem of emergency response with every Chief Officer and
Assistant Chief Officer through collaboration with the Fire Service
College at Moreton-in-Marsh. I am aware that a Home Office working party
is actively reviewing the specification of radiation limits for Fire
Service personnel taking into account advice from the NRPB. In addition a
HAZRAD scheme is being discussed which follows the successful principles
used in the well known HAZCHEM system and all of these discussions take
full account of the lessons learned at Chernobyl, bearing in mind that a
similar accident is not possible in the UK. The January/February 1987
edition of the journal of the Institution of Nuclear Engineers, THE
NUCLEAR ENGINEER, contained a relevant series of papers reporting the
IFE/INucE conference held in November 1986.

MR.W.BRACEY, TRANSNUCLEAR INC.
A further comment on Mr.Bindon's first question. Although we have been
invoved in this programme for designing a cask to transport plutonium
contaminated waste, we haven't been directly involved with the training
programmes etc. but the Department of Energy training in the varous
States, through which nearly 500,000 drums of plutonium contaminated
waste will eventually pass, has been an especially important part of

their programme because of the fear of plutonium in the minds of those who will respond to an emergency.

Concerning the difference in regulations I am sorry that there is so little comment made - I am in the business of shipping a lot of LSA material and I experience tremendous difficulty crossing borders and having to comply with different regulations throughout Europe and I do hope that this will be dealt with properly in the EEC Commission.

LARGE TRANSPORT AND DISPOSAL PACKAGES FOR
NUCLEAR POWER STATION DECOMMISSIONING WASTE

G. Holt and D. L. Brown

Nuclear Waste Technology Branch
Generation Development & Construction Division
Central Electricity Generating Board
Barnett way
Barnwood
Gloucester GL4 7RS
England

ABSTRACT

The decommissioning of nuclear power stations will generate
radioactive waste components that are different in size and nature to the
waste generated during the operation of their facilities. The packaging
developed for operational waste is not appropriate to decommissioning
waste and it is therefore necessary to develop alternative packaging.
Packages for decommissioning waste will need to be larger in size and
suitable for waste loading, handling, transport and direct disposal at a
deep waste repository.

Consideration is given to the activities undertaken during
decommissioning and the waste that this generates. The possible options
for the transport of this material to a waste repository are identified
as road, rail, or sea, or a combination of these. The restrictions on
package design imposed by these transport modes and associated
regulations are identified along with other design requirements.
Consideration of these leads to the conclusion that the most appropriate
package design is a top opening box 4m long, 2.4m wide and 1.8m high.
This could be made from either steel or reinforced concrete and would
have an internal volume of approximately 12m^3 and a maximum weight of
65t. This box would be capable of transport by road, rail or sea.

INTRODUCTION

Radioactive waste materials are routinely transported and the
associated packaging and transport systems are well developed. The waste
material that is presently handled results from the operation of nuclear
facilities and is generally of a size and nature that allows it to be
packaged in small containers such as the standard 200 litre cylindrical
drum. However, the decommissioning of nuclear facilities such as power
stations will generate significant quantities of heavy and bulky waste
items that will need to be packaged and transported to a waste
repository. The existing operational waste packages are not appropriate

for this type of material because it would require extensive volume reduction of the waste which would impose significant penaltiies, e.g. increased radiation dose commitment. It is therefore necessary to develop larger packages, and the associated transport systems, that would be more appropriate to this type of waste.

The first power stations to become due for decommissioning will be the steel pressure vessel Magnox stations, and Stage 2 decommissioning (see section on Decommissioning Stages) will be the first stage at which the bulky waste items will arise. The paper therefore concentrates on this type of station and this decommissioning stage although the principles of package design and the transport system will be the same for the decommissioning waste from other reactor types (e.g. AGR and PWR) and also Stage 3 decommissioning.

DECOMMISSIONING STAGES

The decommissioning of a nuclear power station will be carried out in three stages. Stage 1 decommissioning starts immediately after the shutdown of the station and consists of the removal of the fuel from the reactors and its despatch off-site for reprocessing. The waste materials that are produced during this stage are identical to operational waste and will be packaged and transported off-site in the same manner.

Stage 2 decommissioning will follow directly after Stage 1. This stage involves the dismantling of all radioactive and non-radioactive plant and buildings to the reactor biological shield. The radioactive waste items will generally be lightly contaminated and can be classified as low level waste (LLW). Typical waste items will be refuelling machines, fuel storage pond equipment and boilers from the steel pressure vessel Magnox power stations.

Stage 3 decommissioning consists of removing all remaining structures from the site. This invloves the dismantling of the reactor structure and its internals which will be radioactive due to activation as well as contamination. A proportion of this waste will be classified as intermediate level waste (ILW) but the extent will depend on when Stage 3 is carried out. It would be possible to start Stage 3 immediately after Stage 2 but at this time the radioactivity levels will still be relatively high and would require extensive remote dismantling and handling operations. If Stage 3 is delayed, possibly by up to 100 years, benefits will be gained due to radioactive decay, thus reducing worker dose commitment and simplifying dismantling techniques.

DECOMMISSIONING WASTES

Reference has been made in the previous section to low level and intermediate level waste (LLW and ILW). The official definition of LLW is that waste containing radioactive materials other than those acceptable for dustbin disposal but not exceeding 4GBq/t alpha 12GBq/t beta/gamma. ILW is defined as waste with radioactivity exceeding the boundaries for LLW but which does not require nuclear heating to be taken into account in the design of storage or disposal facilities. A simple rule-of-thumb by which the two waste categories can be identified is that LLW requires no, or minimal, radiation shielding during handling and transport whereas ILW does require radiation shielding.

Stage 2 decommissioning of a steel pressure vessel Magnox station could provide approximately 10,000t of steel and 900t of concrete, all of which is classified as LLW. The principal waste items are the main boilers of which there are typically 4 per reactor, each being 7m diameter, 22m long and weighing 850t. Other waste plant items are much smaller in size and are generally sized to meet access and maintenance requirements. Waste concrete arises from contaminated civil structures such as fuel cooling ponds.

Stage 3 decommissioning wastes will consist principally of steel, graphite and concrete. From a typical steel pressure vessel Magnox station the quantities of waste could be approximately 5,000t steel, 4,400t graphite and 17,000t concrete. Of the total waste, about 40% would be ILW if Stage 3 follows directly on from Stage 2. If Stage 3 is delayed by 100 years only about 20% would be ILW.

The size of the plant items and structures that will become waste as a result of decommissioning vary significantly. However, when consideration is given to the methods that will be employed for dismantling and handling the pieces, it becomes apparent that cut sections of steelwork or individual items that will require to be packaged will not weigh more than a few tonnes, or exceed 3 to 4 metres in length.

PACKAGE DESIGN REQUIREMENTS

It is necessary to design the decommissioning waste package to satisfy, in an economic manner, the varied requirements of waste loading, waste encapsulation, handling, transport and disposal. Consideration of these requirements leads to the identification of the following basic design criteria:

(a) The largest practicable package size should be used so as to minimise the extent to to which items need to be dismantled or size reduced.
(b) The package shape should be such as to maximise the amount of waste per unit volume that can be loaded.
(c) The package lid or opening should be sized and located so as to interface with a dedicated waste packaging plant and allow easy loading of the waste.
(d) The package design should allow the waste and internals to be cement grouted so as to minimise the voidage within the package and to stabilise the load.
(e) The package size shape and weight should be such as to allow it to be handled and lifted with conventional equipment, possibly remotely.
(f) The package should be suitable for transport by road, rail or sea.
(g) The package should be suitablefor direct emplacement in a waste repository.

A fundamental requirement associated with a number of the above items is that the package needs to maintain containment of the radioactive materials. This is particularly relevant to package handling, transport and disposal. For example, following disposal the package will be a principal component of the engineered barriers within the waste repository which will be required to maintain containment of the short half life ($<$30 years) radionuclides until they have decayed to acceptably low levels.

The waste arising from power station decommissioning is destined for the deep geological waste repository to be developed by UK Nirex Ltd. This repository will also be used for decommissioning waste from other nuclear facilities as well as the waste arising from their operation. To simplify the design and operation of the waste repository it is the intention of Nirex that the minimum number of different waste package designs and sizes are used. It is therefore required that the packages developed for the wastes from decommissioning of power stations are suitable for decommissioning waste from other nuclear facilities. Also, as both low level and intermediate level waste are to be placed in the repository it is preferable if the package for both low and intermediate level decommissioning wastes are physically similar externally.

TRANSPORT OPTIONS AND REGULATIONS

The existing CEGB nuclear power stations are located on ten sites which are geographically dispersed around England and in Wales. Nine of the sites are coastal but one, Trawsfynydd in Wales, is located inland. All of these sites are serviced by road and for most of them road transport is the sole method of delivery and despatch of goods both large and small. Two of the sites are directly connected to the rail system with a railhead inside the site boundary. All other sites do have access to railheads but these are located away from the site and up to 24 kilometres away.

The rail system is used for all movements of spent fuel from the stations to the reprocessing plant at Sellafield. The fuel is transported in flasks weighing approximately 50t and for those sites with remote railheads the movement of the flasks to the railhead is by road. One transport mode that is not used regularly is sea transport. With all but one site being coastal, sea transport may have some merit and should not be discounted.

It is considered appropriate at this stage of development of decommissioning plans that all three modes of transport be kept open. This is also sensible if other non-CEGB nuclear facilities are considered as they are also a mixture of coastal and inland locations served to varying degrees by road, rail, and potentially by sea. The decommissioning waste package should therefore be suitable for road, rail, or sea transport. Each of these transport modes has different limitations which could have an impact on the package design. The following sections address these points.

The transport of radioactive materials is governed by a variety of regulations which place various restrictions and requirements on transport activities and package designs. These are also discussed in a following section.

Transport by Road

The principal statutory instruments covering transport by road, which govern the size and weight of loads and other operational controls, are the Construction and Use Regulations (Statutory Instrument 1017, 1978) and Special Types General Order (Statutory Instrument 1198, 1979), and subsequent amendments to these regulations.

The Construction and Use Regulations govern the transport of normal loads by road. The principal restriction relevant to the carriage of decommissioning waste packages is the maximum allowable gross vehicle weight of 38t. If the typical weight of a tractor and semi-trailer combination is deducted this gives a maximum payload of approximately 25t. There are also controls on the distribution of weight but this is not considered to be limiting in this application.

The other relevant sections of the Construction and Use Regulations are those specifying maximum overall dimensions. The most important dimensional limit is that of width, which for a tractor and trailer is 2.5m maximum. The maximum length and overall height of a semi-trailer is 12m and 4.2m respectively. The maximum allowable speeds for a vehicle complying with these regulations is 40mph on single carriageway roads, 50mph on dual carriageways and 60mph on motorways.

Vehicles that exceed the allowable weights and dimensions of the Construction and Use Regulations are governed by the Special Types General Order. Under this regulation abnormal indivisible loads can be carried on special type vehicles. The special type vehicle plus load should generally not exceed 4.3m width, 27.4m length and 152.4t gross vehicle weight. There are also other controls on axle loading and spacing. The specified gross vehicle weight would allow a maximum payload of approximately 100t. Until recently the most restrictive limitation on the use of special type vehicles was a maximum speed limit of 12mph.

A recent amendment (Statutory Instrument 1327, 1987) to the Special Types Order, applicable as from 1st January 1988, has varied some of the limits mentioned above. Of particular importance is the relaxing of speed restrictions. For gross vehicle weights not exceeding 150t the limits are 30mph on motorways, 25mph on dual carriageways and 20mph on other roads.

The use of special type vehicles to carry abnormal indivisible loads requires that notice of journeys and an indemnity be given to Highway and Bridge Authorities, and, in certain circumstances, notice has also to be given to the police.

In comparison with other modes of transport, road transport is considered to be the most flexible and economic system. However, for frequent regular journeys (such as the movement of decommissioning waste packages to a repository) it is preferable to comply with the Construction and Use Regulations which do impose certain weight and dimensional limits. However, should only short journeys be required, for example to a local railhead or dock, then, as with the present movement of fuel flasks, the use of special type vehicles could well be quite acceptable. The major benefit from the use of special type vehicles is the substantial increase in allowable payload. The associated increases in allowable dimensions are unlikely to be of benefit due to the possible restrictions of existing roads and bridges and also the increased disruption to other traffic they would cause.

Transport by Rail

When consideration is given to the use of rail transport for decommissioning waste packages, the most important aspects to consider are the dimensional and payload restrictions of the UK rail network. The dimensions of the rail wagon and hence the size of load it is able to carry is controlled by the B.R W5 Loading Gauge. This gauge specifies the maximum allowable widths at various heights above the rail level, and these widths vary according to the overall length of wagon and bogie spacing. Typically, for standard 2 or 4 axle wagons a load width of 2.7.m and height of 2m would be acceptable. If the width was smaller a taller height would be allowable.

Rail wagons can be fitted with 2, 4 or 8 axles. The maximum allowable axle loading is presently 25.5t although some routes are still restricted to 22.5t. It is therefore prudent to assume the lower axle loading which would give payloads of approximately 33t,68t and 120t for wagons with 2, 4 and 8 axles respectively. The most commonly used and standard wagons on the rail network are either of 2 or 4 axles. Wagons with 8 axles are only used for special applications and tend to be longer and hence have more restrictive limits on the overall width of load due to the potential for overthrow on curved rail sections. The 8 axle wagons, because of their specialised nature, tend to be more expensive.

The use of rail transport, in comparison with road transport, is less flexible and more expensive. If the same practice is applied as for fuel flask rail traffic, charges are on a per train load basis. It is therefore advantageous to operate as near to the maximum load per train as possible. If, as a planning basis, a 1,000t gross trailing load per train is assumed this would allow up to 20, 10 or 5 wagons of 2, 4 or 8 axle types respectively, per train. However, it should be noted that the capacity of the existing railheads that serve the power station is generally such that only 2 or 4 wagons can be accommodated at any one time. This implies that reduced capacity trains could have to be used or marshalling away from the railheads undertaken. Also, the railheads are commonly located away from the power station sites and hence an element of road transport, between the station and the railhead, would be necessary.

Although there are apparent disadvantages in using rail transport, there are advantages to using rail for the regular movement of heavy loads over long distances between two designated points, such as decommissioning power station and a waste repository. Rail transport in comparison with road transport is better able, without undue restriction, to handle heavy loads, it does not disrupt other users of the transport system and the probability of accidents is significantly lower.

Transport by Sea

Sea transport is occasionally used for deliveries to and from power stations. Some large loads for stations under construction were carried by sea. For example, boilers for Berkeley Power Station were transported by this method. Other large loads such as transformers are occasionally transported by sea for part of their journey. As 9 out of 10 of the nuclear power station sites are coastal, it is worth considering the possibility of sea transport. Also, it must be recognised that of the options being considered for a future deep waste repository, access via an off-shore structure or island is a possibility and hence sea transport could be a necessary element of the overall transport system.

There are two basic options for sea transport and that is to use either existing ports or to develop suitable berthing facilities on the power station sites. The use of existing ports would be the least expensive in terms of capital investment but this option would have to involve an element of road transport, and the restrictions this imposes, as suitable ports are some distance away from the power station sites. It would be feasible to construct berthing facilities at the power station sites, which would eliminate the need for a road transport element, but it is uncertain whether this would be an economic proposition.

A number of options exist for suitable vessels, such as barges and roll-on/roll-off or lift-on/lift-off ships. The load carrying capability of these vessels would place no restrictions on the maximum size and weight of load that can be carried. In fact, studies have shown that it would be possible to handle whole boilers partially filled with grout weighing about 2,000t and transport them by sea. However, this would require direct access to the sea from the decommissioning power station site as it would not be possible to transport this type of load by road.

<u>Transport Regulations</u>

There are a number of regulations covering the transport of radioactive materials by road, rail and sea. However, they are generally based on the transport regulations published by the International Atomic Energy Agency (IAEA). There are two editions of these regulations, the 1973 (As Amended) Edition (IAEA Series 6, 1973) and the 1985 Edition (IAEA Series 6, 1985), which are currently recognised by law. The 1985 Edition will come fully into force in 1990, prior to the decommissioning of nuclear power stations, and hence this is the only edition that is considered in the development of the decommissioning waste package design.

The IAEA transport regulations classify radioactive materials into certain categories depending upon their radioactive content and nature. Each category of radioactive material is then allocated to particular packaging design standards and requirements. For example, packages can be required to withstand certain types of tests which simulate either normal or accident conditions of transport.

An assessment of the radioactive inventory of nuclear power station decommissioning waste shows it can normally be classified as either Surface Contaminated Objects (SCO) or Low Specific Activity (LSA). This is certainly the case for the low level waste from Stage 2 and Stage 3 decommissioning and the majority of the Stage 3 intermediate level waste. It is possible that some of the Stage 3 ILW will be outside the SCO or LSA categories. The package appropriate to SCO and LSA is the Industrial Package Type 2 (IP-2).

The IAEA transport regulations specify the design and performance requirements of the IP-2 package. The most significant requirements are that following tests to simulate normal conditions of transport there should be no loss or dispersal of the radioactive contents, that any loss in radiation shielding integrity should not result in more than a 20% increase in the radiation level at any external surface of the package.

The specific tests to demonstrate the ability of the package to withstand normal conditions of transport are a free drop test and a stacking test. For a package of mass greater than 15t the free drop test is from a height of 0.3m on to an unyielding surface in such a way as to impose maximum damage. The package should also withstand the equivalent of stacking six high. It will be easier to ensure the package will withstand the drop test without damage or increase in package surface dose rate if the waste items within the package are restrained from moving. This is one of the reasons why it is intended that the package be filled with grout around the waste items.

As well as ensuring the package will withstand the regulatory test conditions, it is necessary to consider also any other non-transport situations or faults that might impose a higher design standard. For example, consideration will have to be given to the handling methods and lifting heights that are likely to occur on the decommissioning site or at a repository and the probability and consequences of a fault situation.

Another important aspect of the IAEA transport regulations to consider is the allowable radiation levels on packages. The maximum allowable surface dose rate on a large package should be taken as 2mSv/h, and 0.1mSv/h at 2m from the package surface. However, when consideration is given to minimizing the dose commitment to decommissioning and transport workers, it is expected that lower dose rate limits will be required. Even so, for low level decommissioning waste it is unlikely that shielding, other than that provided by the structural walls of package, will be necessary. For intermediate level waste however, additional shielding in the form of thicker package walls could be necessary.

PACKAGE DESIGN

The preceding sections have identified the package design requirements and the constraints imposed by the various transport options. It is now necessary to access these varying factors and identify an outline design that is best suited to the purpose.

Waste Loading and Handling

From an assessment of the decommissioning waste inventory it is apparent that the waste packages will be required to hold large amounts of steel plate, boiler tubing, blocks of concrete and graphite and a miscellany of other items. The simplest package form to accommodate this variety of shapes and sizes is a box of rectangular profile and the easiest way of loading it with waste will be through a top opening lid.

When waste is loaded into the box it will be necessary to ensure that it is distributed in such a way as to ensure that its overall weight, centre of gravity, specific radioactivity and the resulting external dose rates are controlled within acceptable limits. This will require the waste to be assayed prior to it being lowered into the box and possibly for the waste to be placed into a basket which could require internal guides and locating features in the box. This suggested arrangement should allow the complete grouting around and within the waste components.

A top opening box could readily be sealed against a posting port to allow the loading of waste under contamination controlled conditions. A box would be readily suited to handling, stacking and transport. For example, a handling or lifting system using twistlock features set in the box could easily be accommodated.

Overall Dimensions and Weight

In principle the box should be as large as possible to minimise the extent to which it is necessary to size reduce the waste items. However, when the decommissioning techniques are considered it is likely in the majority of instances for individual waste items to be no more than a few tonnes in weight or exceed 3 or 4m in length.

The dimensions of a box will be controlled by the limitations imposed by the transport mode. For rail a width of 2.7m could be possible but this depends on the box length and weight. For road a width of 2.5m would be quite feasible, or 4.3m if a special type vehicle was used. At this stage of development of the box design and decommissioning plans it is not appropriate to design up to the maximum limit. Because of this, and to allow some margin for any tie-downs or covers a width of 2.4m has been adopted. This dimension is compatible with the size and shape of the waste items.

With respect to the height of the box, it is necessary not only to consider the limits imposed by the transport mode, but also to assess the implications with respect to waste loading systems. Taking these points into account and with a knowledge of the waste items, a judgement has been made that a box height of 1.8m would be appropriate.

The choice of overall length is unlikely to be dicated by what will fit on the transport vehicle, as length is not particularly restricted. The controlling factor will be the overall weight and this will depend on the quantity of waste that can be packed into a box. Random packing of steel items into a box is likely to achieve a packing factor of 10%, i.e. a completed box would contain 10% by volume of steel and 90% by volume of grout. With careful packing of waste a maximum packing factor of 50% is unlikely to be exceeded. Figure 1 indicates how box weight will increase with box length for these two packing factors, assuming a box width and height of 2.4m and 1.8m respectively and making an allowances for the thickness of the packaging itself.

The maximum load that can be carried by road and rail without significant restriction is approximately 100t by either a special type road vehicle or 8 axle rail wagon. This is equivalent to a package length of 6 metres or more. There are unlikely to be any individual waste items that would justify this length of package and, recognising the complications of transporting such a large package, it does not appear to be a practical proposition. There would also be potential problems with achieving a uniformly distributed load and an even centre of gravity for such a long box loaded with a number of waste items.

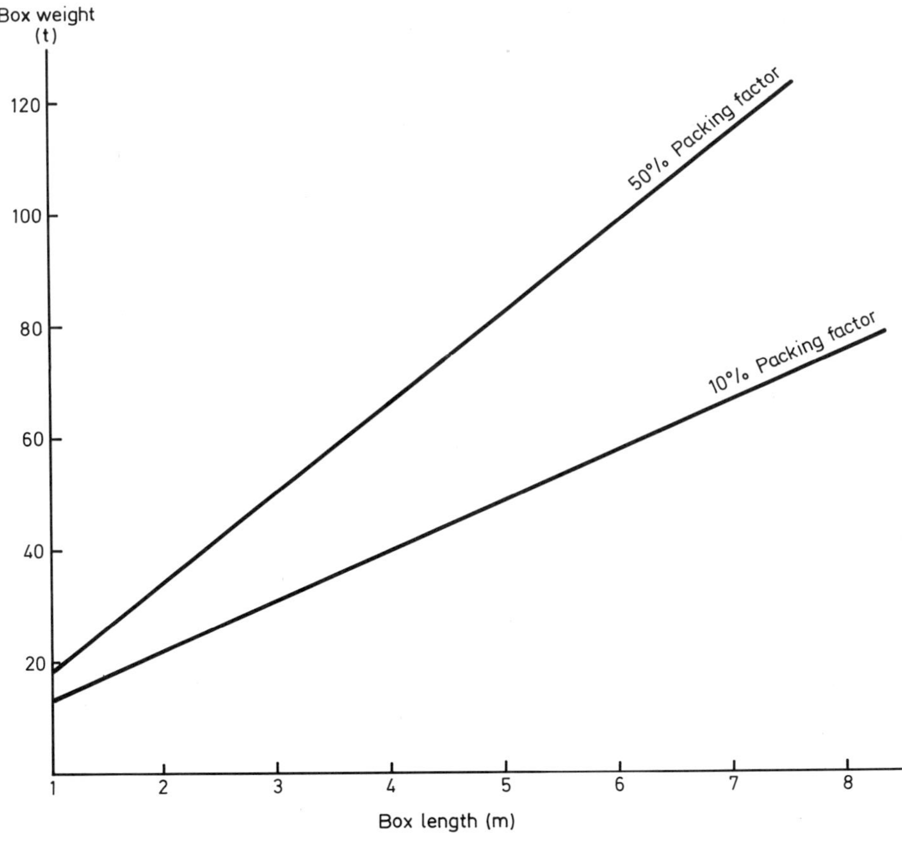

Fig. 1. The variation of box weight with length.

165

If this large box is discounted then the next size to consider would be one suitable for transport by rail on standard 4 axle wagons rather than the special 8 axle wagon. The load capability of a 4 axle wagon is about 65t which equates to a box length of 4m. This size appears well suited to the expected maximum length of waste items, could be loaded more evenly and handled more readily. Road transport and a special type vehicle would still be required. However, this weight is not too dissimilar to a fuel flask whose transport by road over the short distance to existing railheads is a well proven system. A box length of 4m has therefore been adopted giving the overall box dimensions of 4m x 2.4m x 1.8m high and an approximate internal volume of $12m^3$ assuming no additional shielding. A box of such dimensions and weight is considered to be compatible with existing deep waste repository concepts.

Other dimensions of boxes could also be chosen and shown to be suited to particular decommissioning applications. For example, a smaller box that could be transported without restriction by road may have an application. In this case, the weight limit would be 25t which is equivalent to a box 2m x 2.4m x 1.8m.

Materials of Construction

The options for the materials of construction of a box are either steel or reinforced concrete, each of which has advantages and disadvantages. For example, a steel box would require thinner walls for structural strength thus giving a slightly larger usable internal volume. However, a reinforced concrete box could be cheaper to manufacture, although this advantage would be reduced if it is found necessary to add steel cladding to the box to act as an engineered containment barrier for disposal purposes. A box of either material is expected to withstand all the regulatory testing that would be required.

Possible box configurations using both steel and reinforced concrete are shown in Figs. 2 and 3. A box of the proposed external dimensions could be used for both low and intermediate level waste. In the case of intermediate level waste this could require the package walls to be made thicker for shielding purposes. This shielding could be provided by either additional steel or concrete thickness.

CONCLUSIONS

Following consideration of the nature of decommissioning waste and the transport options, it is concluded that a steel or reinforced concrete box, of dimensions 4m x 2.4m x 1.8m, giving an internal capacity of approximately $12m^3$, should be adopted as the standard package to be used for the loading, grout encapsulation, handling, transport and disposal of both low and intermediate level radioactive decommissioning waste.

ACKNOWLEDGEMENTS

This paper is published with the permission of the Central Electricity Generating Board.

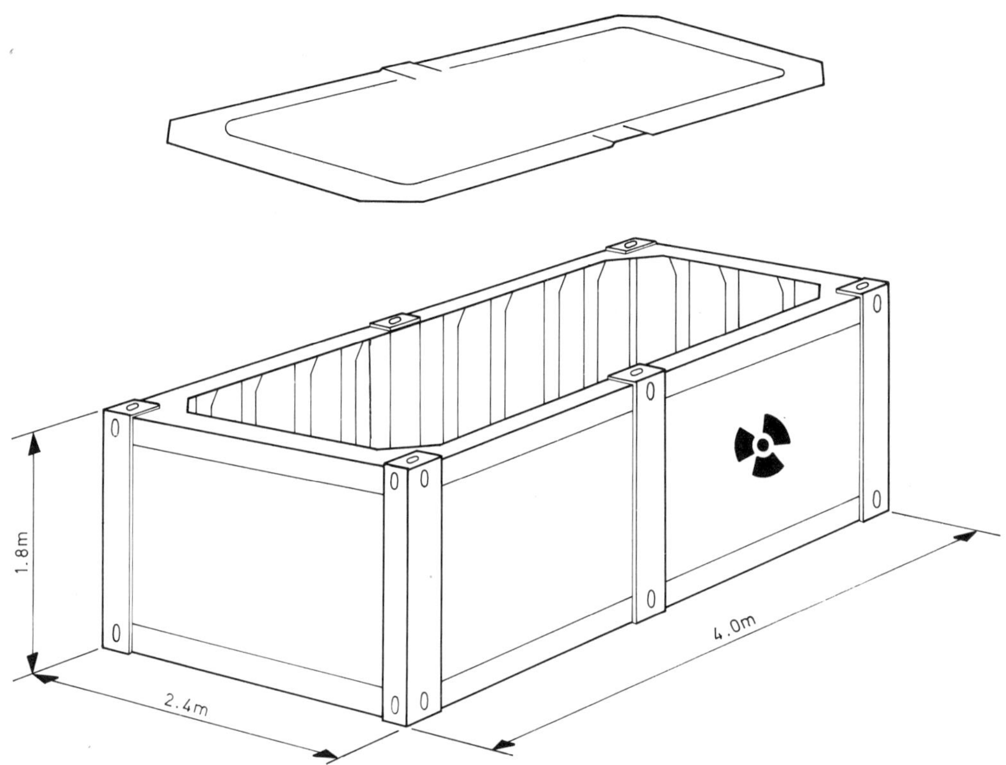

Fig. 2. Steel 12m^3 decommissioning waste box.

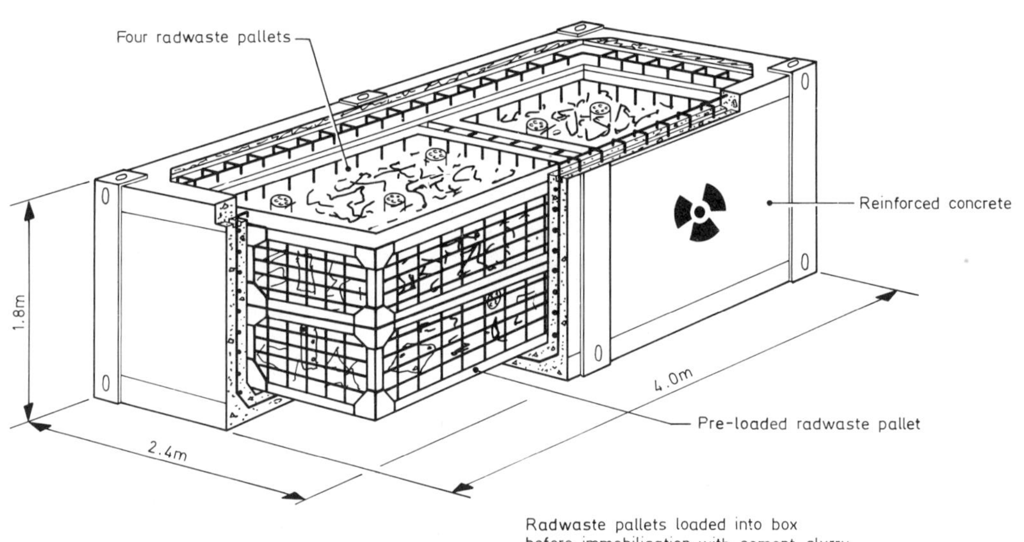

Radwaste pallets loaded into box
before immobilisation with cement slurry

Fig. 3. Concrete 12m^3 decommissioning waste box.

REFERENCES

International Atomic Energy Agency Safety Series 6, 1973, Revised Edition (As Amended). "Regulations for the Safe Transport of Radioactive Materials"

International Atomic Energy Agency Safety Series 6, 1985, "Regulations for the Safe Transport of Radioactive Materials"

Statutory Instrument No. 1017, 1978, "The Motor Vehicles (Construction and Use) Regulations, 1978".

Statutory Instrument No. 1198, 1979, "The Motor Vehicles (Authorisation of Special Types) General Order 1979".

Statutory Instrument No. 1327, 1978, "The Motor Vehicles (Authorisation of Special Types) (Amendment) Order 1978".

A TRANSPORT PACKAGING FOR THE CARRIAGE
OF RADIOACTIVE ION-EXCHANGE RESIN

J.   Armitage

Rolls-Royce And Associates Limited
P.O.Box 31
Derby
DE2 8BJ

ABSTRACT

An assessment leading to the specification and ultimate design of a transport packaging to carry shielded ion-exchange resin storage containers and contents is outlined, including a description of the important physical and legislative parameters and the methods used to satisfy them. The ion-exchange resin was held with its carrier water in portable storage containers and it was in this form, with contents and container complete, that the carriage was to be effected. A future generation of storage container was planned and the transport packaging was expected to include an allowance for size, weight and shielding to cater for this potentially larger design. The ion-exchange resin was assessed against the Low Specific Activity description, and the aim was for a generalised requirement of unrestricted transport by UK road, UK rail, and sea. The transport packaging was to be in the form of a sealed containment, designated IP-2, to retain the liquid content if it leaked from the storage container. A specification, against which commerical designs were produced, is also descibed.

INTRODUCTION AND BACKGROUND

Radioactive ion-exchange resin, when discharged from water clean-up and purification systems, requires storage in a safe, sealed and shielded containment until such time as it can be extracted and conditioned for disposal.

This containment may take the form of large permanent vaults, but in the case considered here, it was in the form of smaller moveable containers in order to accommodate periodic changes in site storage planning. The design is shown in Figure 1(a).

When new, and in good condition, they had been assessed and approved as Type-A standard. However, over a long lifetime, nominally 15 years, some deterioration had occurred and, whilst there was no evidence of loss of integrity, it was difficult to maintain the original safety justification.

One solution was to carry out specific container examinations and reassessment, which would be costly and dose intensive with no guarantee

of success. Alternatively, a transport packaging could be produced to
carry the ion-exchange resin and storage container, and it was this
concept which was investigated.

A secondary transport packaging was designed, into which the storage
container could be inserted, and which would be sealed and prepared in
accordance with the regulations for use with normal transport systems.
All modes, except air, were considered.

It was necessary also to consider the long-term suitability of the
transport packaging as future requirements for waste management suggested
that the current size of storage container was not necessarily the
optimum. A larger, but unspecified size would likely be more
appropriate.

It was decided, therefore, that consideration should be given to a
transport packaging which would cater for the known storage container
size envelope and a potential increase for a future design, with added
guides and fixtures to accommmodate both.

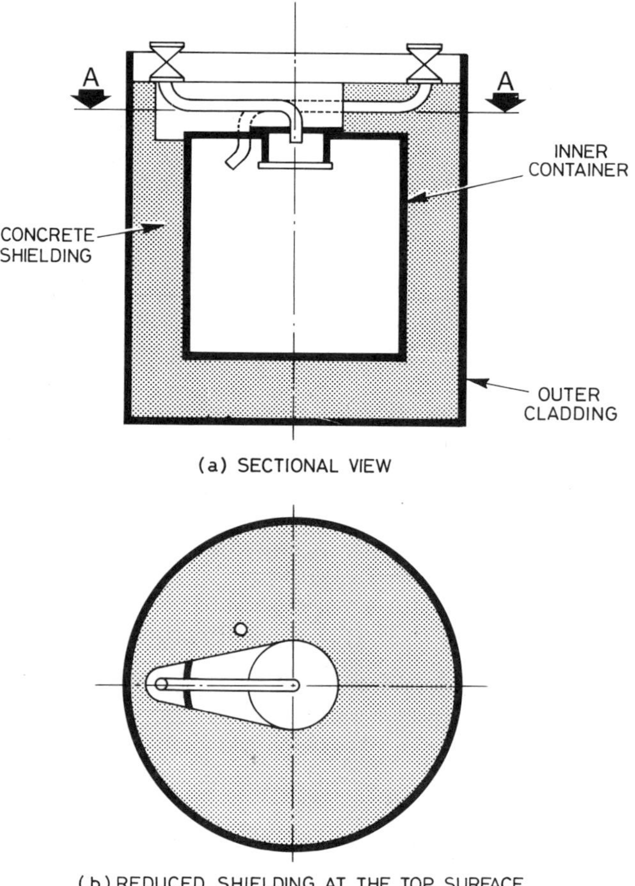

Fig. 1. Ion exchange resin storage container.

TASK DEFINITION

The aim, therefore, was to:-

(i) design a transport packaging which would enable loaded and sealed ion-exchange resin storage containers to be transported in compliance with the appropriate regulatory requirements by all modes of UK road, UK rail and sea transport such that it would enable unrestricted use;

(ii) ensure a suitable packaging size to accommodate future, and larger, generations of storage container in addition to that used currently.

STORAGE CONTAINER/CONTENTS DESCRIPTION

The current ion-exchange resin storage containers are cylindrical double-skinned steel and concrete-lined (Figure 1a).

Nominal dimensions	$1.52m$ diameter, $1.91m$ height
Capacity	$0.71m^3$
Weight	7 tonne

The contents were mixed-bed organic ion-exchange resin, based on gel-type polystyrene beads, and carrier water.

The radioactive inventory comprised that held by the beads themselves, with a small proportion present as particulate matter filtered out during operation.

The circumferential and base shielding, afforded by the concrete lining, was complete but the shielding at the top surface was not. This area of the container housed the operational pipework etc. and it was necessary during operation to retain partial shielding only and defer full capping (Figure 1b).

TRANSPORT CATEGORY OPTIONS AND PRELIMINARY ASSESSMENT

<u>Definitions</u>

In accordance with the requirements of the International Transport Regulations (IAEA 1985, IAEA 1986), the following category definitions were reviewed against the material description to determine the most appropriate mode under which it could be transported.

<u>Low Specific Activity (LSA)</u> Material was that which by its nature would have a limited specific activity.

None of the LSA-I descriptions were satisfied as they covered ores, unirradiated natural uranium, depleted natural uranium, natural thorium and materials with unlimited A2 levels.

LSA-II, which covered material in which the radioactivity was uniformly distributed throughout, was considered to be possible provided that the solid and liquid radioactivity levels could be satisfied.

None of the LSA-III descriptions were satisfied as they referred to consolidated waste in a solid matrix.

<u>Type A</u> material was that which would have a maximum radioactivity

content not exceeding A2Bq, or A1Bq if in special form, and was not related to specific activity as with the previous case.

It was required to consider transport of material, not in special form, and with a radioactivity content in excess of A2Bq. The type A category was inappropriate and discounted.

Type B material was that which would have a maximum radioactivity content exceeding A2Bq, or A1Bq if in special form, and would cover, without reservation, this particular case.

Categories were identified, therefore, as Type B and potentially LSA-II, which required further evaluation, and Type B. The decision to opt directly for a Type B packaging design was not made, deciding initially to carry out a detailed review of the LSA concept and the potential for using the IP packaging.

FURTHER CONSIDERATION FOR LSA MATERIAL

LSA material was limited by the regulations to that in which the specific activity was so low as to become inconceivable that, under any circumstance associated with transport, a sufficient mass could be taken into the body and give rise to a significant radiation hazard.

Two assumptions were implicit in the derivation of the LSA limits, and which would restrict the type of materials considered, namely:

(i)   the radioactivity would be uniformly distributed throughout the material;

(ii) there would be no conceivable mechanism by which its specific activity would be increased during transport.

There was also the additional requirement that the quantity of LSA material would be so restricted in a single package that the external radiation level at 3m from the unshielded material would not exceed $10mSvh^{-1}$.

A number of isotopes were present; the dominant being Cobalt 60 for which $A2 = 0.4TBq$.

In the derivation of an acceptability limit it was recognised that the contents presented as a mixture, could be seen as different states and, therefore, assessed in different ways.

The mixture of ion-exchange resin and carrier water would appear as a settled solid/interstitial liquid layer with a supernatant liquid layer above.

For the Solid Material

Firstly, the mixture could be considered as presented and assessed as the whole.

The specific gravity of the mixture was estimated and the specific activity maximum allowable limit was determined:-

$$10^{-4} A2.g^{-1} \equiv 40 \ MBq.g^{-1} \equiv 43/46 \ GBq.\ell^{1}$$

Secondly, the mixture could be drained leaving a damp ion-exchange resin settled layer. The specific gravity of this state was estimated and the specific activity maximum allowable limit was determined:-

$$10^{-4} \; A2.g^{-g} \equiv 40 \; MBq.g^{-1} \equiv 44/48 \; GBq. \; \ell^{-1}$$

Finally the drained ion-exchange resin could be dried and all non-chemically-held water evaporated. The specific gravity of this state was estimated and the specific activity maximum allowable limit was determined:-

$$10^{-4} \; A2g^{-1} \equiv 40 \; MBq.g^{-1} \equiv 51/63 \; GBq. \; \ell^{-1}$$

The actual specific activities of known contents, in addition to a future anticipated maximum, were obtained and compared. In all cases, the specific activity levels did not exceed the limiting values quoted above.

For the Liquid

The radioactivity level of the carrier water and suspended solids was assessed and two conclusions were reached :-

- The water itself was non-radioactive;
- In the worst case, the water could be evaporated down to leave a solid residue.

The residue would comprise particles of radioactive matter, retained by the resin acting as a filter. Its specific activity was estimated and compared with the maximum allowable limit of:-

$$10^{-4} \; A2.g^{-1} \equiv 40 \; MBq.g^{-1}$$

The limit was not exceeded.

For the Radiation Limit

The anticipated contents, in terms of a working range of radioactivity levels, were considered suitable for classification as LSA-II material in accordance with the limits discussed above.

However, the single package requirement based on radiation level was not so clear. The requirement stated simply that the radiation levels should be from the unshielded material which, for a solid content, would be relatively straightforward. For the type of content under discussion here, a mobile solid/water mixture, shape was variable.

The as-carried cylindrical shape could be considered as also could a right cylinder, sphere, and hemisphere.

The first three options were considered as typical mathematical/geometric shapes only, whereas the hemisphere could be seen as representative of a settled unrestricted mass.

Each was analysed accordingly and the radiation levels for different parts of the surfaces were calculated and presented, in Figure. 2, in terms of limiting specific activity against the particular geomentry.

The maximum activity level estimated for the contents was lower than the minimum allowable level shown in the figure, confirming the selected category of LSA-II for the ion-exchange resin/water consignments to be transported.

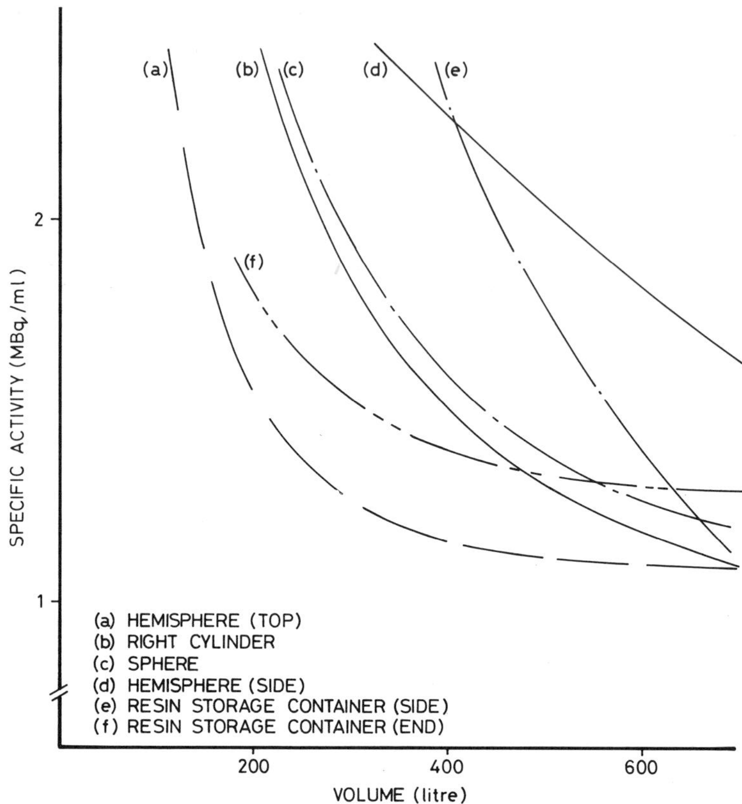

Fig. 2. Specific activity of unshielded ion-exchange resin to produce
10mSv/hr at 3m distance.

PACKAGING REQUIREMENTS

The regulatory requirements (Table 1) for LSA-II material, solid and
liquid, stated that it may be carried in a packaging designated
Industrial Package-2 (IP-2) if carried under Exclusive Use.

Table 1. Industrial Package Integrity Requirements for LSA Material.
(Note: Under certain conditions LSA-I material may be
transported unpackaged).

	Industrial Package Type	
Contents	Exclusive Use	Not Under Exclusive Use
LSA-I		
Solid	IP-1	IP-1
Liquid	IP-1	IP-2
LSA-II		
Solid	IP-2	IP-2
Liquid	IP-2	IP-3
LSA-III	IP-2	IP-3

LSA-II solid material could be carried in an Industrial Package - 2 (IP-2), with or without Exclusive Use declaration, and LSA-II liquid could be carried in IP-2, if Exclusive Use, but would require IP-3 if not Exclusive Use.

The Exclusive Use clause was examined from two points of view:-

(i) Allowable radiation limit could be relaxed from:

$2mSv.h^{-1}$ to $10mSv.h^{-1}$

However, in reality this could not be achieved without infringing other radiation distance limits, eg. vehicle boundary limit, packaging plus 1m limit, vehicle boundary plus 2m limit etc.

(ii) Relaxation from the $2mSv.h^{-1}$ criterion imposed transport limitations if transported by sea, where a package could not be removed from its transport vehicle for separate storage while on board. Exclusive Use packages transported by vessel required special arrangement approval.

In view of this it was considered unlikely that Exclusive Use relaxations could be accepted, although it was anticipated that Exclusive Use transport, in the sense of sole user, would be the normal practice.

The use of Industrial Package-2 specifications would ensure that, under conditions likely to be encountered in normal transport, there would be no escape of contents from the package and there would be limited loss of shielding effect afforded by the packaging.

In addition to the general requirements for all packaging, The Industrial Package Type 2 (IP-2) required that if the package was subjected to the two tests, listed below, it would prevent:-

-    the loss and dispersal of the radioactive contents

-    the loss of shielding integrity which would result in more than a 20% increase in radiation level at the external surface.

The tests were:-

(i) Free Drop Test - a drop on to a target so as to suffer maximum damage.

The height would be function of weight and, in this case for a total package weighing in excess of 20 tonne, would be 0.3m.

(ii) Stacking Test - a compressive load, for a period of 24 hours, equal to the greater of:

a)   5 times actual package weight (which would be nominally 100-110 tonne force)

b)   13kPa multiplied by the verticallly projected package area (which would be nominally 4-5 tonne force).

ALTERNATIVE CONTENTS

The existing size of storage container had been reviewed against future operational requirements. However, and a new design was being contemplated and the capacity was expected to be considerably higher.

The transport packaging was to be designed to accept this larger, but as yet undesigned container.

In principle, it was decided that the existing style of storage container would be retained, i.e. cylindrical double-skinned steel with concrete infill.

The mode of operation would be similar and necessitate a pipe-work array on the top surface, partially set into the concrete shield.

The size would be increased to allow an internal volume of $2m^3$, but the increase would be accommodated by a change of diameter and not a change in height.

The height parity was maintained because access for large containers such as this, was often limited and dictated by crane height. Retention of the current requirement would be of benefit.

OUTLINE DESIGN

The transport packaging design became part of a chain of activities with an enforced information feed-back, to provide for a new ion-exchange resin storage container after the outer packaging was complete.

The sequence was as shown in Figure 3.

A radiological assessment of the current design of ion-exchange resin storage container was carried out to provide details of radiation levels on, or at a distance from, the surfaces (Figure 4).

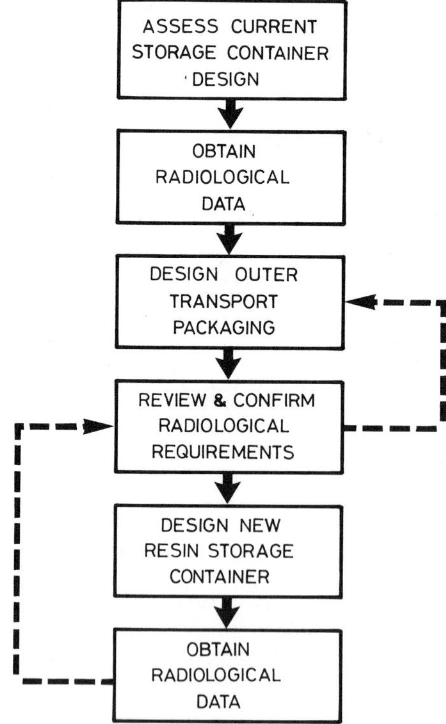

Fig. 3. Storage container/transport packaging design iteration.

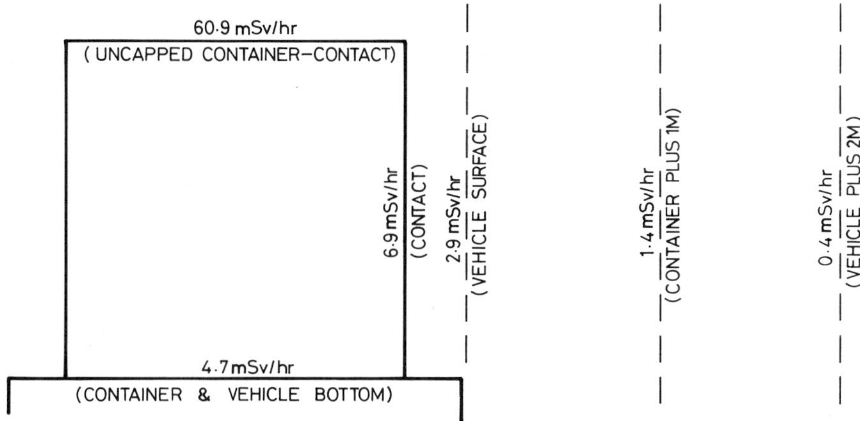

Fig. 4. Ion-exchange resin storage container radiation
levels (current design)

These were based on the maximum allowable radioactive content of
LSA-II level assuming that the container was fully loaded.

The data were used to determine the radiation levels which could be
expected at the transport packaging internal surfaces (Figure 5), with
internal dimensions commensurate with those of the larger proposed
storage container.

The transport packaging shield thicknesses were calculated (Figure
6) and were the maximum thicknesses necessary to ensure that all the
regulatory requirements had been met.

It was possible from this final assessment to estimate the overall
weights, with current and proposed ion-exchange resin storage containers
(fully loaded), which were as follows:-

(i)   Transport packaging - 8/10 tonne
(ii)   Current container plus transport packaging - 16/18 tonne
(iii)   Proposed container plus transport packaging - 20/24 tonne

The design of the future storage container would use this transport
packaging radiation data as prime input. Thus by ensuring appropriate
matching between the content radioactivity level and storage container
shield thicknesses, the levels would not be exceeded.

CONCLUSIONS AND RECOMMENDATIONS

The IP-2 category of packaging was satisfactory for the transport of

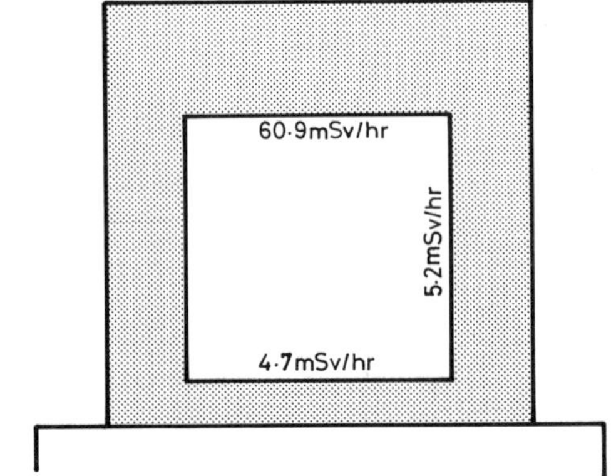

Fig. 5. Transport packaging radiation levels (internal)

the estimated current and future arisings of ion-exchange resin and design should proceed on this basis.

The Type-A category should not be considered further as the estimated content activity levels exceeded those required by the transport regulations.

The Type-B category was more suited to high-level materials and severe accident protection and should not be considered further.

The container design should be adaptable and cater for:-

(i) the current design of resin storage containers and estimated current and future contents
(ii) possible future designs of resin storage containers

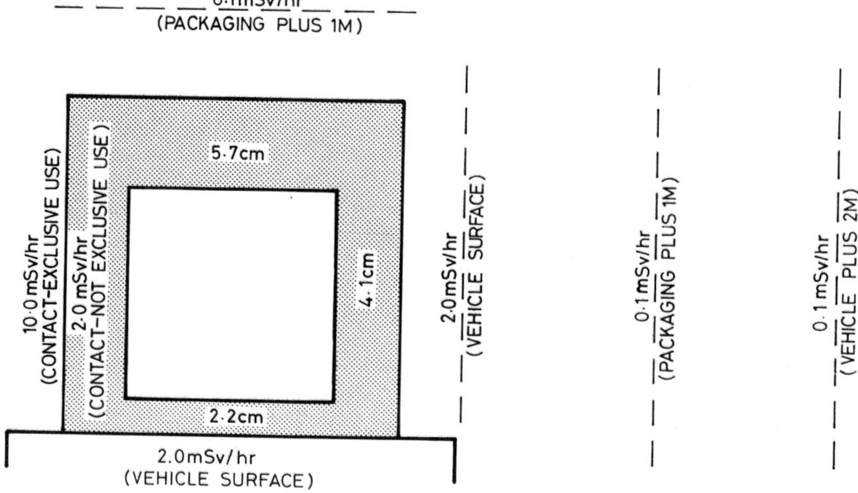

Fig. 6. Transport packaging shield thicknesses.

within reasonable bounds and within the overall envelope of size and
weight allowed by the appropriate current regulations for unrestricted
road transport.

The Statement of Requirements for Invitation to Tender should be
based on the main points identified in Appendix 1.

RESULTS OF THE RECOMMENDATIONS

Production of the transport packaging is proceeding and design of a
new generation of storage container is in hand.

ACKNOWLEDGEMENTS

The author wishes to acknowledge the contribution by Mr.D.Beecroft
of the Shielding Group, Rolls-Royce and Associates Ltd. in undertaking
the radiological assessment work.

REFERENCES

IAEA Safety Series No.6, (1985), Regulations for the Safe Transport of
    Radioactive Material.
IAEA Safety Series No.6, (1985), Suppl.1986, Regulations for the Safe
    Transport of Radioactive Material.

APPENDIX

STATEMENT OF REQUIREMENTS

The requirement would be for a transport packaging which would:-

1. Accept a current ion-exchange resin storage container of stated dimensions:-

   External Diameter   1.52m
   External Height   1.91m
   Overall Weight (loaded) 8 tonne

2. Accept a proposed ion-exchange resin storage container of stated dimensions:-

   External Diameter   1.83m
   External Height   1.91m
   Overall Weight (loaded) 12/14 tonne

3. Be of appropriate quality to meet the definition "Industrial Package-2" in accordance with the appropriate transport regulations.

4. Be constructed of such materials as considered necessary to meet the regulatory requirements for regular and long-term use.

5. Include a bolt-down facility to use during transportation.

6. Be designed to appropriate lifting standards including the required snatch-lift allowance.

7. Provide adequate radiation shielding, including additional allowance in way of the top cover, to meet the requirements of the appropriate transport regulations.

8. Maintain a size envelope to allow unrestricted transport on standard vehicle and wagons by UK road, UK rail and sea.

9. Provide containment for the ion-exchange resin carrier water, assumed to be non-radioactive.

10. Assume exclusive use, in accordance with the appropriate transport regulations, in terms of sole user vehicles; but sea transport with non-vehicular loading and non-Special-Arrangement would be retained.

PROGRAMME TO DEVELOP A LARGE TRANSPORT CONTAINER FOR TRANSPORTATION OF
LARGE PIECES OF CONTAMINATED EQUIPMENT AND OF MEDIUM LEVEL WASTE

G.Chevalier[1], C.Phalippou[1], L.Tanguy[1], C. Ringot[1],
H.Libon[2], J.Draulans[2], I.Lafontaine[2]

[1]Commissariat a l'Energie Atomique (CEA)
31-33 Rue de la Federation
75752 Paris Cedex 15
France

[2]Transnubel
Gravenstraat 73
B-2480 Dessel
Belgium

ABSTRACT

A development programme, sponsored by the Commission of European
Communities was carried out jointly by Transnubel SA Belgium and by CEA
France, with the aim of developing a very large package that complies
with IAEA regulations for the safe transport of radioactive materials.
The packaging will be used for the transportation of contaminated
equipment or waste from various types of nuclear installations. The main
difficulty encountered in designing such packaging consists in ensuring
the leaktightness of the containment system following a drop from a
height of 1 m on to a rigid punch. Most severe damage undoubtedly occurs
when a large surface hits the punch after the drop. Most of the drop
energy is absorbed by the beam grid structure (245,000 Joules for a 1 m
drop).

As a first approach, structural deformation was studied in
reduced-scale tests. This demonstrated the necessity of reinforcing the
beam frame. Work is now in progress to demonstrate that a design
incorporating suitable insulating material can also comply with other
IAEA regulations, such as that governing a 9 m high drop on an edge
followed by a fire at 800°C.

SCOPE OF THE STUDY

A research programme, sponsored by the Commission of the European
Communities was carried out jointly by Transnubel SA Belgium and by CEA
France to develop a transport container that complies with IAEA
regulations for the safe transport of radioactive materials. At present
no such large package is available in Europe. As a result, following
dismantling operations, transportation of most of the large-dimension
contaminated equipment, such as glove-boxes, ventilation systems and
filters requires the setting up of special arrangements, approved by the
competent authorities. Furthermore, in the near future, an increasing

quantity of medium-level waste will have to be transported from
reprocessing plants to final storage areas. A suitable transport
container is thus required for transportation of both the dismantled
parts and this waste. The resulting transport container will be adapted
for use in transportation by both road and rail and will be submitted to
the competent authorities for type B approval.

COLLECTION OF DATA AND REQUIREMENTS OF THE STUDY

On the basis of the answers obtained from an enquiry, sent to the
main European nuclear installations, the following tentative parameters
have been defined:

(1)   Maximum gross weight and ISO standard dimensions: The gross weight
       of the package cannot exceed 25 T due to the 38 T weight limit for
       road vehicles, operating under normal transport conditions. The
       dimensions of a 20 feet ISO-container have been adopted.
(2)   Package containment: A separate containment system, independent of
       the package's external steel frame and of its contents, has been
       chosen. It can be manufactured in various versions, adapted to the
       waste categories to be transported. For example, for 200-250 l
       drums, which, according to an enquiry, seem to be the most widely
       used, the containment can be a vertical vessel, held in a standard
       rack.
(3)   Radiological protection: In the same way, the characteristics of the
       required radiological protection can be specified in terms of the
       contents to be transported. The containment will be fixed to the
       drum and will protect the handling personnel, not only during
       transportation, but also during loading operations and leakage
       controls.

RESEARCH FOR MATERIALS MEETING IAEA REGULATIONS

Following initial experiments, it seems clear that the 1 m high drop
test represents the most difficult problem to be solved. Indeed the
external container shell must withstand appreciable load. Deformation of
the shock-absorbing materials must be limited to avoid any damage to the
containment system, and too great a reduction in the usable volume.

Three punch tests were performed on various types of strengthened
composite materials (C.M):

(a)   C.M.1, corrugated iron sheet, scale 1/1.
       Dimensions: 2,325mm x 1,560mm x 3mm.

(b)   C.M.2, scale 1/3 (Figure 1).
       Dimensions: 840 x 428 x 50mm.
       Components: corrugated iron sheet: 1mm
                   Kevlar   : 5mm
                   Tubes E24.2   : 20 x 20 x 1 mm
                   steel sheet   : 1 mm

(c)   C.M.3, scale 1/3 (Figure 2).
       Dimensions: 840 x 428 x 50 mm
       Components: steel sheet   : 1 mm
                   aluminium honeycomb   : 20.8 mm
                   Kevlar   : 6 mm
                   Tubes E24.2   : 20 x 20 x 1 mm

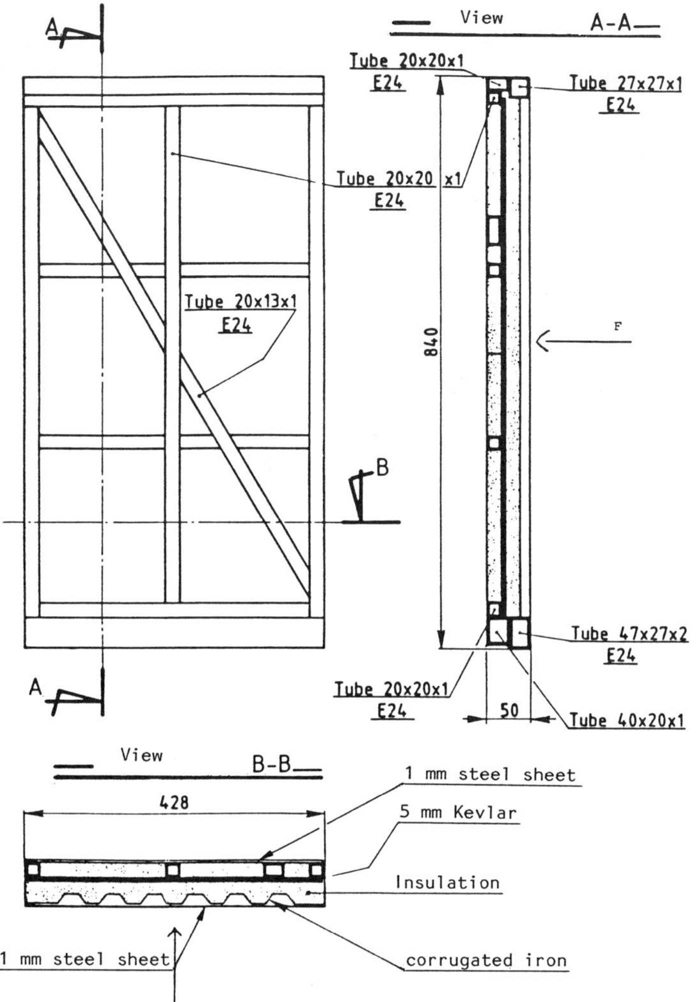

Fig. 1. C.M.2 in static and dynamic tests.

The results for the 3 composite materials are compared with the reference value on a load-displacement diagram (Figure 3). The experimental values are well below the reference value. Strength must be increased fivefold.

Comparison between static and dynamic crush tests confirmed the validity of the static experiments. Indeed, the results show that for similar absorbed energies, buckling is identical.

From Figure 3, it would appear that in comparison with C.M.2 the honeycomb contribution to sheet element C.M.3 increases the stiffness of the structure.

Taking into account the compressive values of high-density honeycomb, it is to be expected that a 75 mm thick honeycomb (650 kg/m^3), stuck between 2 steel sheets 4mm thick, could support the 200 tonne load, necessary for energy absorption.

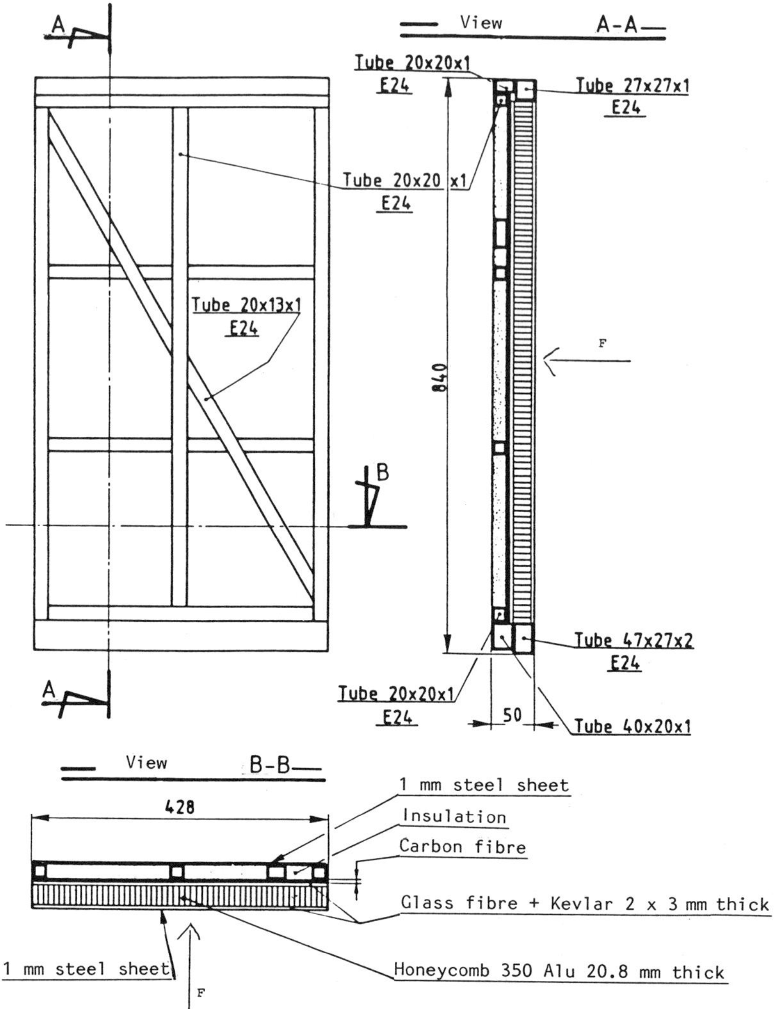

Fig. 2. C.M.3 in static test

PRELIMINARY CALCULATIONS AND TESTING OF THE BEAM STRUCTURE IN ACCIDENTAL CONDITIONS

The main part of the energy liberated in a 1m high drop (245,000 joules) is absorbed by the beam grid structure of the packaging. This grid supports a composite material.

A first approach studied reduced-scale structure crush tests. It demonstrated the necessity for a strengthened beam frame and showed that the energy absorbed by a three dimensional structure is greater than that absorbed by a two-dimensional structure by a factor of 5 (Figure 4). A non-linear finite element code was used for numerical simulation of the collapse load. It takes account of large displacements and instability problems.

The orthotropic grid under the transverse loading is shown in Figure 5. The beam is 160 x 60 x 6mm in section and has an inter-axis of 400mm; the corner beams have a section of 160 x 160 x 9mm.

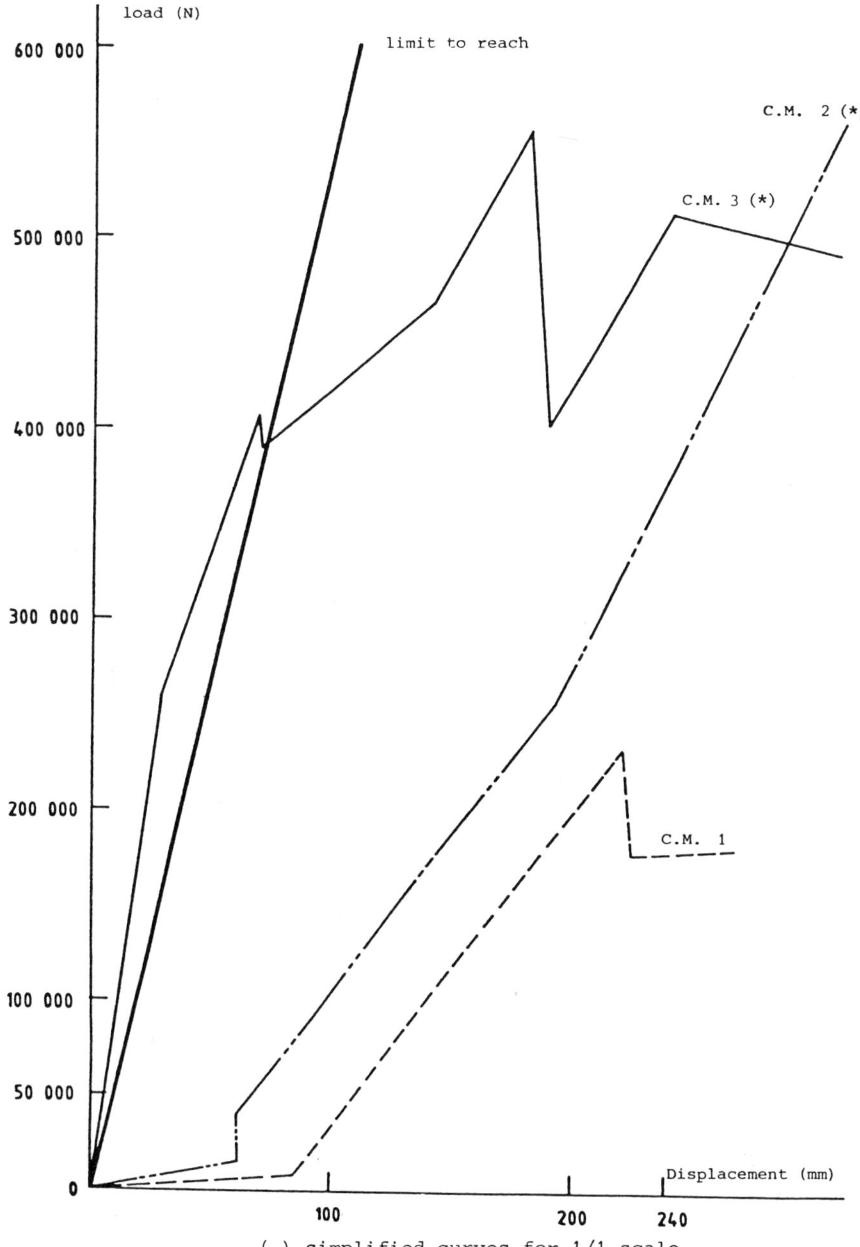

( ) simplified curves for 1/1 scale

Fig. 3. Comparison of static test results.

In order to simplify the calculations, the horizontal cross bars on the lateral sides will not be discounted. For reasons of symmetry purposes, only a quarter of the package has been analysed. The ultimate loading for a maximal displacement of 20cm is obtained at mode 72:
$P_{max} \simeq 4 \times 500$ KN $= 2,000$ KN (Figure 6)

The bending energy reaches more than 259,000 Joules. Stress remains at an acceptable level (5% maximum), but induces the general collapse. Figure 7 shows the bending plastic flow distribution of the grid.

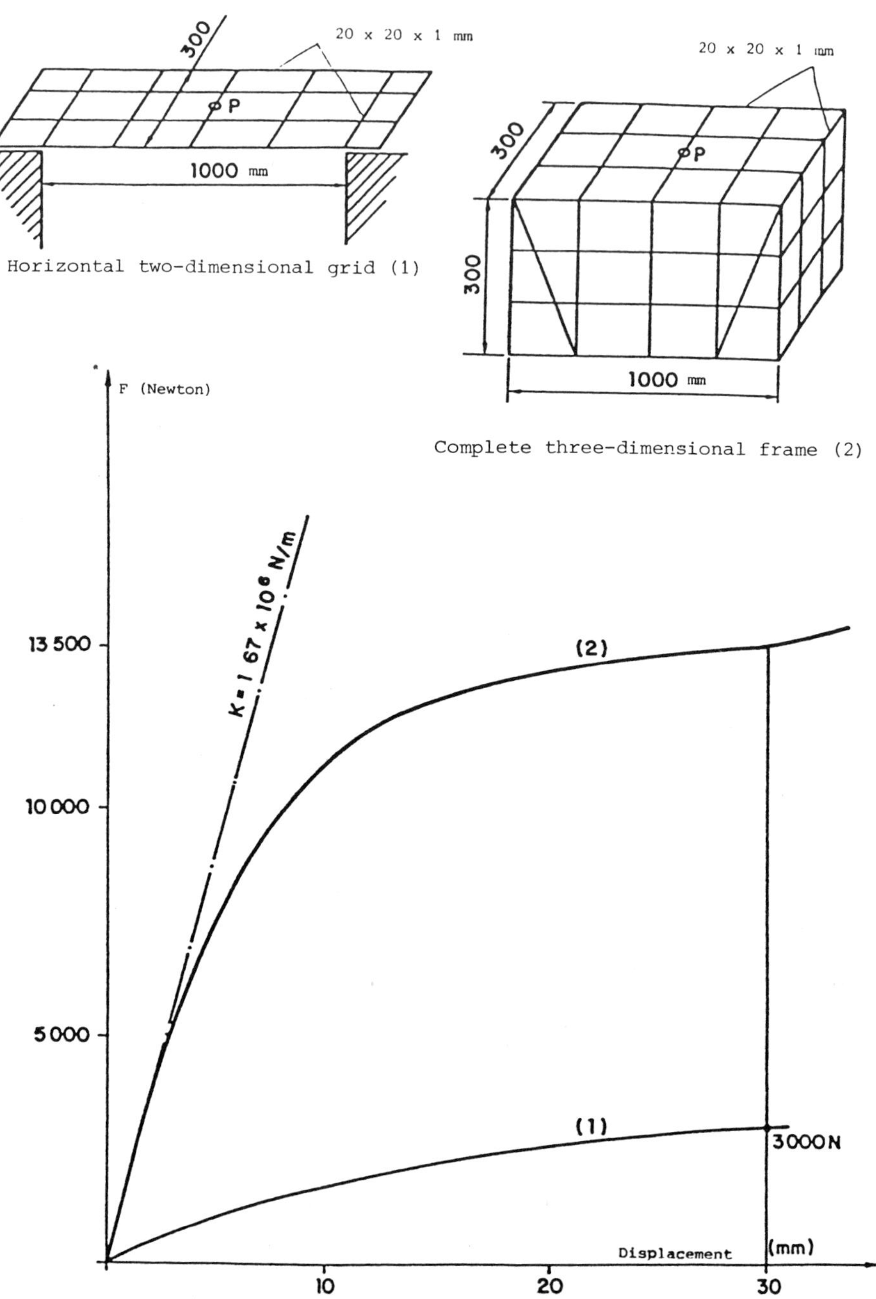

Fig. 4. Energy absorbed by two- and three-dimensional structures.

ANALYSIS OF THE STRUCTURAL BEHAVIOUR OF
PACKAGING SUBJECT TO A FIRE AT 800°C

General Characteristics

The structure evaluated is slightly different from the preceding one. It has the following characteristics:
It constitutes a portion of the 2570 x 2500 x 160mm lining (Figure 8) and has:

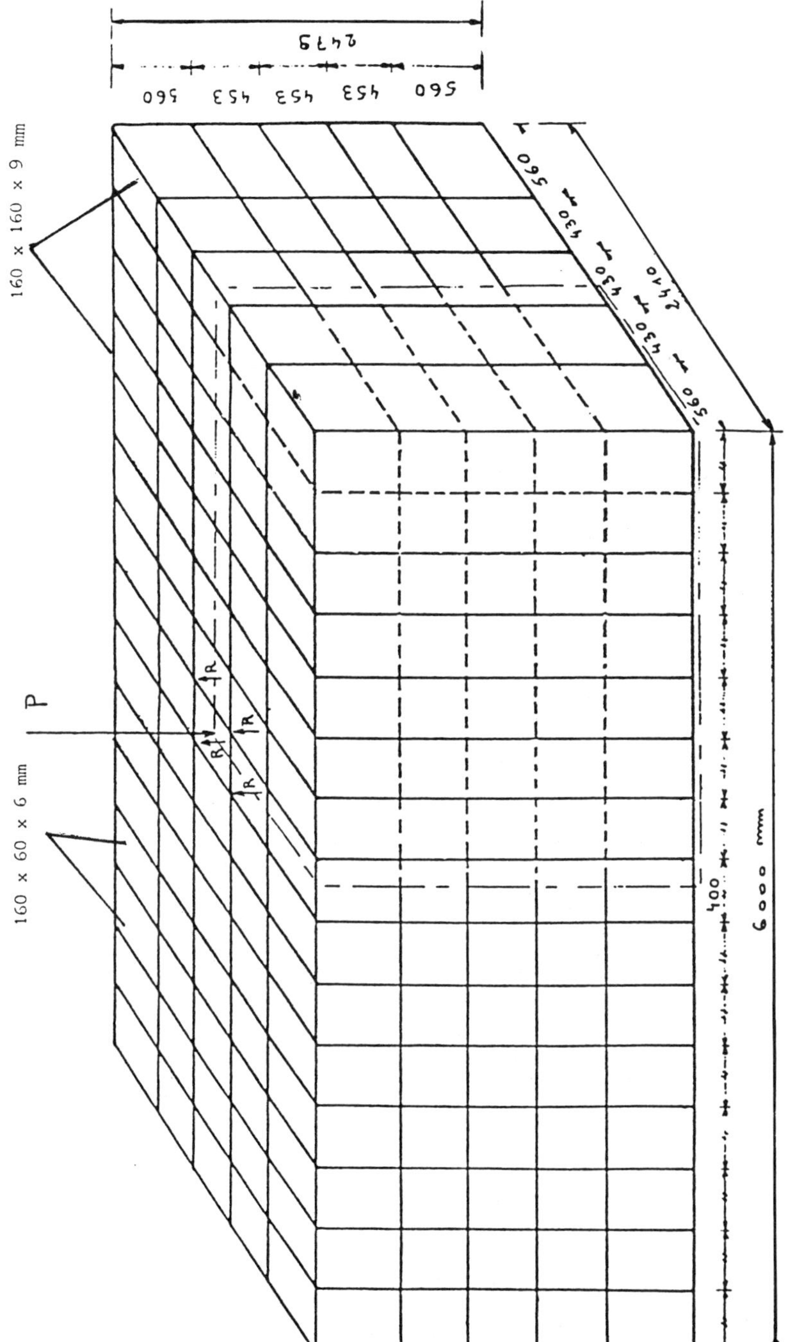

Fig. 5. Finite element calculation.

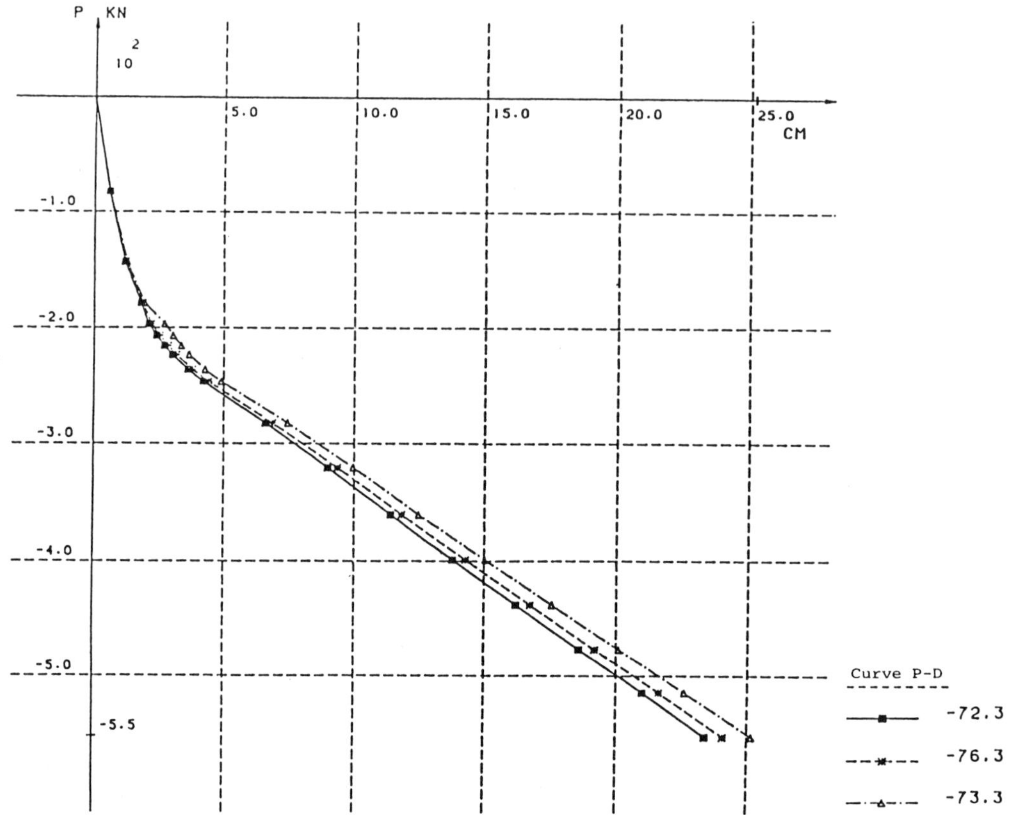

Fig. 6. Study of beam grid.

- A beam frame, 120 x 60 x 3mm and 60 x 60 x 3mm in profile,
- A 2mm thick external lining,
- A 3mm thick internal lining,
- Insulation consisting of 60mm thick rockwool,
- Local damage induced by a punch test (Figure 9),
- Drums enclosed in an open-ended steel caisson.

<u>Simulation Method</u>

So as to evaluate on the one hand the maximal damage induced during thermal testing of drums and, on the other the protective role of the lining, a caisson was built and used to contain and position the drums. As a result, thermal leaks from the cold face of the lining and around drums are minimized and can be considered as a conservative assumption (Figure 9).

<u>Results</u>

Temperature variations at various points in the caisson were monitored by 23 thermocouples. The maximum temperature reached at the punch test location was 110°C. The waste in the drums remained undamaged (Figure 10). It can be concluded that the results indicate satisfactory margins. When the final design is considered to meet punch test requirements, thermal control calculation or model testing will be performed.

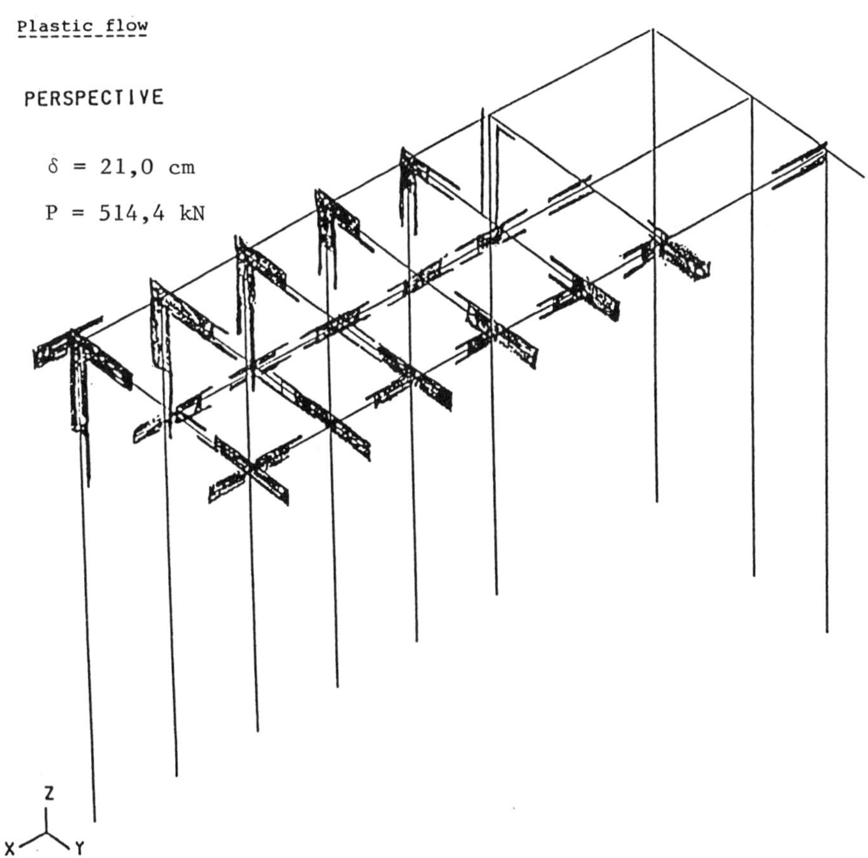

Fig. 7. Study of a beam grid.

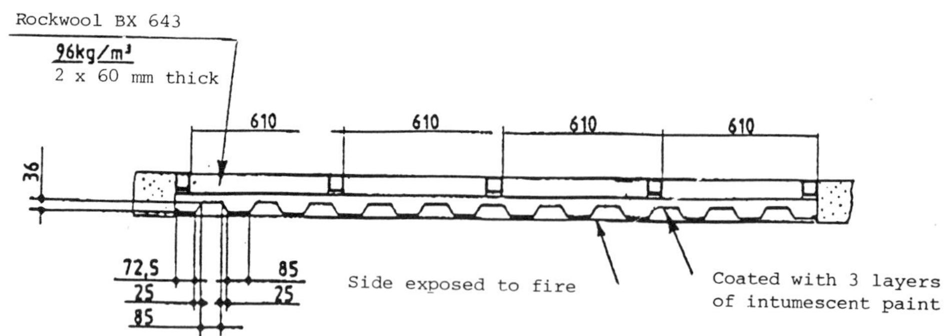

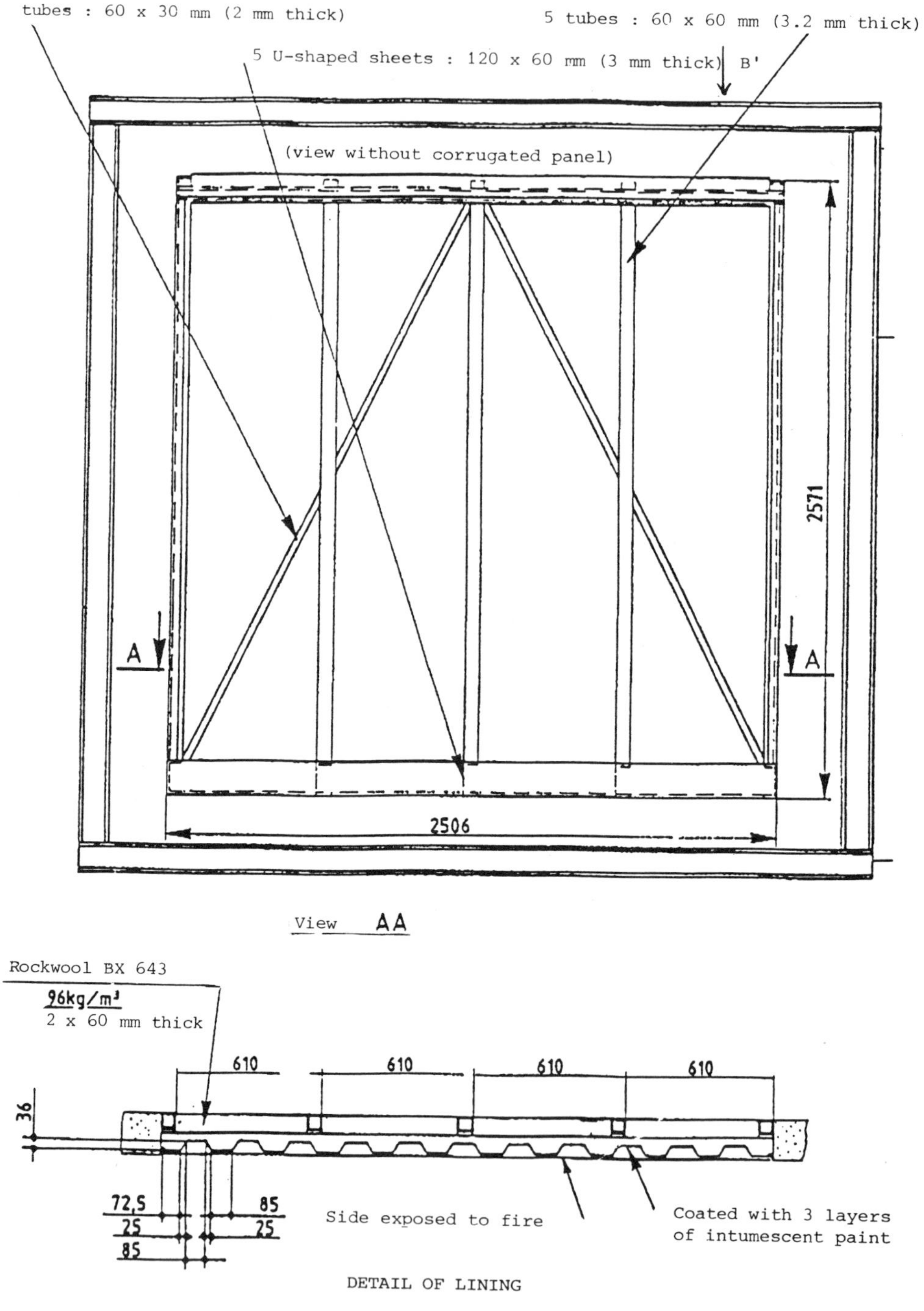

Fig. 8. Detail of lining.

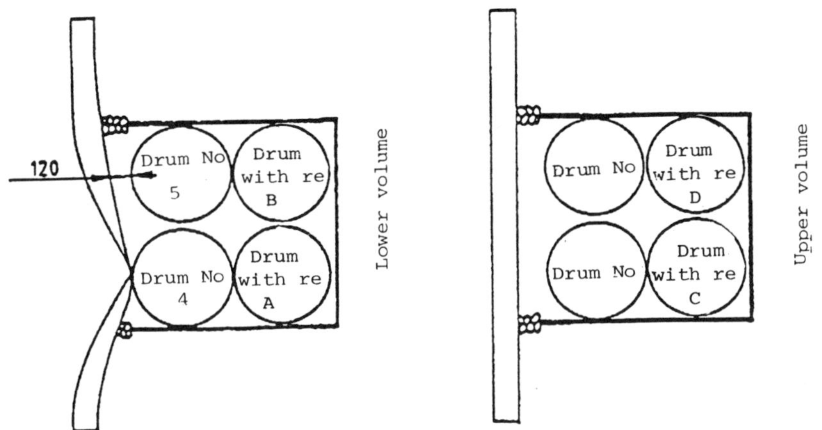

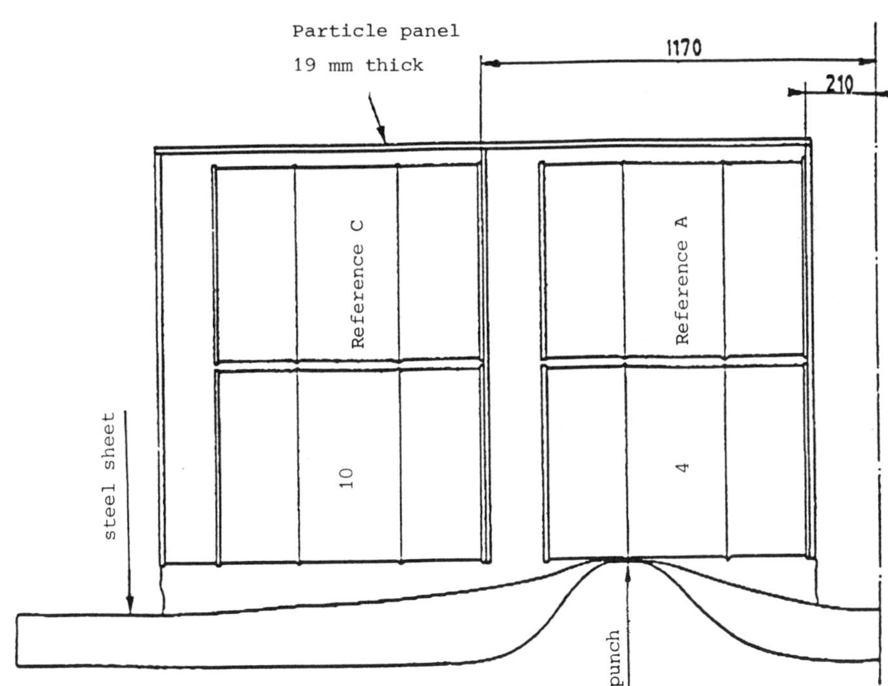

Fig. 9. Position of the drums near the lining.

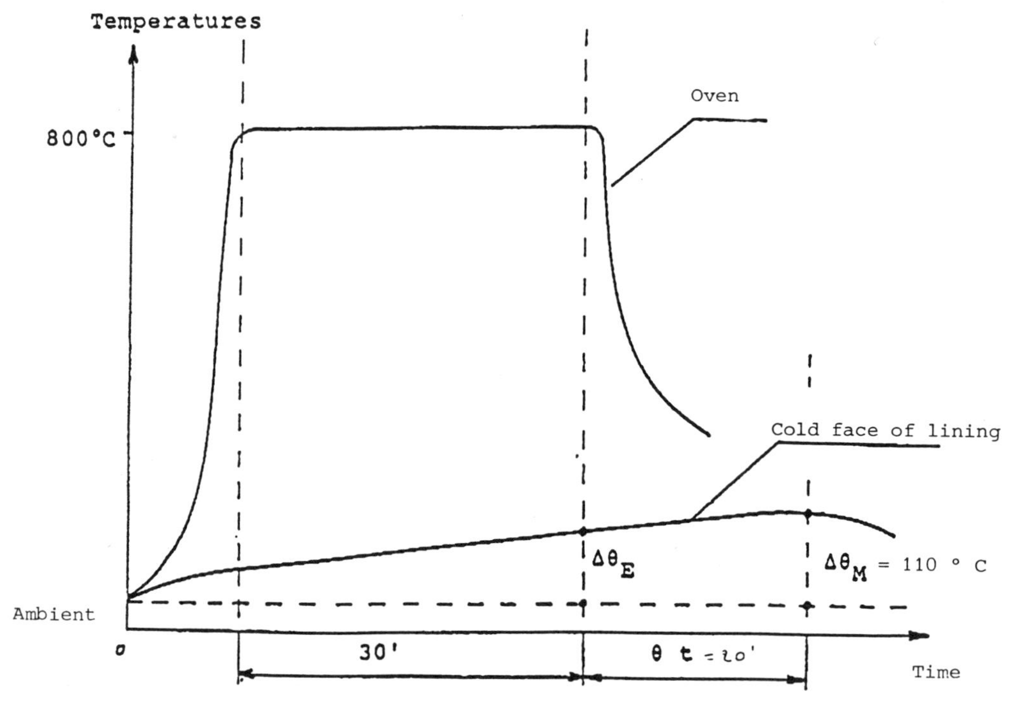

Fig. 10. Curves of temperature.

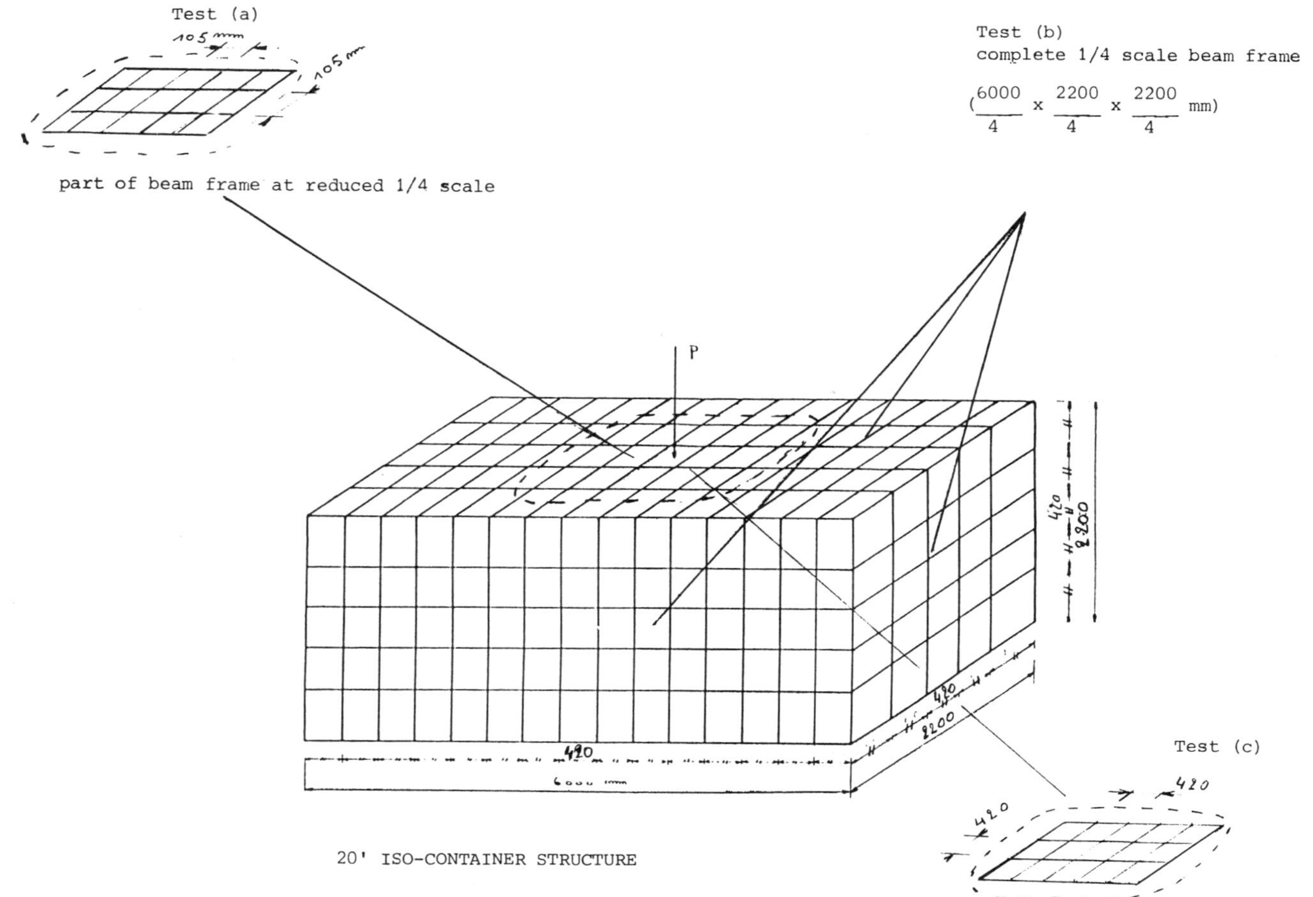

Fig. 11. 20' ISO-container structure.

REDUCED SCALE MODEL TEST PROJECT

In order to verify that a high-density honeycomb composite material
can withstand a load of 2000 KN, since it is very difficult to simulate
the characteristics with reduced scale models, the problem will be solved
using a three-part structure (Figure 11):

Test (a)   A horizontal grid composed of two sets of crosswise tubes,
           identical to those of the beam frame, on a 1/4 scale

Test (b)   A complete 1/4 scale beam frame simulating the frame of the
           package

Test (c)   A portion of the lining including shock absorbing materials.

This three-step procedure represents a more economic solution than
the manufacture of a full-scale prototype.

    This procedure will nevertheless enable design revisions to be made
in the case of errors in the steel frame dimensions or plate thickness.

Test (a)	Test (c)
Horizontal 1/4 scale grid	**Portion of lining**
1) test	
2) calculation	full scale
--------> 3) transposition scale 1	
Test (b)	
	comparison
Complete 1/4 scale steel frame	1) test
a) test	2) deduced container behaviour
b) calculation	
--------> 3) transposition to full scale	

Comparison between the complete steel frame test (a) and that on the
horizontal grid (b) gives an idea of the displacement which will occur in
a similar full-scale frame.

CONCLUSIONS

This first step leads to establishing a preliminary procedure for
reduced scale tests. If the satisfactory progress on the project is
maintained, future work will be required on economic optimization of the
materials.

DISCUSSION FOLLOWING SESSION 2: Papers 5 - 7

DR. R. S. PECKOVER, UKAEA for Papers 5 - 7:
I would like to ask all three speakers whether the containers and
packages they have described were designed with the idea of either
disposal or re-use in mind, because it seems to me that those factors are
the issues and not simply the matter of transport?

MR.G.HOLT, CEGB GDCD:
In my case yes it was designed for re-use. The whole point of the
exercise was that the containers which we use to store the material
needed to be moved and once moved and relocated then the outer packaging
could be shipped back to source and used again.

MR.J.ARMITAGE, ROLLS-ROYCE & ASSOCIATES LTD:
The decommissioning waste package was designed for disposal and not
re-use - one movement only. Though we didn't talk about it we did
consider the disposal requirements for the near field safety case in a
deep repository

MR.H.LIBON:
This container is meant for waste and dismantled parts, and indeed we
intend of course to re-use this container and to use it for waste
transport, that means continually.

DR.J.WILLIAMS, UKAEA Harwell for Paper 2:5
In Mr.Holt's presentation he listed all the wastes to be disposed of from
the Magnox reactors and the largest waste was from the boilers. Are the
CEGB intending to dismantle the boilers and put them into waste
containers of the type described, or are they considering alternatives
such as trying to dispose of them in one piece?

MR.G.HOLT:
It is certainly feasible to dismantle and package them as described but
at present we are not decommissioning any power stations - we are doing
studies on what we need to do to decommission them. Obviously the timing
of decommissioning and the availability of a disposal site is going to be
very important to the final decisions made. So at the moment we are
trying to keep our options open. One of the options is to dismantle and
put into boxes but we do certainly recognize that because of their
physical size, though not highly radioactive, there is going to be quite
a significant dose commitment to the workers in doing that. That is a
problem. We still have the option (though Nirex has abandoned
near-surface disposal for operational waste), as explicitly mentioned in
the agreement between Nirex and the Department of the Environment, that
for very large objects, specifically boilers, we could go for surface
disposal. That again is a possible option. Sea disposal has always been a
possible option, but we have to see what disposal routes at the end of
the day will be available to us.

MR.L.BAKER, (FORMERLY CEGB) PROJECT MANAGEMENT SERVICES:
Could I just reinforce a comment made by Mr.Holt that studies have been
carried out by the Board for the handling and disposal of a complete heat
exchanger. Much of the problem has been in stabilizing the internal
pipework which could not have been fixed due to its thermal cycling it is
proposed to accomplish this by the use of light weight aggregate concrete
filling the heat exchanger. The problem here is that the weight then
increases from 800 tonnes to something like 1800. A scheme has been
prepared for handling these weights and I hope this reinforces Mr.Holt's
remarks.

MR.J.HIGSON, UKAEA RISLEY for Paper 2:5
Why when firming down on the box size, did you not decide on the ISO
container dimensions?

MR.HOLT:
We did consider the ISO size and we are not actually that far away from
it. Maybe what I should say is that we have today fixed on those proposed
dimensions and there is a lot more work to do, as I am sure you probably
gathered from the depth of my paper. It may be as we do more work and as
we engineer the structures a lot better these dimensions will change
either because of external reasons, such as wanting to match ISO sizes or
for internal reasons. Some of the wastes that we put in it might pack
better in a different size box, so that is ultimately still open. We were
a little reluctant to choose the ISO size, in one sense, because ISO
containers are in comparison light weight, about 20 tonnes. We are
talking about 60 tonnes and we did not want to get into the situation
where someone would be tempted to use an ISO container handling frame for
a 60 tonne box. It is still open.

MR.M.J.S. SMITH, UK NIREX LIMITED:
The dimensions are provisional, and Mr.Holt has given some of the
reasons. We certainly do see a box, very close to those dimensions as
being the one that would be used. There is little that I can add. Boxes
with nominal capacities of $12m^3$ and maximum weights of 60 tonnes have
been identified by Nirex as the standard containers for large items of
low and intermediate-level decommissioning wastes. The external
dimensions of 2.4m x 1.85m x 4.0m are provisional but have been arrived
at after consideration of the factors discussed in Mr.Holt's paper,
likely limitations on access to the proposed deep repository, and the
advantages and disadvantages of conforming to ISO dimensions.

THE ROLE OF WINFRITH IN RADIOACTIVE MATERIALS TRANSPORT TECHNOLOGY

M. H. Burgess, M. S. T. Price

United Kingdom Atomic Energy Authority
Winfrith Atomic Energy Establishment
Winfrith
Dorchester
Dorset DT2 8DH

ABSTRACT

Although transport of radioactive materials (RAM) has proved to be a safe operation, with negligible hazard to the public and the environment, there is a need to develop more efficient operations and new aspects of the business. Within the United Kingdom Atomic Energy Authority (UKAEA) there is considerable expertise on RAM transport and related topics and the main UKAEA centre for work in this area is at Winfrith where a large research and development programme over a range of subjects is under way. The programme, carried out for the Department of Energy, is of a generic nature but is usefully augmented in the practical field by specific experiments for commercial customers and by the need to operate radioactive facilities such as the 100 MW(e) Steam Generating Heavy water Reactor. To focus the commercial work both at Winfrith and more broadly within the UKAEA, a Transport Technology Business Centre has recently been established.

The paper gives an overview of the facilities and experience available at Winfrith to study shielding, criticality safety, thermal performance under normal and fire accident conditions, flask decontamination and impact behaviour. Companion papers at this Conference discuss some aspects of the work, e.g. the calculational methods to be applied to the design of shielding, the validation of finite element computer codes to assess impacts, and studies of the sub-criticality of storage and transport arrangements for irradiated fuel.

One of the important recent changes in emphasis has been the increased attention to radioactive waste handling and transport.

In common with other nuclear sites, Winfrith is having to provide new facilities for the storage of intermediate level waste (ILW) pending the availabilty, post-2000, of a NIREX disposal site. Their preferred type of ILW package is an unshielded 500 litre drum which, during transport, will require the shielding and impact protection of an overpack. A programme to examine the performance of the Winfrith drum design under potential handling accident conditions is described. Winfrith has also carried out a range of other waste container impact tests including dropping of 40 Te simulated WAGR decommissioning waste packages.

INTRODUCTION

Winfrith, one of seven establishments of the United Kingdom Atomic Energy Authority, is a main centre for research and development in support of the UK nuclear power programme. The Establishment was founded in 1957 with the aim of providing design data for the growing nuclear power industry; the green field site in Dorset was laid out to provide sites for experimental and prototype reactors such as the 20 MW(Th) Dragon Reactor and the 100 MW(e) Steam Generating Heavy Water Reactor (SGHWR).

The current broadly-based programme covers reactor physics, radiation shielding, criticality assessment, heat transfer, fluid dynamics, control engineering, and instrumentation with a strong bias towards safety-related studies. Winfrith's expertise in chemistry, materials science and engineering fields is applied to the post-irradiation examination (PIE) of AGR fuel and to plant development work in radioactive waste processing. Thus the scientific disciplines and expertise available at Winfrith encompass virtually all those necessary to support an extensive research and development programme on radioactive materials transport.

The specialist areas include:
- shielding
- criticality
- impact testing and package assessment by computer analysis
- thermal performance optimisation and fire testing
- immersion and leak testing
- design safety appraisal

By virtue of the Atomic Energy Authority Act 1986, the Authority became a Trading Fund with the requirement to operate on a fully commercial basis. To meet the Department of Energy's requirements for research and development in the nuclear field, the Authority is required to put forward, for agreement with the Department, a programme letter for each area of declared interest. The programme letters include specific objectives and programmes of work, resource requirements, timescales and priorities.

Although essentially a research and development organisation, Winfrith has the considerable additional advantage of practical experience in handling of radioactive materials because of operation of the SGHWR, from receipt and despatch of irradiated AGR fuel for the PIE programme and from radioactive waste transport to Drigg and Sellafield.

The business opportunities which are inherent in this wide-ranging capability are now coordinated by a recently established Transport Technology Business Centre which is officially launched at this Conference.

Radioactive materials transport is a routine and safe operation. Research and development continues, however, in order to support the evolution of more efficient and economic operations and involves coordination of specialist advice from the scientific and engineering staff augmented by Winfrith's extensive experimental facilities.

This paper reviews these facilities, highlighting aspects of the work carried out in them as well as desribing current and projected radwaste transport operations, thereby demonstrating the comprehensive nature of the radioactive materials transport expertise which is available at Winfrith and more generally within the UKAEA.

WINFRITH FACILITIES AND CAPABILITY

Design

The Engineering Division at Winfrith employs experienced design
engineers who have been responsible for studies of flasks to pressure
vessel standards to cope with potentially high vapour pressures
experienced during a fire test. As an example, a range of flasks has
been designed to carry commercial fast reactor fuel, cooled either by
sodium or helium. A smaller, general purpose flask has been designed and
licensed to an outside organisation.

Manufacture

The Winfrith workshops have established a high standard of
workmanship consistent with the demands of zero-power and
electricity-generating nuclear reactors. These skills and the workshop
facilities are ideally suited to the manufacture of flask components up
to a unit mass of about 25 tonnes, as has recently been demonstrated by
the completion of two modular flasks with full Quality Assurance to
standards accepted by the Department of Transport. The quality built
into these flasks has resulted in very high levels of leak tightness and
good shielding as shown by scintillation tests (Pullen, 1983).

The workshops incorporate an Inspection Department with the latest
instruments and techniques. Non-destructive testing by X-rays and
gamma-rays, and by ultrasound, complements the more usual visual and
dye-penetrant inspection techniques applied to welding. Magnetic
particle detection and eddy current methods are also available.

Shielding and criticality

Flask design inevitably required an input from shielding assessments
and often from criticality calculations. The Shielding Group at Winfrith
is responsible for the development of computer codes and data sets for
the UK and validating these against experiments. This work is aided by a
variety of experimental facilities, ranging from an MTR reactor (NESTOR),
used as a source of neutron and gamma radiation, to radioactive sources
and detectors which can be used to test full-sized flasks or components.
The codes used have been included in the ANSWERS service (Packwood, 1988)
available on subscription to members. They include sophisticated Monte
Carlo codes, capable of high accuracy and detailed geometrical
representations, and highly efficient codes for scoping studies where
accuracy and detail are less important.

Criticality calculations are based on the MONK (Rushton, 1978) and
WIMS (Halsall, 1987) codes which are also offered as part of the ANSWERS
service. The former is a Monte Carlo code used with point or group cross
sections and capable of representing geometries of arbitrary complexity.
WIMS is a collection of modular codes from which various selections of
solution strategy can be made. Neutron flux distributions and
K-effective can be calculated using diffusion theory, collision
probability, Monte Carlo or discrete ordinates methods. Both codes are
well validated (Walker, 1976; Halsall, 1988) and subject to Quality
Assurance as part of the ANSWERS service.

While the criticality codes have been validated against a wide
variety of critical experiments, few of these have included materials and
geometries found in transport flasks. The DIMPLE reactor at Winfrith has
been arranged to simulate AGR fuel assemblies in a boronated steel skip
as used for transport. This work will enable the codes to be compared

Fig. 1. AGR Skip with asymmetric AGR fuel clusters in dimple.

with perturbations to this transport situation, for example when fuel clusters move close to each other within the constraints of the square cavities, see Figure 1.

THERMAL ANALYSIS AND FIRE TESTS

The extensive reactor and heat transfer rigs at Winfrith have needed the development and support of skills and facilities which are also directly applicable to flask design, assessment and manufacture. Flasks designed to carry irradiated fuel must dissipate the heat generated by the fuel and also survive a 30 minute fire test without significant loss of shielding and containment functions. Because of temperature limits set by the contents or the flask seals, the heat dissipation capability can limit the payload of a flask with economic consequences. Careful assessment and/or practical simulations are needed to derive temperatures of important components. Winfrith staff have performed many heat dissipation tests on full-size flasks and flask components for commercial customers. As a result, computer code assessments can be used with confidence for similar situations. Codes developed and used for this purpose include TAU (Johnson, 1981), FLUFF (Fry, 1985) and RIGG (MacGregor, 1981).

Full-scale flask tests at Winfrith have been reported at PATRAM meetings (Middleton and Livesey, 1986) as have analyses of such experiments (Fry et al, 1983).

The new Winfrith Pool Fire Test Facility (PFTF, Figure 4) is capable of accommodating the largest transport flasks in an area 9.5 m x 6.5 m. Commissioning tests are complete and a programme of commercial tests and generic studies of finned and plane surfaces in pool fires, coupled with instrument development, is under way. This work extends the analytical studies reported at PATRAM '86 (Burgess, 1986).

IMPACT TESTS AND ANALYSIS

The Transport Regulations (IAEA, 1985) require Type B packages (flasks) to undergo tests for accident situations. These include a nine metre drop test, a penetration test and a crush test with limits on the post-test leak rates. There are several facilities at Winfrith available for the impact tests. The external Winfrith Drop Test Facility (WDTF) (Figure 2) incorporates a 150 tonne crane and a 700 tonne target buried in the ground. The protruding surface of the target is faced with 150 mm steel plate 3 m x 3 m in area. Bomb releases are available to drop packages or models in pre-arranged attitudes from various heights, e.g:

-   44 tonnes from up to 65 m (using jib extensions)
-   93 tonnes from up to 30 m

A guidance system is being developed to control the attitude at impact.

There are two drop test facilities within existing buildings which can be used to test scale models or smaller packages (up to 500 kg) with drop heights up to 24 m.

The major new Horizontal Impact Facility (Cooper and Wicks, 1987) has been designed to accelerate flask models for impacts onto a 1000 tonne pre-stressed concrete target faced with steel.

Fig. 2. 40 Te prototype WAGR decommissioning waste package, raised prior to a bottom face impact.

This is an indoor facility (Figure 3) with complete attitude control and which is independent of weather conditions. It is capable of accelerating 2 tonne models to 45 m/s (100 mph) or smaller masses (50 Kg) to near sonic velocities, subject to a maximum kinetic energy of 2 MJ. It will thus allow quarter scale models of 100 tonne flasks to be impacted at 100 mph. One of three barrels of 2.0, 1.0 and 0.5 m diameter respectively can be selected for use in conjunction with the compressed air reservoir. This facility extends the impact capability to that required by several national authorities for air-transportable packages.

These rigs, and a 300 MN hydraulic press, are being used to measure material properties and plastic deformations during impacts to validate calculations with computer codes. The codes include HONDO II (Key et al, 1978) DYNA 3D (Hallquist, 1982) and ABAQUS (Hibbitt, 1984) each mounted on a Cray computer. In addition, the I-DEAS (CAE International, 1986) interactive mesh generation package is mounted on a local VAX 11/750 computer. These codes are, of course, used as design tools for package development.

Strain gauges and accelerometers can be attached to packages and more than 100 channels of data collected and analysed. Extensive facilities for examining the package for damage and leak-tightness exist at Winfrith. These include sophisticated NDT equipment including helium mass spectrometry leak test equipment. Suitable techniques are selected for each application and a comprehensive report on the tests and conclusions is offered.

IMMERSION, PENETRATION AND CRUSH TESTS

The WDTF also incorporates a water immersion tank and a selection of masses suitable for dynamic crush tests. By use of the large crane and the construction of suitable structures, the engineering support organistion can arrange crush tests in any attitude. The immersion tank is 4.3 m deep and so can provide the 0.9 m head of water for packages up to 3.4 m in length. The 15 m and 200 m immersion tests can be carried out using off-site facilities. Spear-type penetration tests, as requested by some authorities, can be accomplished by dropping the spear from the crane jib or by means of the HIF.

DESIGN SAFETY REPORTS

Information derived from design, assessments and testing is collated into a design safety report for submission to the Competent Authority responsible for approving flask designs and quality assurance programmes.

The assessments include studies of irradiated fuel behaviour and the release of volatile products from damaged fuel pins. The behaviour of radioactive gases and aerosols must be modelled for water and gas spaces to provide estimates of leakage from flasks under normal operation and accident conditions. This modelling has much in common with studies of release from hypothetical reactor core disruption accidents, performed by the Reactor Safety Analysis Division at Winfrith.

BUSINESS CENTRE

A Transport Technology Business Centre has been established to coordinate the exploitation of these wide-ranging activities. It is recognised that virtually all aspects of RAM transport technology are

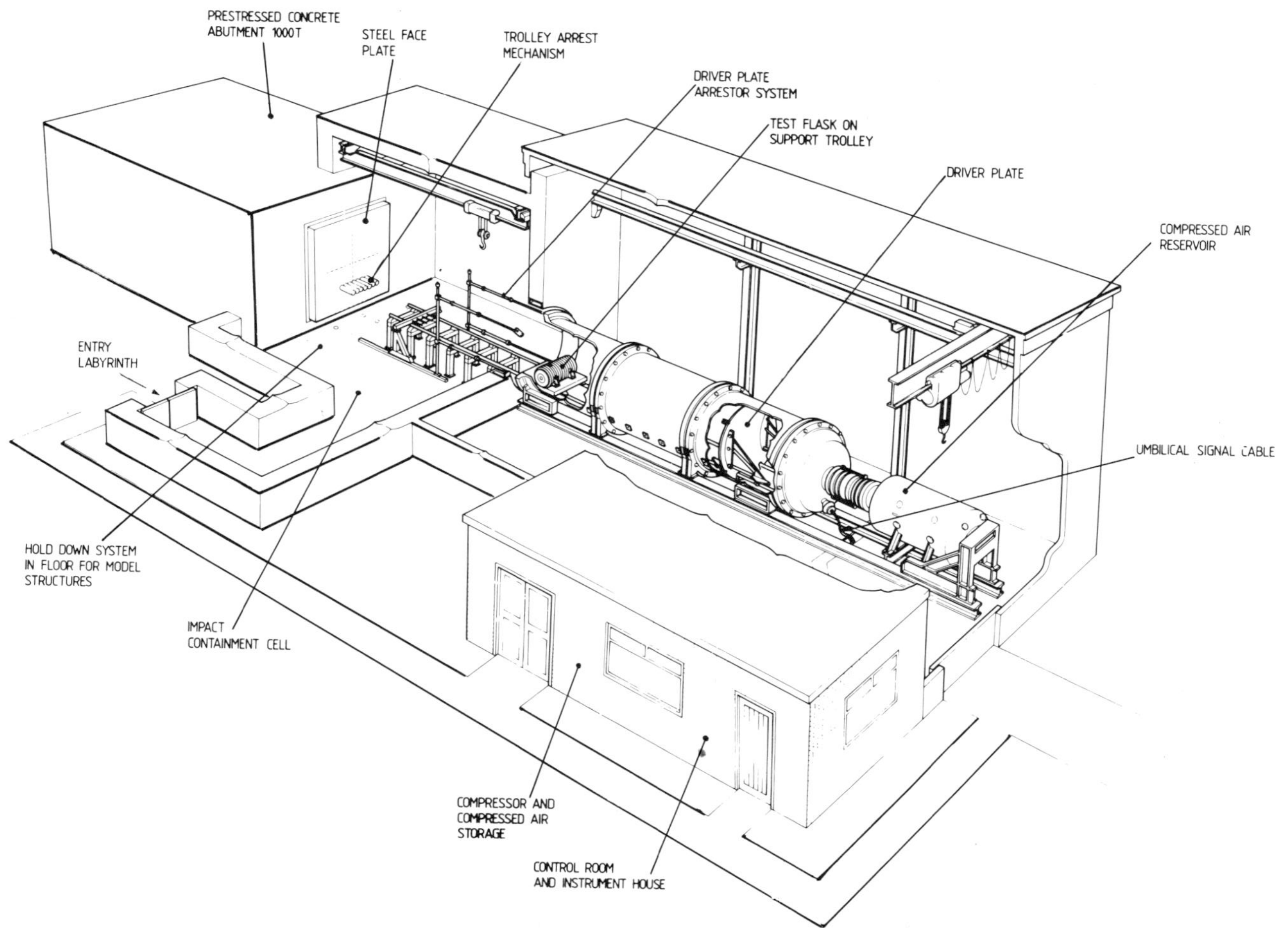

Fig. 3. The horizontal impact facility.

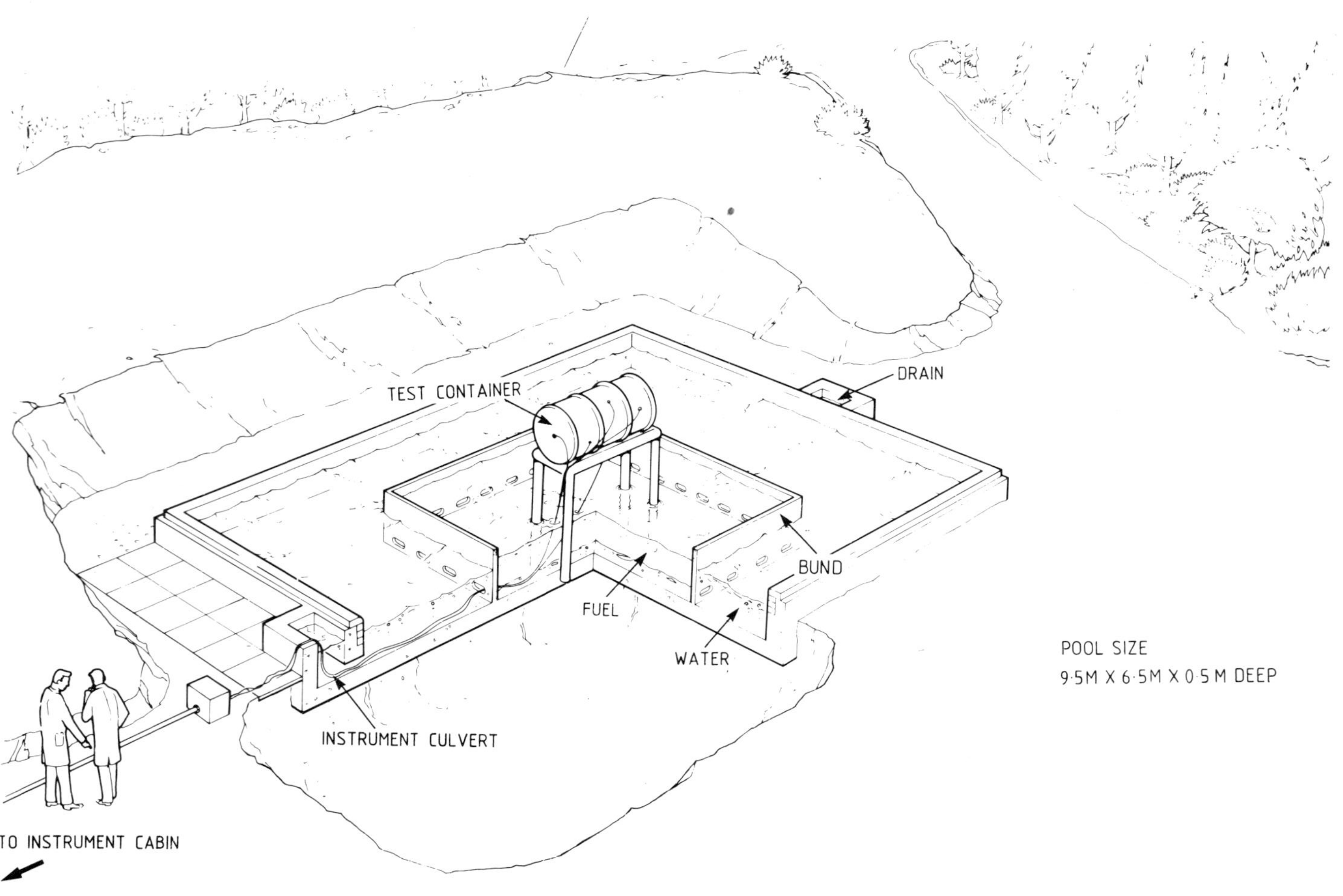

Fig. 4. The Winfrith pool fire test facility.

available at Winfrith, so the Transport Technology Manager provides a single contact for customers from inside and outside the nuclear industry, with the ability to call on the wider resources of the UKAEA. Customers are offered a comprehensive service of design, manufacturing, testing and maintenance for RAM transport packages. While separate requirements are defined for radioactive materials, they are included in the overall definition of dangerous goods and some of these services are required for non-radioactive dangerous goods packages. The stringent conditions for RAM packaging have produced testing and inspection techniques which could, with benefit, be applied to packages for other dangerous goods.

CURRENT AND PROJECTED RADIOACTIVE MATERIALS TRANSPORT

Low Level Wastes are carried to the BNFL disposal site at Drigg in ISO containers (see the section on Low Level Waste). The upper end Intermediate Level Wastes, resulting mainly from PIE work, are moved to BNFL silos in small general purpose flasks manufactured at Winfrith, (see the Section on Intermediate Level Waste – Upper End).

Additionally, there are smaller packages and flasks used for carrying sources and samples to and from Winfrith for a variety of purposes ranging from R & D operations to industrial applications of non-destructive testing with radiographic sources.

LOW LEVEL WASTE

The national site for disposal of solid low level waste (LLW) is at Drigg in Cumbria, a few miles south of Sellafield. Solid LLW arises at Winfrith principally as a result of operation and maintenance of the SGHWR and from PIE of AGR fuel as well as from a variety of research and development activities. Since 1983 compressible LLW has been reduced in volume using an in-drum compactor which applies a force of 3 MN inside a standard 220 litre drum. The compression ratio averaged 2.8:1 last year for an input of 450 m^3. The amount of solid LLW sent to Drigg for disposal is typically in the range 700–1000 m^3/y of which about 73% is drummed, the remainder being parcelled. The LLW is transported in accordance with the Transport Regulations (IAEA, 1979) as Low Level Solid radioactive material (LSA III) in strong industrial packaging. All the waste is carried in standard ISO freight containers. New Quality Assurance (QA) procedures have been introduced in compliance with revised requirements for Drigg disposal. The Winfrith QA plan for solid LLW came into force in October 1987.

From 1 January 1988 all wastes for disposal at Drigg are required to be containerised. This is linked to BNFL's plans to upgrade the standard of waste handling operations at Drigg and the timing is set by practical considerations to preserve remaining trench capacity whilst a new disposal facility (Vault 8) is being constructed.

Following discussions within the industry, BNFL has confirmed that a design based on an ISO freight container will be acceptable for Drigg disposal. The standard ISO freight container has had to be modified to satisfy disposal requirements eg to spread the load in the burial trench and to avoid the normal air-gap under the container. Three concepts are currently being pursued with a) and b) having the higher priority viz:

a)   a 20 ft x 8 ft x 8.5 ft high ISO freight container suitable for carrying drummed waste under the Transport Regulations, (IAEA, 1979)

b)   a 20 ft x 8 ft x 4 ft high ISO freight container suitable for
      carrying loose and wrapped waste under the Transport Regulations
      (IAEA, 1985)
c)   a 20 ft x 8 ft x 8.5 ft high ISO freight container suitable for
      carrying drummed, wrapped and loose waste under the Transport
      Regulations (IAEA, 1985)

Concepts (a) and (c) have to be capable of being reduced in height
to 8 ft since this height variation may be necessary if rail transport
has to be used.

INTERMEDIATE LEVEL WASTE (UPPER END)

About 110 CAGR and 2 SGHWR irradiated fuel assemblies are examined
each year in the PIE caves at Winfrith. These operations produce a
relatively small volume of upper end intermediate level waste (ILW)
(approximately 1 m^3/y) containing activated and contaminated materials
with gamma radiation levels in the range 1-100 Gray/hour at 30 cm. In
former times there were significant arisings of similar waste from the
Dragon Reactor and SGHWR fuel element breakdown. Some material has been
sent to one of the BNFL silos at Sellafield in general purpose flasks,
but about 10 m^3 of such waste has accumulated in temporary storage
facilities at Winfrith. A programme of work to free space in the storage
holes local to the PIE facilities is under way. The waste consists
mainly of fuel element structural material coming from the breakdown of
Dragon and SGHWR fuel elements prior to 1976. The procedure involves
repacking the waste into disposal cans, with size reduction as necessary,
the can and contents being crushed with 15 MN force. By this technique,
the volume of can and contents are reduced to about 60% of the original
volume. This upper end ILW is transported to the Sellafield Silo in
general purpose flasks, e.g. Type 1991.

INTERMEDIATE LEVEL RADIOACTIVE WASTE (LOWER END)

At the inception of NIREX (now UK NIREX Ltd) proposals were made for
the standardisation of packages and for the development of larger
transport modules (Grover and Price, 1984). Because ILW requires to be
shielded it is better to transport unshielded drums in multiples in a
shielded overpack but to place unshielded drums in the disposal
repository. A drum size larger than the 200-220 litre size, which is
standard for low level waste, was considered to be necessary and this led
to standardisation by NIREX on a 500 litre drum size. Most organisations
have tended to use steel for the material of the drum and Winfrith has
chosen Type 316 L stainless steel.

Lower end intermediate level waste at Winfrith consists of sludges
and solids from clean-up of the water circuits and decontamination of the
SGHWR, and solids from a variety of operations. Previously these wastes
would have been packaged for sea disposal. With the cessation of sea
disposal and following the Ministerial Statement of 1 May 1987 this waste
will have to be retained on site until UK NIREX Ltd can accept it for
deep burial, probably no earlier than 2005. The sludge is a mobile waste
and on safety grounds Winfrith wishes to immobilise it in cement at an
early date. A plant is being built which will have the dual purpose of:

a)   mixing SGHWR sludge with cement in a 500 litre drum fitted with an
      in-drum lost paddle
b)   grouting solid items inside a 500 litre drum

The cementation plant does not have significant plant hold-up so that, after manufacture, the unshielded 500 litre drums have to be moved to an on-site long term store. Although the design of the handling equipment for this set of operations has been made very secure against handling accidents an assessment of the consequences of a handling accident is necessary.

Preliminary experiments have therefore been carried out to assess the behaviour of freshly cemented and grouted drums when subjected to an accidental drop from 15 metres. The worst condition for impact occurs in the Treated Radwaste Store when a blade of the paddle is adjacent to the point of impact and when the cement is not set hard. Impact tests on 'green' cement simulant sludge waste in this worst attitude have been carried out and a more extensive set of qualification tests will be carried out in 1988 to support the safety submission for the store. It is clear that these tests are far more severe than any likely to be encountered during the eventual transport to the NIREX repository.

INTERMEDIATE LEVEL DECOMMISSIONING WASTE

The main technical centre for work on decommissioning of nuclear power stations in the UK is at Windscale Nuclear Laboratories. The programme there has been supported by Winfrith having carried out drop tests on 40 Te prototype reinforced concrete packages (Figure 2). In addition Winfrith has coordinated work, funded in part by the Commission of the European Communities, on large transport packages for decommissioning waste (Lafontaine and Price, 1985). This work is being continued in collaboration with Ove Arup and Partners and Windscale Nuclear Laboratories in a follow-up study designed to investigate the constraints of manufacture, handling, transport and disposal on design and due to be completed in 1988.

DECONTAMINATION AND MAINTENANCE

The operational requirements imply regular maintenance of flasks and reusable packages which almost invariably involves decontamination of internal and external surfaces to allow hands-on operations. External decontamination is required to satisfy normal handling and transport limits. Internal decontamination is also required to reduce cross-contamination of internal containers and cave facilities.

This need for operational decontamination has resulted in the development of a variety of techniques (e.g. Bond 1985) which are now available for commercial application. The experience allows Winfrith staff to offer the development of new and existing techniques for specialised applications. Techniques in use range from simple manual washing with mild reagents through high pressure water jetting, Arklone (Freon) jetting, wet and dry grit blasting, vibratory and ultrasonic cleaning, to chemical foams and gels and electrolytic deplating. The techniques originally developed for decontaminating the SGHWR circuits have been successfully applied and incorporated into the design of a transportable decontamination rig (TRANSDEC) which can be brought to a large flask. Waste treatment and disposal facilities are already provided at Winfrith and a full health physics and radiation monitoring service is in operation. The scientific and technical services are coordinated through a Decontamination Business Centre which works closely with the Transport Technology Centre.

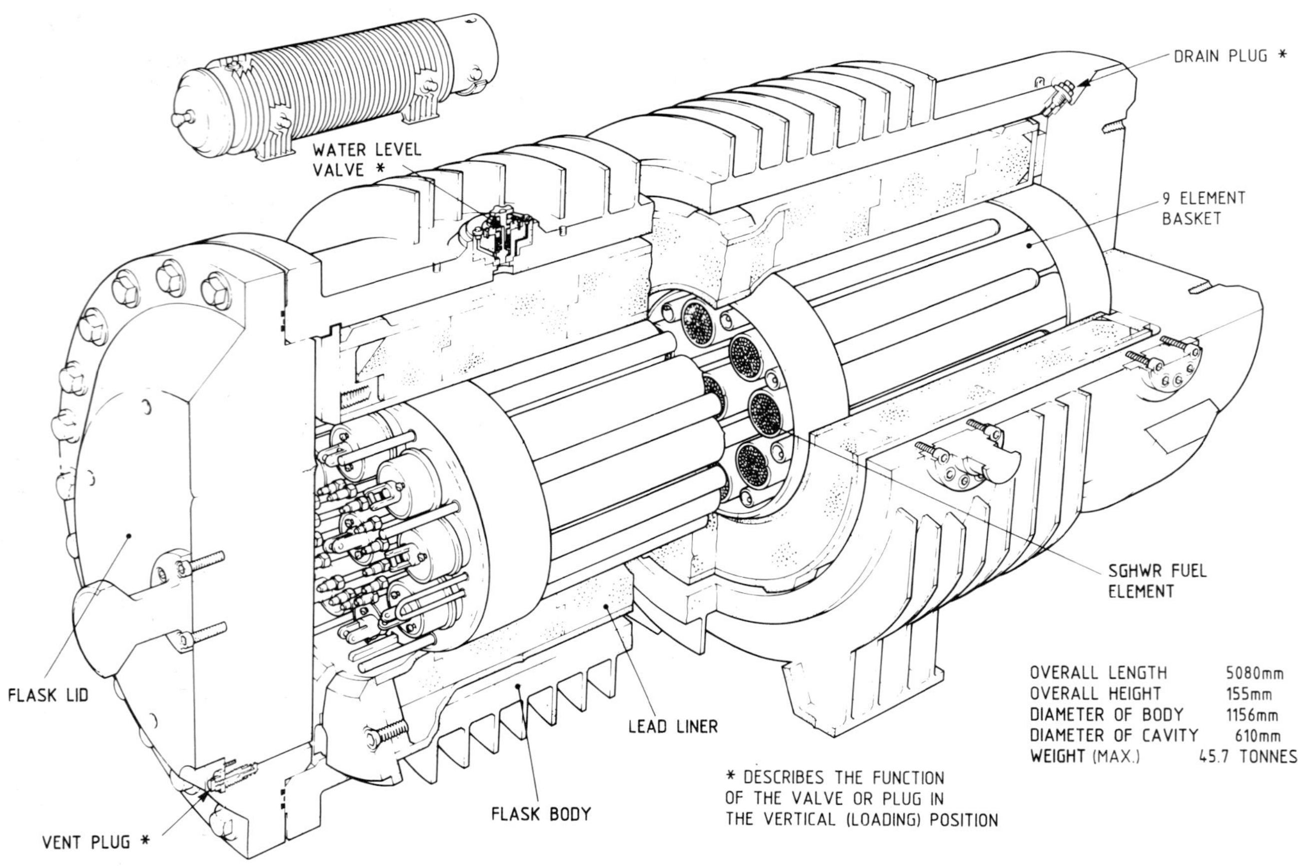

Fig. 5. Type 1120 water-cooled flask for transport of irradiated SGHWR fuel.

CONCLUSIONS

This paper has given an overview of the role of Winfrith in RAM
transport technology thereby providing an insight nto the extensive
technical base which is needed to underpin transport. The facilities and
experience available at Winfrith, which are augmented by support from
other UKAEA sites, have been described with reference, as appropriate,to
recent published work and include:

- design including the preparation of design safety reports
- manufacture
- shielding optimisation
- criticality assessment
- thermal analysis and experimental testing
- impact testing and analysis
- other regulatory testing
- operation
- maintenance
- decontamination

While Winfrith is able to provide all of the above services and
applies them to its own programmes, the larger resources of the UKAEA can
be used to supplement these and to provide others. There is much
practical experience with the operation of flasks, particularly at
Harwell, Windscale and Dounreay. The Safety and Reliabilty Directorate
at Culcheth provides risk assessments based on large files of accident
statistics. The UKAEA can therefore provide a complete service to
transporters of radioactive material.

REFERENCES

Bond, R.D., 1985, Decontamination of a Fuel Transport Flask using
        Chemical Foams, AEEW-R1865
Burgess, M.H., 1986, Heat Transfer Boundary Conditions in Pool Fires,
        in PATRAM '86 Proceedings pp 423-431, IAEA-SM-286/75P
CAE International, 1986, I-DEAS User Manuals (Hitchin, England)
Cooper, C.A. and Wicks, S.J., 1987, Impact Studies at Winfrith,
        AEEW-R2182
Fry, C.J., 1985, The FLUFF Code for Calculating Finned Surface Heat
        Transfer - Description and Users Guide, AEEW-R1918
Fry, C.J., Livesey, E., Spiller, G.T., MacGregor, B.R., 1983, Heat
        Transfer in a Dry, Horizontal LWR Spent Fuel Assembly, in PATRAM
        '83 Proceeding pp 1087-1094
Grover, J.R. and Price, M.S.T., The Packaging and Transport of LOW and
        Intermediate Level Waste, in Proceedings of BNES Conference on
        Radioactive Waste Management, London, 27-29 November 1984, pp
        189-195
Hallquist, J.O., 1982, Theoretical Manual for DYNA3D, UCID-19401
Halsall, M.J., 1987, Future Management of the WIMS Group of Codes at
        Winfrith, NEACRP-A-870
Halsall, M.J., 1988, The Effect of the 1986 WIMS Library on the Analysis
        of Benchmark Lattice Experiments, AEEW-R2262
Hibbitt, H.D., 1984, ABAQUS-EPGEN; A General Purpose Finite Element Code
        with Emphasis on Non-Linear Applications, Nucl Engng Des 77, pp
        271-297 (1984)
IAEA, International Atomic Energy Agency, 1973, Regulations for the Safe
        Transport of Radioactive Materials, Revised Edition (As amended)
        (IAEA: Vienna), 1979
IAEA, International Atomic Energy Agency, 1985, Regulations for the Safe
        Transport of Radioactive Materials, Edition (IAEA: Vienna), 1985

Johnson, D., 1981, TAU: A Computer Program for the Analysis of Temperature in Two- and Three- Dimensional Structures using the UNCLE Finite Element Scheme, ND-R-218(R)

Key, S.W., Beisinger, Z.E. and Krieg, R.D., 1978, HONDO -II; A Finite Element Computer Program for Large Deformation Dynamic Response of Axisymmetric Solids, SAND-78-0422

Lafontaine, I and Price, M.S.T., 1985, System of Large Transport Containers for Waste from Dismantling Light Water and Gas-cooled Nuclear Reactors (Volumes 1 and 2), EUR-10232

MacGregor, B.R., 1981, RIGG - A Computer Code for Calculating Transient Temperature Distributions in Pin Bundles in a Gas or Vacuum - Description and Users' Guide, ND-R-609(D)

Middleton, J.E. and Livesey, E., 1986, A Full-Scale Experiment to Determine the Thermal Response Following Loss of Coolant Water from a Flask Containing Irradiated LWR Fuel, PATRAM '86 Proceedings pp 445-455, IAEA-SM-286-86P

Packwood, A., 1988, The ANSWERS Subscription Service, Brochure Pack, AEE Winfrith

Pullen, D.A.W., 1988, Gamma-Scintillation Testing of Shielding Integrity in Nuclear Materials Containers in British Journal of NDT January 1983, pp 17-20

Rushton, K.G., 1978, The Monte Carlo Code MONK - A Guide to its Use of Criticality Calculations, SRD R88

Walker, G., 1976, The Monte Carlo Code MONK - Validity for Use in Criticality Calculations, SRD M85

SATELLITE TRACKING OF RADIOACTIVE SHIPMENTS -
HIGH TECHNOLOGY SOLUTION TO TOUGH INSTITUTIONAL PROBLEMS

P. D. Grimm

United States Department of Energy (DOE)
Washington DC 20585
U.S.A.

ABSTRACT

Four troublesome institutional issues face every large-quantity radioactive materials shipment:

- Routing: Where is the shipment going and how is it getting there?
- Pre-notification: States want to know what is being shipped and when?
- Emergency Response: What kind of response to accidents is needed for this shipment and who will respond?
- Transportation Management: Optimizing use of the vehicle on packages transporting materials.

The Transportation Communication System (TRANSCOM), which is under development by DOE, is based on a rapidly developing technology which determines geographical loction using geo-positioning satellite systems. This technology will be used to track unclassified radioactive materials shipments in real-time. It puts those charged with monitoring transportation status in contact with every shipment. In addition to practical benefits in the areas of logistics planning and execution, its demonstrations of emergency preparedness have indeed been considered and close monitoring is possible. This paper will describe TRANSCOM in its technical detail and DOE plans and policy for its implementation. The state of satellite positioning technology and its business future will also be discussed.

BACKGROUND

The Atomic Energy Act, Hazardous Materials Transportation Act, and the Nuclear Waste Policy Act (NWPA) require the Department of Energy (DOE) to provide for the safe, secure, and efficient transportation of radioactive materials. The DOE, in carrying out these requirements, needs to monitor shipments of nuclear material in support of emergency response and other activities.

Recent experience in major shipping campaigns has shown increasing concerns on the part of State and local governments over radioactive material shipments moving through their jurisdictions. The concerns will become even more intense when transuranic wastes begin moving to the WIPP site and commercial spent fuel is shipped in volume for either monitored retrievable or repository storage.

The States have expressed the need for obtaining more information on radioactive material shipments to improve their ability to prepare for and respond to various transportation contingencies. In particular, they want to know the exact date and time that each radioactive shipment will enter their State and the precise route it will follow in order to be fully prepared for an emergency.

Partly because of the relative lack of such information, many State and local governments have enacted laws and ordinances banning or restricting shipments of radioactive matrials through their jurisdictions. The resulting patchwork of laws and ordinances places conflicting and route-restricting demands on DOE, whose primary task is to safely transport radioactive materials from origin to destination in the most direct and efficient manner. Three transportation issues at the top of DOE's need-to-solve list are prenotification, routing, and emergency response.

TRANSPORTATION COMMUNICATION SYSTEM-TRANSCOM

DOE has developed a prototype transportation tracking, communication, and information management system (TRANSCOM) that will enhance DOE's oversight and response capability as well as respond to State and local concerns. TRANSCOM combines a data base management system with communications technology which will provide DOE Headquarters and field offices with instant information about the location and status of their shipments. For the first time in the Department's history, Headquarters, the eight DOE operations offices, and their contract carriers, as well as key State and local officials, will be linked together, in near real-time, by a system to share information.

IMMEDIATE ACCESS TO CRITICAL INFORMATION

The system will provide a real-time map location as well as detailed bill of lading and routing information on all shipments. Access to this information significantly enhances emergency preparedness posture for both DOE and State and local officials. It also allows DOE to improve the dispatching and efficient use of its packaging fleet.

Vehicle Location

The TRANSCOM system uses advanced technology to track the location and status of DOE's unclassified nuclear shipments. This technology will make use of satellites to determine the latitude and longitude coordinates of shipments. Satellite will also serve as two-way communication links between the vehicle and the shipper.

Tracking the location and status of nuclear shipments is the central purpose of TRANSCOM. The scope of the tracking capability is the continental United States. A map of the 48 States projected on a computer monitor uses color-coded icons to provide a general overview of all shipments underway. The TRANSCOM user will also be able to view separate screens. A map of the U.S. railway system displays on-going DOE rail shipments, and a map of the U.S. highway system displays truck shipments. The tracking function will be able to zoom in on three levels of maps: U.S., State, and county. Interstate highways, State level roads, and major rail links will be identified.

Bill of Lading Information

Data will be entered to provide designated officials with information about current and upcoming nuclear shipments. For each planned shipment, specific data such as the schedule, planned routes, and type of radioactive material will be available. All the information carried on the truck or rail manifest and required by the Department of Transportation (DOT) regulations will be available.

DOT Emergency Response Procedures

DOE users will be able to select emergency response information from the DOT.

Guidebook for Hazardous Materials Incidents (1987)

The scope of this file includes all the regulatory categories of nuclear materials that can be expected to be shipped in highway route-controlled quantities. DOE or States can use this function in an accident to quickly determine what response measures are needed for the shipment in question.

Federal/State Points of Contact

The emergency information includes the key emergency response contacts at DOE Headquarters, the originating field office, and the shipper. The contacts for the State and the field office closest to the accident will also be shown.

Shipment Schedules

A separate screen will display an accouting of planned shipment activities. This will allow states to more efficiently monitor shipments relative to their geographic boundaries.

READILY-AVAILABLE HARDWARE

The TRANSCOM system will support a fully-implemented system with many remote users. DOE HQ will operate and maintain the TRANSCOM Central Computer (TCC). The central computer receives the shipment location information from the ground station and maintains and updates the principal TRANSCOM data base. The remote user computers can access the central computer via telephone lines to obtain current information concerning shipment locations, bills of lading, and emergency response data. The TCC will provide a central point to ensure program quality assurance in both the system hardware and software.

Transceiver

Each identified shipment would have a transceiver, which will automatically calculate and send location information to the satellite evey 15 minutes. This data will be used to automatically track the progress of the shipments on the display maps. Alternatively, or for back-up purposes, a driver can enter data as he passes specific checkpoints. The driver will be given computer-generated checkpoint lists at the originating facility for his preapproved route and or any preapproved alternate routes. These checkpoint lists will include the route and descriptions of each of the checkpoints, including the checkpoint number and a description of the highway intersection or State border crossing corresponding to it. The driver uses the keyboard on the

transceiver to enter the number of the checkpoint and presses the enter
key when he passes a designated checkpoint. If the signal has been
transmitted, the on-board system will indicate a successful transmission.
If the driver does not receive this kind of message, he will immediately
phone his dispatcher. His dispatcher can then track the shipment
manually. The same procedure could be used if the driver must use an
alternate route which has not been preapproved.

Microcomputer Configurations

The TRANSCOM central computer is designed to handle multi-tasking
and multi-user operations. It must be a relatively well-equipped,
high-speed microcomputer such as an IBM PC/AT or compatible. In the
fully-implemented system, the AT will be linked to up to six slave PC's.
Each slave PC will have a multiport serial communications board for
BELL-212 data transmission protocol, along with a set of four Hayes
compatible smart modems of 1200 baud rate capability for supporting
remote users.

The remote user terminals are single user computers such as an IBM
PC/XT, or compatible, equipped with a 20-megabyte internal fixed disk for
using the TRANSCOM database and an enhanced color graphics monitor for
map displays. In addition, each XT will be equipped with an internal
1200 baud Hayes compatible smart modem for communicating with the central
computer. Each remote user's system will have already stored the mapping
file to improve system response time. This will assure a rapid response
even with multiple system inquiries.

Data Security

Although the information in TRANSCOM is not classified, the
information will be protected through the following design features:

- user passwords;
- unique transmitter identification codes;
- dedicated phone lines and;
- system encryption devices, if necessary.

Telecommunication Systems

A number of commercial industry options will be thoroughly reviewed
prior to selecting the specific telecommunication system to support the
TRANSCOM project. Each system will be considered by evaluating compatibility, accuracy, reliability, security costs and other program specific
needs. However, most proposed telecommunications suppliers will be
capable of establishing satellite tracking systems, by ultimately
providing a basic latitude and longitude location.

One telecommunication system, which is being considered for the
TRANSCOM project is the Geostar systems, discussed below. It resembles
other commercial systems which could be used to support TRANSCOM.

PROPOSED GEOSTAR SATELLITE COMMUNICATION SERVICE

The GEOSTAR System, proposed as the world's first commercial "Radio
Determination Satellite Service", may provide a unique, patented means of
nationwide two-way communications. Messages will travel in complete
privacy, encrypted if necessary, directly through satellites in a matter
of seconds.

Based on the GEOSTAR's potential for saving lives, reducing crime, and increasing the efficiency of American industry, a formal proposal was published in 1984 by the Federal Communications Commission to allocate the special radio frequencies needed for the new "Radio Determination Satellite Service."

The GEOSTAR satellites will be in "stationary" orbit, at fixed points 23,000 miles above the continental U.S. A computer at GEOSTAR central will transmit general interrogation signals to each transceiver through satellite relays, many times per second. Each transceiver will be identified by a unique "fingerprint". A digital code, built in at the factory, will identify the message it sends and select the messages addressed to it.

The GEOSTAR System will measure the location of each transceiver by combining the time of interrogation, the times at which the response is received through two satellites, and stored digital information on local terrain heights. That location will not be revealed to anyone without authorized access. No one can use the transceiver without first keying a private code. Complete condfidentiality of information is assured.

Because the required number of satellites for determining latitude and longitude is not yet available, TRANSCOM is being configured at two levels: as a one-way identification system for the near term, and as a two-way communications system for the long term (late 1980's when the satellite links will be available).

SATELLITE POSITIONING

The Satellite System is a network that provides radiodetermination, radiolocation, radionavigation, two-way digital communications, and emergency assistance from the same set of user equipment. The elements of the system include a control and central processing center (The Center), two or more geosynchronous relay satellites, a widely spaced grid of "benchmark" transceivers, and the user transceivers. The Central Processing Center can position the user transceivers accurately in the continental United States (CONUS) by measuring the time required for the user to respond to an interrogation signal relayed to the user through a single satellite. Because the response passes through the two satellite relays, this defines two roundtrip ranges to the user transceiver. These two ranges, plus knowledge of the user altitude, provide the necessary information to calculate the user position at the Center. All user transceivers have a unique identification number and can send and receive digital messages. Thus, the Center can relay the calculated position information to the user transceiver using an addressed digital message.

Because the Center has knowledge of all users positions, the GEOSTAR Satellite can provide a variety of additional services. These include navigational guidance, fleet loction to a central dispatcher, and collision warning. In addition, the two-way message capability allows GEOSTAR satellite users to exchange short messages and to interconnect to data bases and to other communication systems.

The Satellite System will be a licensed operator under the new Radio Determination Satellite Service (RDSS) that is approved by the Federal Communications Commission (FCC). The FCC rules define radiodetermination as "the determination of position, or obtaining information relating to position, by means of the propagation of radiowaves". The goal of the RDSS frequency allocation was to provide routing and accurate position to a variety of users including land vehicles, ships, aircraft, and pedestrians. The FCC has allocated several frequencies for RDSS use in the United States. Allocation of frequencies for global service is under consideration by the International Telecommunications Union (ITU) and independently by many countries.

GEOSTAR POSITIONING TECHNIQUE

A general interrogation is sent to all users from the Center through one of the relay satellites. The interrogation and subsequent messages from the Center addressed to users are modulated onto an 8.0 MHz direct pseudonoise (PN) spread sequence. All user transceivers in view of the transmitting satellite remain locked on to this outbound PN stream of interrogations and messages.

Users respond with a short burst. The user's response burst is a spread spectrum signal using a PN code, and contains the user's unique ID and any messages to be sent to or via the Center. To prevent system saturation, the user will only respond to one of many interrogations. The rate of response is determined by the class of user (truck, aircraft etc.), and can be controlled by Center or the user. For example, aircraft in the landing phase will respond to interrogation more frequently than slow moving land vehicles.

The user's response bursting is "bentpiped" through at least two satellites to Center where the roundtrip signal transit time is calculated. The Center measures the roundtrip time by delay locking a local replica of the user PN code with the received code, and measuring the associated time delay. The time delay scaled by the velocity of light is the measurement of the roundtrip range from the Center to the user transceiver.

DIFFERENTIAL OPERATION

An important contributor to Satellite System accuracy is the use of fixed "benchmark" transceivers at known locations. This eliminates many of the sources of ranging bias errors that plague other navigation systems.

The Satellite System was designed from the outset to be a differential system. Because all user positioning calculations are done at the Center, it is a simple task to incorporate differential correction into the user positioning algorithm.

The benchmark corrections can be implemented as either a measurement correction or as a positioning solution correction. Both methods have been successfully demonstrated in other differential navigation systems.

The position solution correction involves using the normal positioning algorithm to calculate the position of the benchmark from uncorrected range measurements. This estimated position is compared with the known position, and the difference can be used to correct the positioning solutions of nearby users.

Satellite Errors

All satellites will be continuously tracked by the Satellite System. Using the smoothed roundtrip ranges to the benchmarks, the satellite positions can be estimated accurately. Normally, any residual satellite estimation error accumulates. However, a large part of this residual

216

error is a correctable range bias and is eliminated by benchmark correction.

If the error is along the line-of-sight from the benchmark to the satellite, then the error is largely a range bias and thereby easily eliminated. If the error places the estimated satellite position "next to" the true satellite positioning (i.e., perpendicular to the line-of-sight), then the error is not as completely cancelled by the benchmark correction. Analysis shows that the uncorrected error is very small even in a worst case situation.

A CONUS user separated by 250 km from a benchmark will see an uncorrected range error of less than 0.3 meters from a worst case 25 meter satellite error in the down range direction.

Uncorrected satellite errors under more normal conditions will be very small (0.1 meters issued in the error budget).

RELATIVE ACCURACY

The majority of users are primarily interested in their positions relative to another point (as opposed to "absolute" accuracy with respect to an arbitrary coordinate frame located, for example, at the center of the earth). Landing an airplane, docking a boat, determining the distance to a truck depot, and directing rescuers to an injured hiker are all 'relative' problems. For these applications, the Satellite System can provide positioning accuracy of the order of the above two-dimensional accuracy. This is accomplished by using a Satellite System generated "time difference" map instead of the digital terrain map.

Points of interest to the satellite systems customers can be "time difference" mapped by recording the roundtrip signal times between the Center and the point to be mapped. Next, this signal time is corrected by a benchmark measurement to yield a "time difference". Two time differences define that location in a unique repeatable manner, and can be used as the basis for positioning nearby users relative to that point. Essentially, this is a way of eliminating the terrain map errors. The Satellite System can map primary highways, waterways, railways, airports, etc., such that even if the absolute position of a user is not known to better than 35 meters, the distance between users and the distance to various land marks can be known to a much higher precision (of the order of meters).

Near-Term System

Figure 1 illustrates the configuration of the TRANSCOM prototype system, which will be tested at DOE HQ and three field offices. The prototype is based on currently available technology and can be implemented as soon as the first telecommunications satellite is launched. A ground-based, low-frequency navigational system known as LORAN-C determines the latitude and longitude of the position of the shipments to within 1/2 mile. The LORAN-C box carried on the vehicle calculates the position of the vehicle on the basis of signals emitted from the nationwide system of LORAN-C towers. Latitude, longitude, status, shipment identification, and optional text messages are transmitted periodically to a telecommunications satellite. Telecommunications relays the information to its ground station near Princeton, New Jersey, from where it is sent to DOE. A driver-initiated status indication can also be transmitted via telecommunications to the central computer.

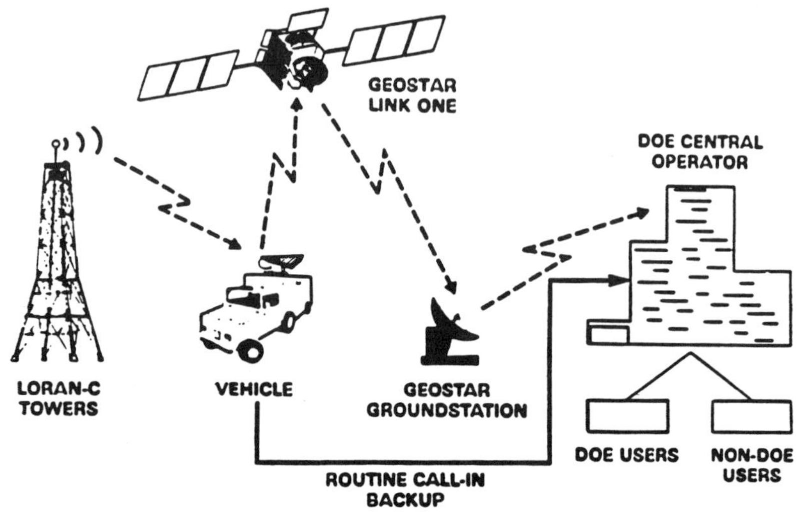

Fig. 1. GEOSTAR LINK ONE service.

Long-Term System

Figure 2 depicts the final full-scale TRANSCOM system. This system will be used by DOE and other Federal agencies. As Fig. 2 indicates, the communications link from the vehicle to the TRANSCOM central computer is the same as is used in the prototype. The full-scale system differs from the prototype; latitude and longitude position of the shipment will be computed more precisely on the basis of satellite signals. The tele-communications satellite positionsing system is expected to be fully operational in the late 1980's. The TRANSCOM prototype is designed so that the improved position location accuracy can readily be integrated into the system with little or no disruption. In the full-scale system, there is an automated status indicator which is supplementary to the driver-initiated status. In addition, this version will have expanded emergency response capability, including automated DOE procedures and checklists.

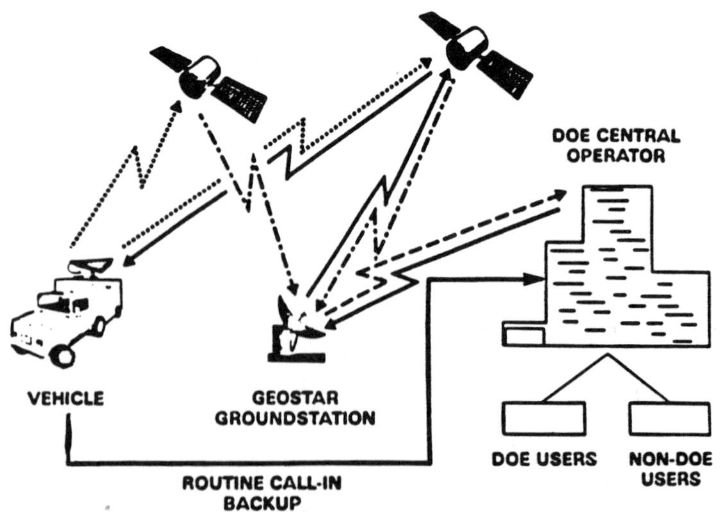

Fig. 2. GEOSTAR prime service.

IMPLEMENTATION PLAN

To date, the TRANSCOM system requirements have been defined and first generation software has been completed. The TRANSCOM software was developed by Systems Research and Application Corporation (SRA) and was formally demonstrated to DOE in January 1987.

The satellite service to be used in the near-term system (LINK ONE) is expected to be offered by Geostar Corporation in early 1988. This date is predicated on a successful satellite launch. Phase over to the longer term satellite system is dependent on subsequent satellite launches, and will be implemented as soon as the two-way communication service is available. It is currently projected that full system use will be available in 1989, including two-way communications capability.

TRANSCOM's initial testing schedule will incorporate a gradually increasing distance check of the accuracy of U.S. Geological survey mapping systems. The initial tests have proved map accuracy to approximately 80 feet in densely populated areas and greater accuracy capabilities with equipment which eliminates transport equipment noises. When all necessary revisions have been incorporated and tested, the system will be made available to all DOE field offices and shippers, and then to State designees. Non-DOE users will only use the system to monitor activities. DOE program and emergency operations office will have the ability to reschedule, change shipping status, or schedule checkpoints during normal use.

SOFTWARE OUTPUT

Each enroute shipment which is being tracked by satellite or by carrier call-ins will be displayed on the map with its corresponding transmitter number. The user has the option to zoom in on a particular State and then to a particular county. All shipments in the lading file which have left their point of origin but have not arrived at their destination will be tracked on the map.

When the tracking function is chosen from the main menu, the user has the choice of viewing all shipments, motor or rail for all of DOE, or motor or rail for a particular field office appears. The all-shipment option enables the user to view a map of the U.S. with the States designated, but without rail or road markings. This map displays all on-going shipments, regardless of whether they are rail or motor shipments. Each shipment will be designated as rail or motor by the shape of the icon. The shipment status is indicated by various colors. In addition to the counrty-wide overview, the user has the option of viewing just rail or just motor shipments for a particular field office or viewing rail or motor for all DOE shipments. DOE Headquarters will alway view all on-going shipments for either rail or motor. Since higher density areas where roads, rail lines, and cities cause the mapping pictures to be overloaded with information, the software system has single function keys to "create the desired picture". Cities, roads, rail lines, or country borders can be added or deleted at the discretion of the user.

The rail maps display all major rail lines for the U.S., States, or counties. The motor maps of the U.S. and the States display all major U.S. interstate and State highway routes. In addition, the U.S. maps show State names and major city names. The state maps show major city names and county names. The county maps show major city names. At each level of the mapping function, the user may choose to investigate the lading information for a particular shipment.

DOE personnel responsible for nuclear shipments will require access to detailed information about each shipment. The lading information file provides this information; it includes all the data elements required by 49 CFR 172 and 173, which is the relevant DOT transportation shipping information, including a listing of all radioisotopes by name and their total activity level. The purpose of this file is to provide a readily-accessible, detailed source of information about the kind of material being transported. The shipment identification code displayed on the tracking map provides the link between the tracking function and the lading file. In shipment status areas, the software will record all previous transmissions by latitude/longitude and alert signals will automatically provide the last vehicle position and speed and the projected location of the vehicle over a specified period of time and route.

Choosing the checklist option from the main menu will place the user in a second menu from which he will be able to execute a checklist in case of an emergency. The user will be prompted for a shipment number, which will be used to locate the emergency contact information for this shipment. The user will then be presented with a menu of hazardous materials. After the user selects the appropriate material, he will be presented with several screens which contain all emergency contact information and recommended emergency procedures.

SUMMARY

DOE's mandate to ship radioactive materials safely and efficiently has been affected by institutional issues arising form serious State and local concerns. These concerns result from the local officials' view that a lack of information prevents them from being on top of the situation and responding to emergencies.

The TRANSCOM system offers a cost-effective solution to DOE's shipment tracking requirement. It enables DOE HQ and field office users to share information in near real-time. If the system is made available to a State or local government, it would provide required prenotification and routing information. The precise time of arrival at a given border can be accurately estimated by monitoring the progress of the shipment as the system tracks its location. Built-in emergency response guidelines will prompt State and local as well as Federal officials with the proper procedures if there is an accident. In short, this high-technology, real-time tracking system solves major institutional as well as operational shipping problems.

QUALITY ASSURANCE IN SPENT FUEL TRANSPORTS

B. Cooke

Nuclear Transport Limited
Delenty Drive
Risley
Warrington
Cheshire

ABSTRACT

The IAEA Regulations, Safety Series No. 6 (1985), specify that transporters shall establish a quality assurance programme.

NTL has been transporting spent fuel from European light water reactors to storage and reprocessing facilities since 1972. NTL's formal application of quality assurance to its transport activities began in 1983. Introduction began with the Group Quality Assurance Manual which stated the company policy, identified areas and extent of application, together with the provision of material support for the company quality assurance programme.

The quality assurance programme for spent fuel transports identifies all the necessary procedures required to comply with the company manual, as well as identifying responsibilities and determining levels of authority. The programme covers all those activities associated with package approvals, transport licences, flask handling, flask and vehicle maintenance, modification, and repair.

The paper discusses a quality assurance programme specifically applied to the business of transporting spent fuel.

INTRODUCTION

Up to 1985 the nuclear transportation industry had no regulatory directive concerning the application of quality assurance. In 1985 the IAEA issued the latest version of Safety Series No. 6 (IAEA 1985), in which the application of quality assurance was defined:

"Quality assurance programmes shall be established for the design, manufacture, testing, documentation, use, maintenance and inspection of all packages and for transport and in-transit storage operations to ensure compliance with the relevant provisions of these Regulations. Where competent authority approval for design of shipment is required, such approval shall take into account and be contingent upon the adequacy of the quality assurance programme. Certification that the design specification has been fully implemented shall be available to the

competent authority. The manufacturer, consignor, or user of any package
design shall be prepared to provide facilities for competent authority
inspection of the packaging during construction and use and to
demonstrate to any cognizant competent authority that:-

(a) The construction methods and materials used for the construction of
     the packaging are in accordance with the approved design
     specifications; and

(b) All packagings built to an approved design are periodically
     inspected and, as necessary, repaired and maintained in good
     condition so that they continue to comply with all relevant
     requirements and specifications, even after repeated use".

However, in Safety Series No. 6 there were no QA guidelines on
reference standards quoted to assist in the formulation of QA systems.

This was overcome in 1987 with the latest issue of the IAEA Safety
Guide No. 37 (IAEA 1985) which contained in appendix IV the details of
recommended quality assurance for the nuclear transportation industry.

It was recognised in IAEA Guide No. 37. that QA had been applied by
the industry prior to the introduction of IAEA Guide No. 37. It was also
recognised that QA may have been applied using a non-transport related
IAEA publication, IAEA Safety Standard 50-C-QA^3, this being the closest
the industry could associate with its own activities.

IAEA Guide No. 37 recognising this fact, detailed a comparison
between the 50-C-QA QA structure and the structure of QA recommended for
the industry in IAEA Guide No. 37.

Also, in 1986 the UK Department of Transport (DTp), issued a
statement concerning quality assurance.

Briefly, the DTp stated that organisations involved in the business
of transportation of nuclear matrials shall be required to have quality
assurance implemented and operating for radioactive material transport
activities by 1st January 1990.

In the statement by the DTp, it was recognised that QA programmes in
accordance with IAEA Guide No. 37 would be acceptable along with other
national and international QA standards.

REQUIREMENTS FOR A QA PROGRAMME FOR NTL

NTL is an international group of companies (See Figure 1) which owns
and operates spent LWR fuel flasks throughout a number of countries in
Western Europe and has offices in the UK, France and Germany. Therefore
the development of a QA programme had to take account of all of the
Company's activities, the countries concerned and the manner in which it
operated in order to achieve the optimum QA programme.

This meant analysing the company activities in terms of QA and non
QA related activities in the various NTL offices and identifying the QA
interfacing and responsibilities of the NTL offices.

In applying QA to NTL's business of transporting spent LWR fuel,
differences of attitudes, approach and account of local and national
regulations regarding QA were required to be borne in mind so that an
international approach to QA could be achieved.

222

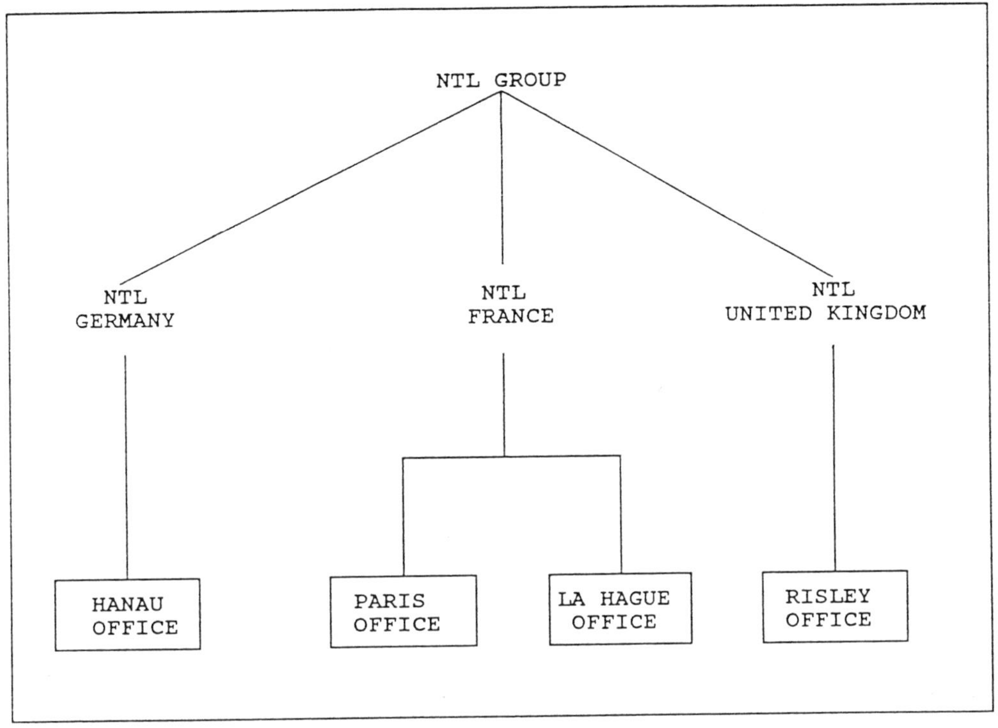

Fig. 1. NTL group of companies.

DESIGN OF THE NTL QA DOCUMENTATION SYSTEM

The NTL QA documentation system is shown in Figure 2. It consists of the following four tiers of documents:-

1ST TIER - GROUP QUALITY ASSURANCE MANUAL
2ND TIER - QUALITY ASSURANCE PROGRAMMES (PAQs)
3RD TIER - QUALITY PLANS
4TH TIER - INSTRUCTIONS, PROCEDURES, REPORTS, ETC.

<u>Group Quality Assurance Manual</u>

In recognising the fact that the business from all four of the NTL offices (Figure 3) was a part of the business of the NTL group it was decided to state NTL's commitment to QA by drawing the business of the NTL together under one QA umbrella - the NTL Group Quality Assurance Manual.

The manual, (written in English - the agreed language within the NTL group) identified the range of the group's QA related activities, the QA hierarchy and responsibilities. Figure 4 shows the NTL QA organisation chart and responsibilities.

Much of the transport business involves the interfacing of the NTL offices, with various responsibilities being delegated between the offices. The manual recognised that more than one NTL office would have responsibilities within one particular company area of business and specified that this would be achieved by the common approach to all QA activities within the NTL group.

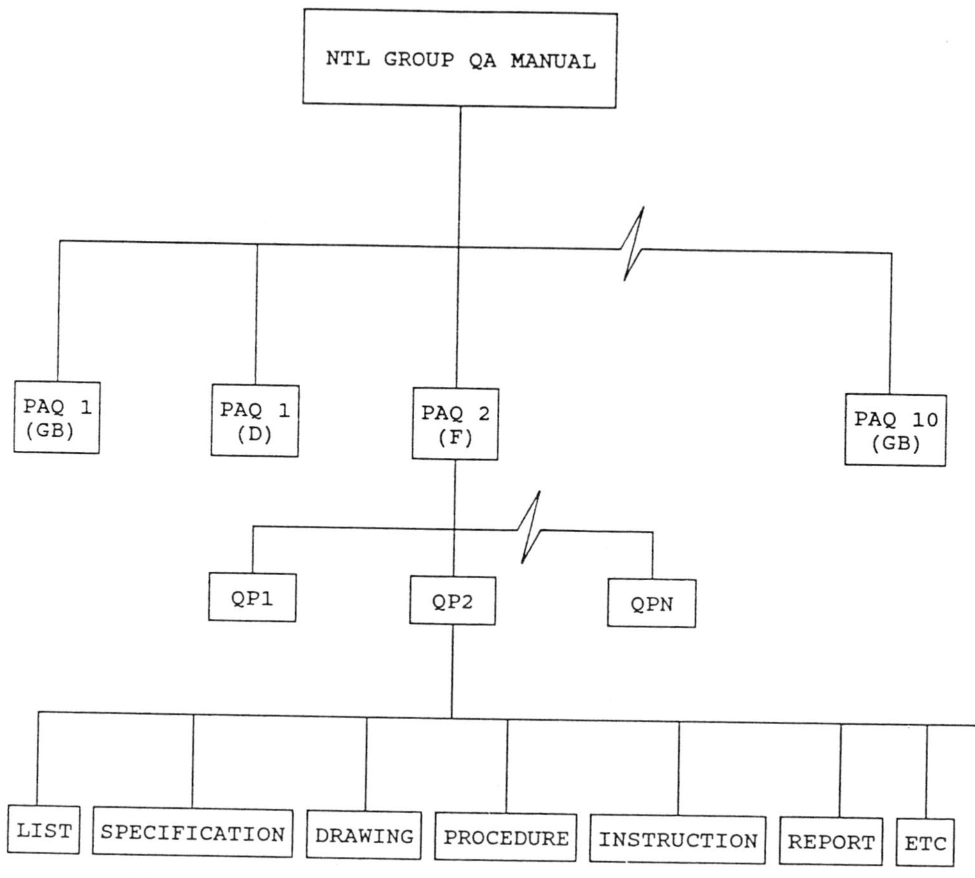

Fig. 2. NTL group quality assurance document system.

It should be noted that Figure 4 shows that in a number of cases, the same type of business is carried out by more than one NTL office. The manner in which this is dealt with is addressed in the second tier of documents - Quality Assurance Programmes (PAQs).

Quality Assurance Programmes (PAQs)

The documents immediately subordinate to the NTL group QA manual in the NTL QA documentation system were deemed to be Quality Assurance Programmes and are referred to in NTL as PAQs (Programme d'Assurance de Qualité).

A PAQ covers the QA activities in one specific NTL area of business.

The PAQ is specific to the QA activities within that area of business only and details the management of quality in general terms. The management of quality includes detailing who is responsible for the various activities within that area of business and any interfaces within the PAQ. The finite detailed requirements of how to apply the PAQs are covered in documents subordinate to the PAQ.

The principal methods in which QA is addressed in each PAQ, follows the pattern shown in IAEA No. 37.

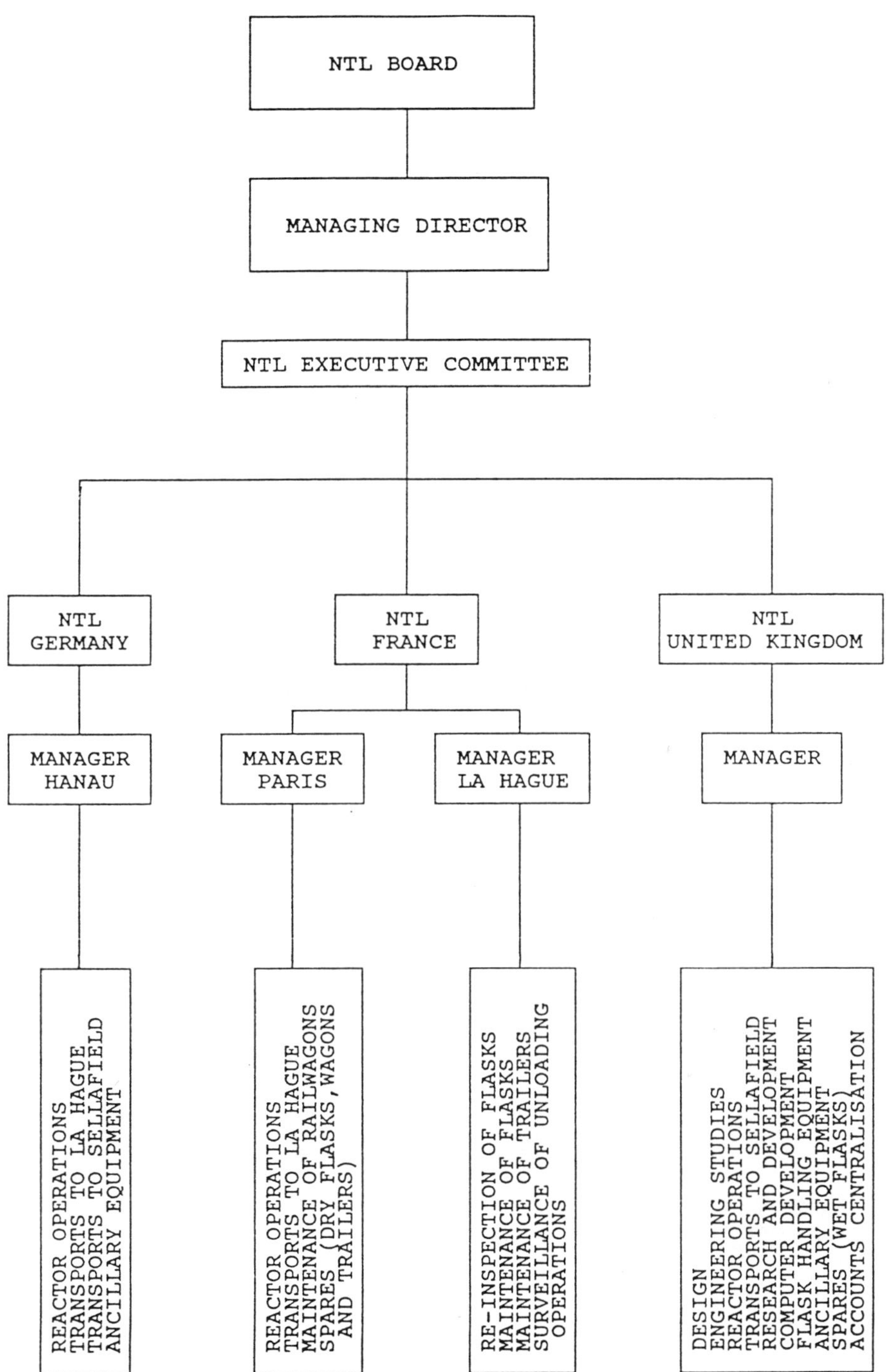

Fig. 3. NTL organizational chart.

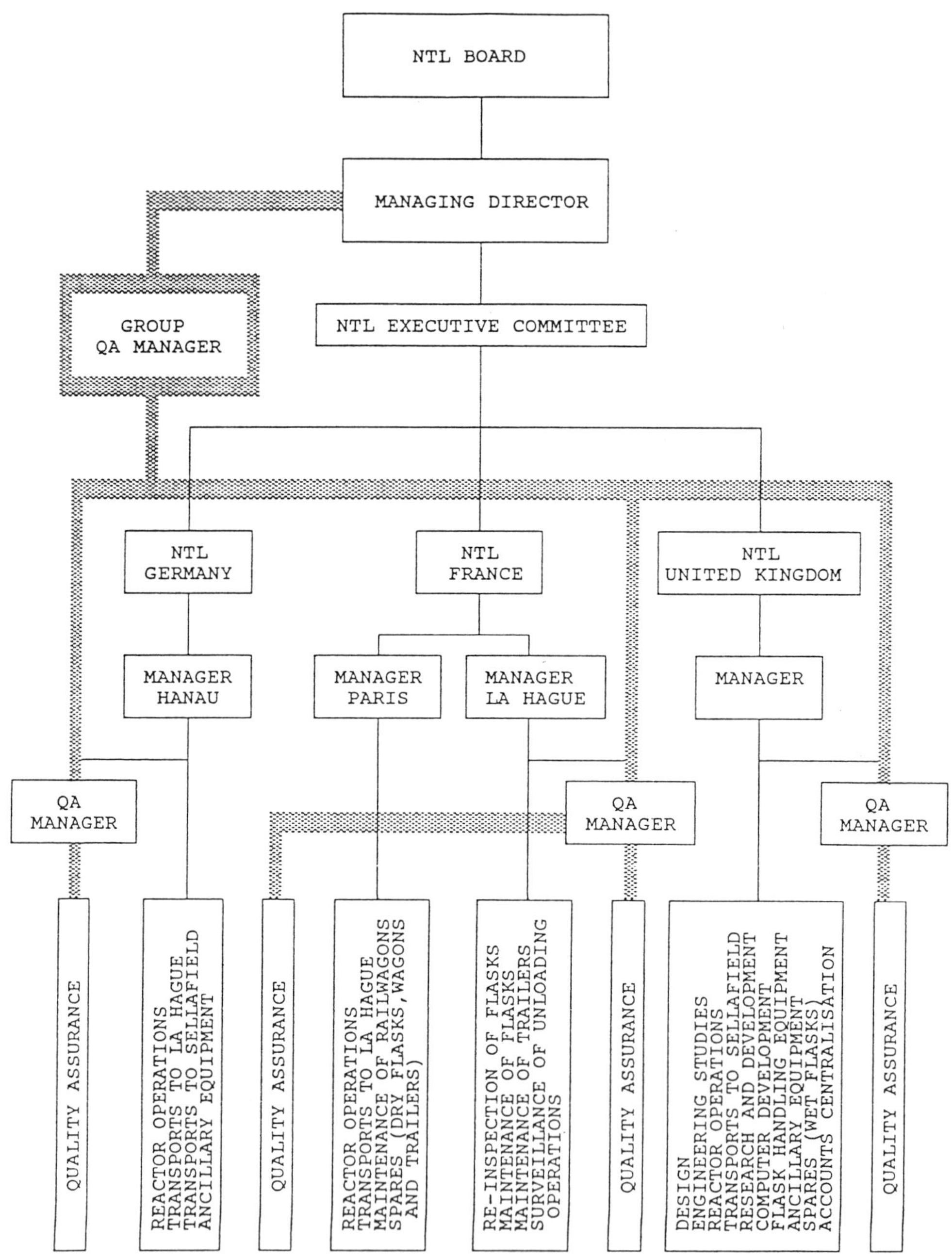

Fig. 4. NTL QA organization.

Following NTL's decision to have one PAQ per company area of business, it was deemed (for transport) that one PAQ was required to cover spent LWR fuel transport to each specific storage/reprocessing site. As NTL carries out transports to both Sellafield in the UK and to La Hague in France it was decided therefore that two PAQs would be required. Local and national requirements, and the attitudes and approaches of the two recipient countries concerned are incorporated into the PAQs as well as the requirements of countries the flasks pass through.

The content and approach within these two PAQs were very carefully considered as the methods of applying QA in two or more different countries had to be incorporated to meet the total QA requirements defined in the NTL group QA manual and not to corrupt any local or national methods and approaches to all of the areas a PAQ covers.

For the non (direct) transport areas of business in NTL where this business is duplicated by different offices, i.e., where NTL's UK and German offices are engaged in design, there is one UK PAQ and one German PAQ each essentially covering the same subject. However, as only the approach to QA is universal, the application of QA can be completely different in different countries. The duplicated PAQs take account of the application and apprach of QA in that country responsible for that area of NTL's business. In some cases the requirements, attitudes and approaches in different countries are the same. In those cases, the appropriate area of the PAQs are also the same.

Quality Plans (QP'S)

Control of the application of the PAQs is carried out using Quality Plans (QPs). QPs are working documents covering one instance of the application of a PAQ i.e., for one spent fuel transport one QP will be used as a 'live' document from commencement of the transport to its completion. The QP will be completed as specific stages of the transport are achieved and the QP will be finally completed when the transport is also complete. The QP will provide a management record of the QA activities of that one particular transport.

The QP contains the responsibilities of all the activities within the QP and details where reports and records are generated. QPs detail the sequential order of carrying out the activities.

Instructions, Procedures, Reports etc

Subordinate to the QPs are the base tier of documents covering the detailed application of the QPs. This tier of documents is made up of all working documents such as procedures, instructions, reports, specifications, drawings etc. etc.

INTRODUCTION AND IMPLEMENTATION OF THE QA PROGRAMME

NTL took its first formal quality assurance steps with the introduction of the company QA manual in 1983. The manual at that time was based upon IAEA 50-C-QA and is now being rewritten using IAEA No. 37 as a basis.

The manual (written in English) identifies the range of the company's QA related activities, the QA hierarchy, organisation and responsibilities.

To date, NTL have identified 10 'areas of business' that require PAQs to be produced. A number of these PAQs are duplicated as previously discussed to cover similar area of business being carried in the different offices.

The PAQs and all supporting documents are written in the language of the country responsible for the particular area of business.

To direct, guide and implement QA throughout the company, regular formal QA meetings are held, at office level and company level. At these meetings the implementation of the various levels od QA is defined and programmed. QA policies are reviewed and iteration is carried out in the light of experience gained throughout the company to ensure the implementation of QA is following and meeting the agreed programmes.

MONITORING THE NTL QA SYSTEM FOR EFFECTIVENESS

Monitoring is carried out on a continuous basis to ensure the effectiveness of the QA system.

All methods of application of QA, and the activities in particular areas of business are liable to change to reflect changes in practice, improved methods of working, etc.

To this end a schedule of audits and reviews is operated in NTL to check the effectiveness of QA applied to specific company activities and pinpoint areas requiring updating and improvement.

At present a part of the monitoring is caried out by the internal auditing of company departments against the activities detailed in the PAQs. (The PAQs are also reviewed but this is carried out separately). This enables the detailed application of QA in one specific company activity, to be both audited in detail and gives the opportunity for inconsistencies, changes, amendments in the PAQs to be updated to be consistent with the current practice in applying QA within that activity. Reviews and audits are, at present, every 6 months. The periodicity was chosen to aid the implementation of QA whilst it is in its 'infancy'. (It is envisaged that the periodicities of audits and reviews will be extended to 12 months at some future date).

A further part of the monitoring is by the auditing/assessment of 'approved suppliers' who supply goods or services to NTL. As NTL is not a large organisation and hence not ordering large quantities of goods or extensive services, the auditing/assessment of approved suppliers has been tailored to meet the company's requirements. One of the methods adopted in this tailoring has been 'project' auditing. This is where NTL has audited suppliers supplying goods or services in support of a specific NTL project. The project auditing enables NTL to audit specific supplier services instead of a total supplier audit.

Reviews have also taken account of the change of NTL's QA reference from IAEA 50-C-QA to IAEA Safety Series No. 37. The original issue of the NTL QA Group manual was based upon IAEA 50-C-QA. This was also the case for some of the earlier PAQs that were written prior to 1987. The manual and the appropriate PAQs are in the course of being rewritten in accordance with IAEA No.37.

REFERENCES

IAEA Safety Standard, Safety Series No. 6, regulations for the Safe Transport of Radioactive material - 1985 Edition.

IAEA Safety Guide, Safety Series No. 37, Advisory Material for the IAEA Regulations for the Safe Transport of Radioactive Material (1985 Edition) Third Edition.

IAEA Safety Standard, Safety Series No. 50-C-QA, Quality Assurance for Safety in Nuclear Power Plants. A code of Practice.

UK Department of Transport letter to NTL, September 1986, titled Compliance and Quality Assurance in the Safe Transport of Radioactive Materials: Department of Transport Policy and Requirements.

TECHNIQUE OF STOWING PACKAGES CONTAINING RADIOACTIVE
MATERIALS DURING MARITIME TRANSPORTATION

G. Ringot[1], G. Chevalier[1], E. Tomachevsky[1],

J. Draulans[2], I. Lafontaine[2]

[1]Commissariat a l'Energie Atomique
31-33 Rue de la Federation
75752 Paris, Cedex 15
France

[2]Transnubel
Gravenstraat 73
B-2480 Dessel
Belgium

ABSTRACT

The Mont Louis accident, in which uranium hexafluoride packages were
involved, alarmed a large number of European competent authorities
directly or indirectly concerned in the accident, including the
Commission of European Communities (CEC).

The CEC sponsored, in 1986-1987, a preliminary research, mainly a
bibliographic data collection, to obtain a first view on the problem.

However, the purpose of the CEC is to produce, by further work, a
code of good practice for stowing, with accident conditions taken into
consideration.

The research work was carried out jointly by Commissariat a
l'Energie Atomique (CEA), Paris, France, the Institut de Protection et de
Sûreté Nucléaire (IPSN) at Fontenay-aux-Roses, France and by
Transnubel S.A. Brussels, Belgium. The final report of these contracts
was completed in November 1987.

The work performed in the scope of these tasks includes:

-   data collection concerning the acceleration to be considered during
    normal and accidental conditions and concerning safe areas in the
    ship to be reserved for the stowing of packages containing
    radioactive materials.
-   data collection concerning accident statistics.
    Lloyds Register of Shipping supplied the basic information needed to
    select reference ship-accident scenarios as a basis for further
    work.

The collected data provide the necessary basis for this further work, aiming to increase the safety of transporting radioactive material by ship.

OBJECTIVE AND SCOPE OF THE STUDY

Definition of the problem

The accident (August 25th 1984 in the North Sea) to the Mont-Louis cargo which transported uranium hexafluoride, alarmed a large number of European competent authorities directly or indirectly concerned in the accident, the Commission of the European Communities included.

During a meeting held on July 11 1985 at the offices of "Le Secrétariat d'Etat à la Mer de la République française" it appeared that stowing prescriptions for packages containing radioactive material, taking accident conditions into consideration, were nearly non existent. Even the rules for normal transport conditions differed, for example longitudinal accelerations between 0, 5 and 2 g were considered.

Aim of the study

The study should collect the different deceleration values used by the transport companies, and should define the accident conditions to be considered.

This work should serve as a basis for later research, to end with the proposal of a code of good practice for stowing.

Tasks included in the scope of a general study

The general study should include :

- a bibliographic study of the work published in this field
- statistical analysis of the accidents
- a theoretical approach of the problem:
  . impact against quay
  . collision
  . contact
- a number of collision tests to define the deceleration values obtained in several collision types with respect to the collision angle, the relative speed and the ship sizes
- the comparison of the test results with the calculated acceleration values
- a proposal for standards to establish the stowing conditions.

Tasks included in this first phase

This first phase includes two main tasks:

- a statistical analysis
- a bibliographic study of ship accidents.

These tasks have been carried out by performing the following studies and inquiries :

- collection of information from British and French ship-registering companies
- discussions with the French Compagnie Générale Maritime
- discussions with the Belgian Compagnie Maritime Belge

232

-       bibliographic study with the help of Organisation Maritime
        Internationale (OMI),
-       analysis of accidents in open sea and in the harbour,
-       investigation of the French standards relating to the stowing of
        trucks on car-ferry decks,
-       investigation of the sea transportation of packages loaded with
        irradiated fuel on board ships built especially for this kind of
        transportation,
-       attempt to define the number of decelerations (g) generated during
        storms and on impact.

DATA COLLECTION

        Forty-eight documents have been consulted to obtain information
concerning, on the one hand, the accelerations to be considered during
normal transport conditions and in contact conditions, the ship area to
be reserved for the storage of packages containing radioactive products
and, on the other hand, the accident statistics. The list of these
documents is given in addendum I.

<u>Data collection concerning the accelerations to be considered during
normal transportation conditions and in contact conditions.</u>

        Some of the documents, listed in addendum I, contain information
concerning the accelerations to be considered. The following values have
been found:

1.   longitudinal accelerations   : between 0,29 and 2 g
2.   transversal accelerations   : between 0,75 and 2 g
3.   vertical accelerations   : between 0,15 and 3,5 g
4.   roll with a maximum angle of 30° and a period varying between 10
        and 50 s
5.   pitch with the maximum angle of 10° and a period varying between 5
        and 15 s

        The accelerations given for collisions differ from the ones
mentioned above:

1.   longitudinal accelerations   : between 0,135 and 1,5g
2.   transversal accelerations   : between 0,11 and 3,65g

<u>Note:</u> Most of the data found concern the ship itself and not the
cargo.

<u>The highest values found are:</u>

1.   longitudinal acceleration   : 2 g
2.   transversal acceleration   : 3,65 g
3.   vertical acceleration   : 3,5 g

<u>Data collection concerning the areas on board ship reserved for the
stowage of packages containing radioactive products</u>

        Document 15 in the list, "American national standard for highway
route controlled quantities of radioactive materials - domestic barge
transport, ANSI N 14.24 - 1985" defines the area <u>on board of barges</u>
where such containers can be stowed as follows:

-       at a distance of minimum of B/5 inboard from the side of the vessel,
        where B is equal to the beam of the barge

- at a distance abaft the forward perpendicular and forward of the
  stern transom of $L^{2/3}/3$ or 14.5 m min. where L is the length of
  the barge.

Data collection concerning accident statistics

Several documents contain information concerning the probability of
a ship having an accident. These collected data are given in Table 1
"Collected data concerning ship accident probability".

Furthermore, Lloyd's Register of Shipping was requested to supply
the following information :

- an analysis of the types of accidents near most of the European
  coasts for 5 different ship size bands (500 to 15,000 tonnes)
- the collection of more detailed information concerning 16 accidents
- an analysis of ship movements for the English Channel and the
  assessment of accident frequencies
- a selection of reference type accidents to be used in further tasks
  (theoretical approach and scale tests).

Summary of the study performed by "Lloyd's Register of Shipping"

Study area. The study area involved includes the Western
Mediterranean and its approaches, Eastern Atlantic Waters, the North Sea
and the Baltic, as well as the Kiel Canal. The area is shown in Figure 1.

Time period considered in the study. The period 1978 to 1986 has
been considered.

Chosen ship types and deadweight size bands. The chosen ships are:

- general dry cargo
- Ro-Ro (cargo and/or passengers)
- container and nuclear fuel carriers
- combinations of these classes

The chosen deadweight size bands in tonnes are:

- between   500 and   1,999
- between   2,000 and   4,999
- between   5,000 and   9,999
- between   10,000 and 14,999
                15,000 and above

Analysis of the accident types. The following averages per year
(over 9 years) have been found:

-   35.2 ships are lost
- 199.4 ships are heavily damaged
- 234.6 ships are either lost or heavily damaged.

The study gives as main causes of accidents to ships carrying
nuclear materials in European Waters the following percentages :

- collision   (16.5 %)
- contact   (6.5 %)
- fire/explosion   (11.2 %)
- foundered   (9.1 %)
- hull/machinery damage (34.5 %)
- wrecked/stranded   (20.5 %)

Table 1. Collected Data Concerning Ship Accident Probabilities

Document title	Le transport par mer des matières radioactives	Safe stowage and securing of cargo on board ships	An engineering assessment of the probability for damage to ram transport casks due to barge collisions
Authors	P. Gilles + C. Ringot	P. Anderson	B.L. Hutchison a.o.
Organization	CEA 83–108	Mariterm AB 1982	Glosten Ass.
PROBABILITIES			
a. Per ship/year			
a.1. Swedish territorial Waters			
– ship lost by collision	2.10^{-4}		
– ship lost by stranding	4.10^{-4}		
– loss of cargo	10^{-5}		
– cargo damaged by collision	10^{-4}		
– fire lasting a long time	10^{-7}		
a.2. Japanese data			
– total loss	$7,6.10^{-7}$		
b. Per voyage			
b.1. French data			
– heavy collision (Channel–North Sea)	10^{-4}		
– ship lost by collision	10^{-5}		
c. Per ship/km			
– collision with tanker (Channel–North Sea)	7.10^{-9}		
d. Per 1.000 ships/year			
– total loss/period 78–80			
. Ro/Ro		4,6	
. tanker		4,2	
. World Fleet		7,2	
e. % of penetration in case of collision			
e.1. In the area defined by ANSI			
– on the Mississippi			25,64
– in the ocean			16,48
e.2. Out of the area defined by ANSI			
– on the Mississippi			74,36
– in the ocean			83,52

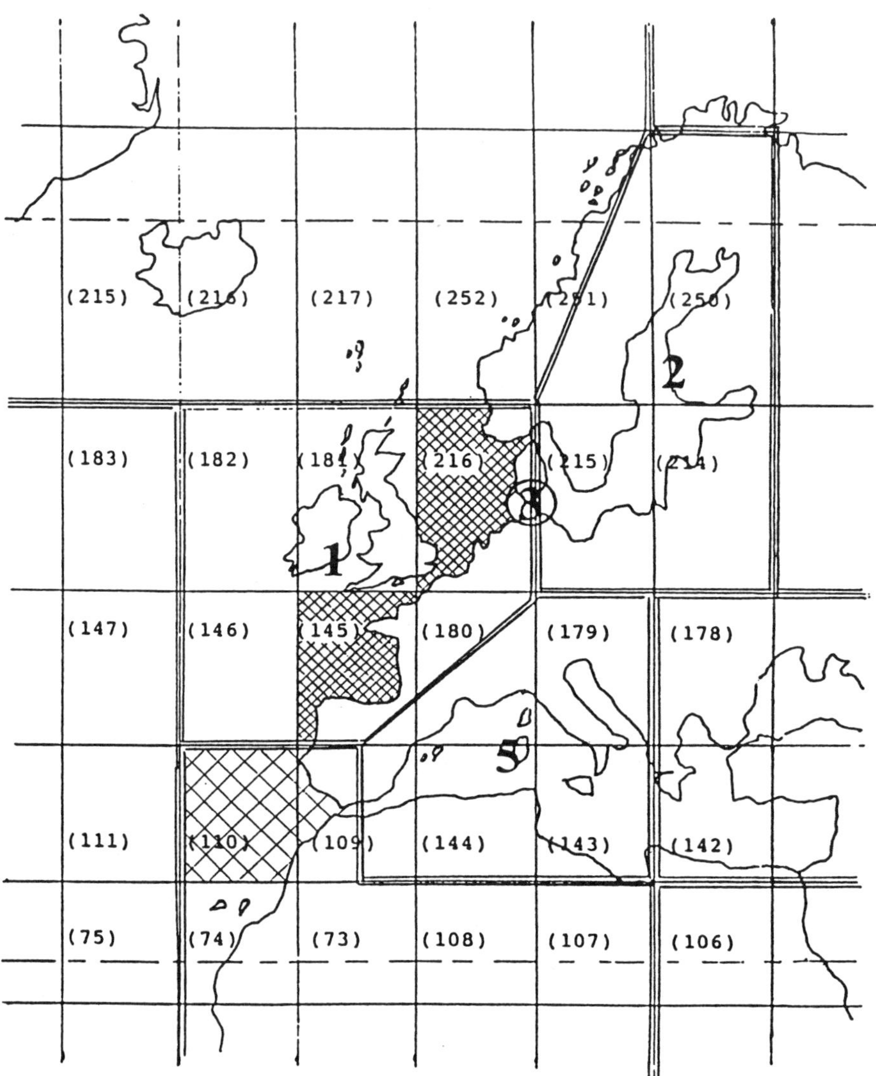

Fig. 1. Considered European waters and their approaches.

In the study, the hull/machinery failure has been omitted since no damage to cargo is likely as a direct result and also foundering where cargo damage is due to water ingress.

<u>More detailed accident investigations</u>. Sixteen accidents have been selected for more detailed investigation. High penetrations were found for Ro-Ro ships (3.9 m - 7.0 m - 4 to 5 m - 4 to 5 m).

<u>Final selection of suggested reference scenarios</u>. Eight suggested reference scenarios have been selected to provide a typical cross-section of "worst-case" accidents for the selected ship types. It is considered that collision is likely to provide the most common scenario incident with respect to acceleration forces on the cargo. Therefore, the selection of the scenarios reflects this. Information concerning these eight accidents is given in Table 2.

Table 2. Information Concerning the Eight Suggested Reference Scenarios

Information \ Scenarios	1°case	2°case	3°case	4°case	5°case	6°case	7°case	8°case
Type of accident	Collision	Collision	Collision	Collision	Collis./ Fire	Collision	collision	Contact
Year	1979	1980	1979	1985	1983	1982	1984	1980
Location	River/Est	Sea	Sea	River/Est	River/Est	Sea	Sea	Restr. Waters
Sea state	Unknown	Unknown	Unknown	/	/			Unknown
Weather	Unknown	Unknown	Unknown	Fog	Dense fog	Heavy wind	Fog	Unknown
Struck ship . type	Cargo	Cargo	Cargo	Tr. cont.	Tr. cont.	RoRo pass.	RoRo	Cargo
. dim. −long. (m) −breadth −draught	106 14,9 7,25	153 23 8,7	149 22 9,1	147 22 8,1	203 30 11,2	133 20 5,8	135 19 6,5	84 13,2 5,2
. deadweight (t)	5.960	13.600	16.500	13.100	28.900	3.950	5.900	2.800
. speed (km)	5,6 à 7,4	25,9	27,8 ·	18,5	16,7	28,1	20,4	18,5
. loaded/ballast conditions	Unknown	loaded cont.	loaded cont.	loaded cont.	loaded cont.	laden vehicles	laden cont/veh.	Loaded cont.
. angle of colli- sion	120°	80°	Unknown	180°	120°	94°	90°	/
. depth of pene- tration in m	2 à 3	/	1 à 2	1 à 2	1,5	3,9	7	2
Striking ship . type	Cargo	RoRo cargo	Cargo	Ore Transp.	Cargo/ tr. cont.	RoRo cargo	RoRo ferry	Nihil
. dim. −long. (m) −breadth −draught	144 20,5 8,8	148 23 6,6	93 13,5 5,5	261 40,6 16	177 23 9,7	142 18,5 4,5	153 24 5,8	
. deadweight (t)	15.200	4.300	3.000	117.600	14.000	5.600	2.900	
. speed (km)	5,6 à 7,4	37	20,4	16,7	16,7	10,4	22,2	

The following penetration depths have been found:

```
- 1° case          2 to 3 m)
- 2° case          -       )
- 3° case          1 to 2 m)
- 4° case          1 to 2 m)   average of the
- 5° case          1.5 m   )   max. values : 2,67 m
- 6° case          3.9 m   )
- 7° case          7.0 m   )
- 8° case          2.0 m   )
```

<u>Movement analysis in the Marsden grids 145 and 216.</u>

A total casualty rate of 3.53×10^{-7} is obtained per ship kilometre sailed in these grids.

This rate implies for a typical 220 km voyage across the North Sea (e.g. U.K. to Rotterdam) a probability of 7.8×10^{-5} for a serious accident per voyage.

Similarly, the probability of a serious collision would be 1.5×10^{-5}.

CONCLUSIONS

The complete study has given the following results :
- the accelerations to be considered for the ship during normal transport-conditions and in collision conditions
- information concerning the areas on board ship to be reserved for the stowing of packages containing radioactive products
- the selection of 8 accident types, providing a typical cross-section of "worst-case" accidents for the selected ship types
- the definition of the probability of a serious accident per voyage and for a serious collision during a typical 220 km voyage in the North Sea.

The collected data supply the necessary basis for further tasks in this field as there is a theoretical approach to the problem and the necessary tests in order to increase the safety of stowing packages containing radioactive products on board ship.

ADDENDUM 1

CONSULTED DOCUMENTS

1.   Vers une harmonisation des règlements.
      Ronald B. Pope.
      AIEA Bulletin, Printemps 1985, 5.1/2 p.

2.   Compte rendu réunion 12 juillet 1985.
      C. Ringot.
      C.E.A. - IPSN, 85-168, Septembre 1985, 4,5 p.

3.   Note de travail sur l'arrimage de colis de matières radioactives.
      P. Gilles.
      C.E.A. - IPSN, 84-182, Décembre 1984, 2 p.

4.   Compte rendu rëunion 4 Octobre 1984.
      Au siège de la Compagnie Générale Maritime.
      P. Gilles.
      C.E.A. - IPSN, Octobre 1984, 1.1/2 p.

5.   Compte rendu de visite du M/S "Borodine".
      30.11.1983
      E.S. Tomachevsky.
      C.E.A. - IPSN, 83-163, Décembre 1983, 3 p.

6.   Le transport par mer des matières radioactives.
      P. Gilles et C. Ringot.
      C.E.A. - IPSN, 83-108, Août 1983, 7 p.

7.   Recent advances in hazardous materials transportation
      research. An international exchange, Transportation Research Board/
      National Research Council, 1986.

8.   Casualty return 1980.
      Statistical summary of merchant ships totally lost, broken up, etc.
      Lloyd's Register of Shipping, 34 p.

9.   Casualty return 1981.
      Statistical summary of merchant ships totally lost, broken up, etc.
      Lloyd's Register of Shipping, 43 p.

10.   Casualty return 1982.
      Statistical summary of merchant ships totally lost, broken up, etc.
      Lloyd's Register of Shipping, 52 p.

11.   Casualty return 1983.
      Statistical summary of merchant ships totally lost, broken up, etc.
      Lloyd's Register of Shipping, 65 p.

12   Casualty return 1984.
      Statistical summary of merchant ships totally lost, broken up, etc.
      Lloyd's Register of Shipping, 104 p.

13.   Casualty return 1895.
      Statistical summary of merchant ships totally lost, broken up, etc.
      Lloyd's Register of Shipping, 120 p.

14.   Arrimage des véhicules routiers utilitaires chargés ou non.
      NF J 39-010, Juin 1984, 7 p.

15. American National Standard.
    Highway route controlled quantities of radioactive materials -
    domestic barge transport.
    ANSI, N 14.24-1985, 28 p.

16. Safe stowage and security of cargo on board ships.
    Research report by : Peter Andersson, Björn Allenström, Martti
    Niileksela.
    Mariterm AB 1982, 160 + 182 p.

17. Cargo securing manual for MS Transaren.
    Enclosure to research report "Safe stowage and securing of cargo on
    board ships". 61 p.

18. Reliability and risk analysis services document.
    For Marine, Offshore and Industrial Projects.
    Lloyd's Register of Shipping, April 1985 issue, 42 p.

19. Particularités de la structure des navires rouliers compte tenu de
    Leur chargement.
    P.J. Latreille, Ingénieur principal, division constructions neuves
    du Bureau Veritas.
    No. 415, Navires, ports et chantiers, Déc. 1984, 4 p.

20. Arrimage des conteneurs à bord des navires.
    Bureau Veritas.
    NI 186 CNI, Juin 1984, 38 p.

21. Dommages par suite de l'échouement d'un vraquier.
    J.C. Alizon.
    Département Recherche Appliquée et Développement, Bureau Veritas,
    Paris.
    Bulletin technique du Bureau Veritas, Juin 1985, 6 p.

22. Study on the structural strength of ships in collision.
    M. Arita, N. Ando, K. Arita.
    International conference of Fracture Mechanics and Technology.
    Hong Kong, March 1977.

23. Défence des réacteurs nucléaires de navire contre les abordages.
    F. Spinelli, professeur de construction navale à l'Université de
    Naples.
    Association Technique Maritime et Aéronautique, Session 1962, 20 p.

24. Protection du compartiment du réacteur nucléaire contre les
    abordages, résultats d'essais sur modèles.
    F. Spinelli, Professeur de construction navale à l'Université de
    Naples.
    Association Technique Maritime et Aéronautique, Session 1964, 25 p.

25. Protection du compartiment du réacteur nucléaire d'un navire
    contre les abordages, résultats de 24 essais sur modèles.
    F. Spinelli et V. Belli.
    Association Technique Maritime et Aéronautique, Session 1971, 9 p.

26. Collision tests with ship models, Topical report. Eur 4560 e, 1971.

27. Conclusions from collision examinations for nuclear merchant ships
    in the FRG.
    G. Woisin, OECD, Paris 1978, pp; 137-147, Symp. on the Safety of
    Nuclear Ships, Hamburg, December 1977.

28. An engineering assessment of the probability for damage to
    radioactive material transport casks due to barge collisions.
    Bruce L. Hutchinson, David L. Gray, Glenn Bauer
    (The Glosten Associates, Inc.) Seattle Washington, USA.
    IAEA-SM-286/164, June 1986, 8 p.

29. Les collisions.
    Y. Loisance.
    Bulletin Technique du Bureau Veritas, Mai 1978, 5 p.

30. An analysis of ship collisions with reference to protection of
    nuclear power plants.
    V.U. Minorsky, Journal of Ship Research, October 1959, 4 p.

31. Equivalent added mass of ships in collisions.
    S. Motora, M. Sugiura, M. Fujino, M. Sugita.
    J.S.N.A. Japan, Vol. 126, Dec. 1969, 10 p.

32. Critical evaluation of low-energy ship collision-damage theories and
    design methodologies. Volume I : Evaluation and recommendations.
    P.R. Van Mater, Ir/J.G. Giannotti.
    SSC-284, Ship Structure Committee 1979, 90 p.

33. A literature survey on the collision and grounding protection of
    ships.
    Norman Jones, Massachusetts Institute of Technology.
    SSC-283, Ship Structure Committee 1979, 55 p.

34. Ship collision with bridges and offshore structures.
    IABSE Reports, Volume-Band 41, Copenhagen 1983, 20 p.

35. Ship collision dynamics and the prediction of the shock environment
    for colliding ships.
    Pakstys Michaël Peter, 1977, University of Rhode Island, USA, 159p.

36. Barge collisions, rammings and groundings.
    An Engineering Assessment of the Potential for Damage to Radioactive
    Material Transport Casks.
    B.L. Hutchinson, The Glosten Associates, Inc.
    Contractor Report, SAND85-7165, TTC-05212, January 1986, 239 p.

37. Lloyd's Register of Shipping Forces on container vessels due to
    collision.
    A.J. Williams/L.R. London.

38. Large shielded containers for transport of reactor waste to the
    final repository in Sweden.
    S. Pettersson (Vattenfall, Swedish State Power Board).
    B. Gustafsson (Swedish Nuclear Fuel and Waste Management Co).

39. Prediction of structural damage, penetration and cargo spillage due
    to ship collisions with icebergs. 15 p.
    D.S. Aldwinckle, B Sc, Ph D, C Eng, FRINA and
    K.J. Lewis, B Sc, C Eng, M Inst.E.

40. Transportation system for spent nuclear fuel.
    Preliminary Safety Report.
    Prepared by: AB Kärnkraft, AKK, Salen Technologies AB, Saltech.
    Svensk Kärnbränsleförsörjning AB, SKBF.
    Swedish Nuclear Fuel Suppy Co, March 1980, 9 p.

41. Evaluation des contraintes auxquelles sont soumis les emballages
    durant le transport maritime.
    P. André et J. Victor.
    LNE 15.212.10 octobre 1980, 33 p.

42. Stowage of goods in freight containers.
    British Standards Institution
    BS 5073 : 1982, 10 p.

43. Cost 301. Marine traffic casualties in the cost 301 area 1978-1982 -
    Final Report.
    Capt. R. Tresfon, C.E.C., Directorate-General for Transport,
    DG VII/A-3, July 1985.

44. Cost 301. Comparison of traffic methods in the Mediterranean
    - Final Report.
    M. Monica, C.C. Glausdorp, C.E.C. Directorate-General for Transport,
    DG VII/A-3, July 1985.

45. Cost 301. Ship casualties in the Greek seas - Final Report.
    Const. A. Philippou, Th. E. Frangoulakis, Alex. C. Philippou.
    C.E.C. Directorate-General for Transport, DG VII/A-3, July 1985.

46. Etudes de collision - Quelques références.
    Institution de recherches de la construction navale. 19.06.87, 20 p.

47. Cost 301. Validation, marine casualty figures.
    (Greek seas) + addendum (Sept '86).
    C.A. Philippou, A.C. Philippou, T.E. Frangoulakis.
    C.E.C. Directorate-General for Transport, DG VII B-3, March 86.

48. Method for estimating ship collision damage.
    Kikuo Arita, S.B., Yokohama National University, 1961.

DISCUSSION FOLLOWING SESSION 3: Papers 1 - 4

MR.F.NUHAAN, WESTINGHOUSE INTERNATIONAL, BRUSSELS for Paper 3:2
I see in the example that the Bill of Lading indicates shipment of
limited quantity. I wonder whether it is really interesting to have the
tracking system also include this category of shipment or would you not
rather limit this to more hazardous types of shipment?

MR.P.D.GRIMM, UNITED STATES DEPARTMENT OF ENERGY:
Yes, as a matter of fact we do, the intent of the tracking system was not
to track all nuclear material shipments, principally those that are
identified as highway route controlled or the larger quantity radioactive
material shipments. That was unfortunately just one of the demonstrations
of the current work. But it is true we would be trying to track the
larger quantity shipments.

DR.B.J.TYMONS, W.S.ATKINS ENGINEERING for Paper 3:2
Looking more closely at the question of security - if the transceiver is
moved to another vehicle, does your system know this has happened. In
other words does your system really track the vehicle or just the
transceiver?

MR.P.D.GRIMM:
As a matter of fact I have heard this question raised on numerous
occasions. The point to recognise is that the system is not intended to
be a tracking system of weapons type "classified" shipments. In fact I
would answer to those that would say you don't want those shipments being
traced, by saying that there is a better network for people who track
those kinds of shipments than anything else in the United States. The
type of shipments we are talking about and the security that can be
provided is in fact in place but it is not flawless; for instance taking
a piece of transceiver out of the vehicle and putting it in another is
certainly conceivable but for the most part these are unclassified
shipments and we are only trying to identify the location. If a piece of
equipment is removed then that would send a signal identifying that there
is something wrong with the system. It is not infallible, people can
certainly break access codes to systems, but it at least provides a
measure of security that we feel is appropriate.

MR.G.LUCK, CEGB for Paper 3:2
One of the claimed benefits of satellite tracking system is ability to
direct emergency services. In view of this, what requirement is there for
the transceiver to withstand an accident, particularly bearing in mind it
only transmits every fifteen minutes?

MR.P.D.GRIMM:
Yes, it is our concern that if the piece of equipment which is placed in
a vehicle or on a train is involved in an accident, and it is destroyed,

that it will send a signal that it is no longer communicating and
obviously stimulate a status change to an emergency condition. But you
bring up a good point, we never intend that the transceiver device
becomes an integral part of a piece of equipment, for instance, a packed
containment system, and be able to withstand the types of accidents and
areas that type B packages can withstand. That is not the intent of this
system. It is strictly a system to deal with those institutional problems
such as identifying emergency response, making sure equipment is where it
is supposed to be essentially, etc.

MR.R.GELDER, NATIONAL RADIOLOGICAL PROTECTION BOARD for Paper 3:4
We investigated accidents and incidents, including events at sea, so this
is a question for Mr.Draulans. We have information on some four events,
three with sealed sources lost from rig supply vessels, one with
uranium-hexafluoride cylinders torn from mountings, does Mr.Draulans know
of any other information on shipping events, apart from the Mont Louis?

MR.J.DRAULANS, TRANSNUBEL SA:
The statistics given are based on accidents with ships which could carry
radioactive materials, but it was not specified whether they contained
radioactive materials or not, only the size of the ships determined
whether they were able to carry radioactive materials.

MR.A.HANSON, TRANSNUCLEAR INC.
In response to the last question, about two weeks ago in a hurricane, a
cylinder containing UF6 went overboard from a vessel off Charleston on
the east coast of the United States, and was lost. That is the most
recent incident we know of with regard to a sea related transport of
nuclear material. This particular accident did not affect the public in
any way and has in fact had very little coverage in the trade press.

MR.A.HARRISON, FORGEMASTERS ENGINEERING for Paper 3:2
Does the use of a satellite tracking system decrease public opposition to
transport of radioactive materials in the USA, especially spent fuel, and
is its use likely to bring forward the date of shipment of spent fuel
from the power stations to a central repository?

MR.P.D.GRIMM:
Certainly the satellite system is designed to give a feeling of, let's
say, comfort to those police and fire emergency officials who are the
responsible local officials for any given incident in their jurisdiction
and that is what we attempted to address. There is only so much education
you can do to give that local representative who is first on the scene
the feeling that they have the capability to do something or control the
situation and that is perhaps the most important thing. To my mind, there
will never be a complete removal of public apprehension but at least this
minimizes the concern from the officials that are important unseen
representatives and, of course, they influence the political leaders to
feel that there is a comfort factor associated with these shipments. In
relation to your second question I think you are talking about the
commercial waste repository which is planned under the Nuclear Waste
Policy Act, is it that you are questioning? (Yes). This would be utilized
for that particular programme as well, but I don't think there was any
drive as far as the satellite system was concerned in bringing forward
the opening date of that facility. That is entirely dependent upon
acceptability in Congress.

MR.H.GEISER, GNS for Paper 3:2
Can you precisely define again how you are going to use your satellite
tracking system, for what kind of transportation? As far as I know
security transportation is now escorted by police, so you do not need to

use a communication system by satellite to inform the police.

MR.P.D.GRIMM:
There are many different aspects, but I guess I would categorize three
major components of transportation of larger quantities of radioactive
materials in the United States. One is our Department of Energy's
shipments under weapons type activities, which is conducted entirely in a
classified mode by department equipment and that is not what we are
talking about here, it is a completely separate segment of
transportation. Another part is the commercial transportation business
handled by commercial licensees as we referred to those entities that
would move fuel on behalf of commercial power reactor operators and be
regulatory by the Nuclear Regulatory Commission, and that is not what we
are referring to here either. The last category is the component that I
am referring to and that is strictly where the DoEnergy is a shipper and
in fact transporting large qantities of radioactive material shipments.
That is what we would be utilizing the system for. For large quantities
of high level waste or spent fuel and in addition those shipments we
would like to track, for instance, the trans-uranic waste moving to the
waste isolation pilot plant will be utilizing this system. We can
designate work where we want to attach a satellite system - it is
limitless as far as the amount of units that can be tracked - and in the
Department, I presume it is only limited by what types of programs we
want to try to track. For the commercial industry if you were an entity
trying to transport fuel on behalf of a commercial power reactor and you
wanted to utilize the system, we have made provisions to allow them to
use the system. It has not been worked out in detail, but we contemplate
allowing commercial users to utilize the system as well.

MR.F.J.L.BINDON, INSTITUTION OF NUCLEAR ENGINEERS:
I would like to ask our American colleagues about the transportation of
radioactive materials, and I don't mean just spent fuel, across the U.S.
from state to state. What are the prohibitions in crossing state
barriers? Are there some states which will not have any radioactive
materials across their boundaries and so necessitate detours for long
distances in order to overcome that problem? Finally, I am not too sure
whether Mr.Grimm took this point in in his paper about satellite
tracking, but how is he allowed to track the radioactive shipments
outside the boundaries of the United States and say, any movements
between his country and Canada?

MR.P.D.GRIMM:
Concerning prohibitions, there are incredible amounts of ordinances and
statutes that have been enacted to try to prohibit shipments across
certain jurisdictions - states, counties, municipalities. It is these
particular statutes that cause the impediments we are trying to address
with this system. What specific types? We have had anything from 'time of
day' restrictions to escorting with certain security types of equipment,
to general bans, absolutely no transportation through this area.
Certainly we have had to challenge many of these restrictions with what
we refer to as inconsistency rulings with federal regulations and/or
DoTransportation regulations (which would allow you to transport the
shipments) but that is a very slow process and it causes a lot of concern
and animosity from the local officials. Just to tell a quick story,
involving the movement of the Surrey fuel from Surrey, Virginia to Idaho,
practically ten states' worth of transit. The manufacturers had built a
cask that was a very short trailer and violated perhaps seven or eight
states' weight restrictions not as far as overall weight, 80,000 lb, but
as far as bridge weight formulae, axle differential and weights, and
consequently we had to apply for permits in all those states. Well you
have one of those "you can't get there from here" stories. One state

would make you wait until 9.00 am, then by 3.00 you were no longer
allowed to transport but had to get to another state to lay idle. It
became an incredible process. Eventually the route had to be changed,
adding an additional four to five hundred extra miles, and go a
completely different way, so there are shipping impediments, and some we
really need to address up front, both institutionally as well as
technically.

For the second part of the questions, at this point I would say that
there is no intention of utilizing this system for anything other than
shipments within the United States. The system would allow you to track
overland and within the navigational boundaries around the United States,
but we do not intend to track shipments into Mexico or Canada. Once it
crosses the international boundary we would stop at that point. That is
not to say that there couldn't be a co-operative effort to continue the
tracking. The technical capability is there but it as far as the policy
and the programme goes at this point. We haven't expanded it that far.

MR.F.J.L.BINDON:
That does not seem a very satisfactory system to me, beause I am talking
about international, legal, regulations. How do we in some other country
know what you are tracking, where is the openness of this syctem to the
public at large in the world?

MR.P.D.GRIMM:
I am not sure I understand your question. Do you mean, would anybody be
given access to the system or would it be from a shipper's...?

MR.F.J.L.BINDON:
Well you are saying, I think, that you do not think you would do this or
that, what are the legal restraints on your satellite programme, do you
have inspectors from say, an international agency who oversee what you
do?

MR.P.D.GRIMM:
No we do not.

CHAIRMAN:
There may be some misunderstanding. The vehicles need transcievers on
them, I don't think the system is capable monitoring anybody's vehicle
anywhere.

MR.P.D.GRIMM:
Maybe I brushed over this too quickly. Certainly the equipment has to be
purchased and the DoEnergy is funding the Transcom Central — the location
that would process the information to give you a latitude/longitude fix
and ultimately identify that on a computer screen. So while we would be
funding that effort, a user that would want to get on to the system would
still have to purchase a transceiver, be trained in its use, as well as
be given access codes and I guess that is the control we would have over
it. There is no intent at this point, however, to provide access
indiscriminately to anyone.

CHAIRMAN:
I would like to open up the question of quality assurance. Mr.Blackburn
presented Mr. Cooke's paper on this and I know that it is a topic within
the UK which has attracted considerable interest, I know that the
Government's Advisory Committee on the Transport of Radioactive Material
(ACTRAM) has spent happy hours addressing the concerns of quality and
compliance assurance, and indeed I think its document is due to be
published shortly and I assure you it makes absolutely riveting reading.

I would like to ask Mr. Blackman how the Department of Transport sees the QA/CA scene evolving in this country?

MR.D.BLACKMAN, DEPARTMENT OF TRANSPORT:
Thank you Mr. Chairman, I had intended to put my hand up just to give Mr.Grimm a rest. I'm pretty sure that most people in the room know by now that that date of 1st January 1990 is the target date set by the International Atomic Energy Agency for member states to adopt the 1985 regulations, as well as all the international modal organizations. Everything is aimed at that particular date, and this is why it was stated in the paper, it wasn't just a date picked out of the air. Yes, in fact as you so rightly say, Mr. Chairman, ACTRAM's report on compliance assurance and quality assurance will be published very shortly and a very great deal of work has gone into it - certainly it places a very great deal of emphasis on quality assurance compared with the situation under the previous set of regulations and as far as the government is concerned we are going to have to keep quite a wary eye on what is going on throughout the country, a lot of responsibility is being placed on us for compliance assurance. Paragraph 2.09 states that quality assurance programmes must be developed by transporters, para 2.10 states that each competent authority is responsible for assuring compliance with these regulations. I would like to point out that my own department is not an inspectorate as such, we do not have armies of people trailing round to every transporter of radioactive materials. What we are doing in fact, is sending out a questionnaire asking about the general pattern of radioactive materials transport. This is one thing we are trying to do as a preliminary to try to ensure nationwide compliance with the regulations. We are trying to establish the overall pattern of radioactive materials transport so we know who is moving the stuff and where. This will enable us to target our efforts towards those transporters who are moving lots of material about and who will be the focus of our attention.

I have in fact just had a note passed to me that the ACTRAM report has just been published and can be obtained from Her Majesty's Stationery Office (HMSO), at £5.80. Very many copies are being distributed free, but how many members in this room will be on the lucky distribution list I am not sure. For those who are not, it does make extremely intersting reading and represents a very considerable amount of effort on the part of Study Group II. The Chairman of this session was the chairman of the group which put all this together. Was that the sort of thing you were hoping for? Perhaps not the commercial at the end!

CHAIRMAN:
I was anxious that people understood the programme that the Department was working to and how they were trying to reassure the public that not only do operators have effective quality assurance systems in place but that the regulatory and compliance inspectorial systems are also sensibly in place to audit those arrangements.

MR.BLACKMAN:
Yes, I can assure that we in the DoTransport are very conscious of our responsibilities in this field.

MR.J.FIELDS, BRITISH NUCLEAR FUELS PLC, SAFETY DEPARTMENT for Paper 3:4
Some of the frequencies quoted in that presentation regarding accidents at sea were quite close to the numbers we in BNFL would get for site extreme natural events, such as earthquakes and really bad climatic conditions. I should therefore like to ask a general question - are these sort of events - earthquakes, hurricanes, volcanoes, etc. given due consideration in assessment of transport safety?

MR.J.DRAULANS, TRANSNUBEL:

In the values given by Lloyds Register of Shipping, all natural phenomenon have been taken into consideration as far as in the last eight years, when they consider these phenomenon could have happened.

MR.J.FIELDS:

The problem with collecting statistical data of that type is that earthquakes happen very infrequently and we have not had any of significance in the last eight years to my knowledge in this area, but nevertheless we are talking in terms of frequencies of something like 10^{-5} per shipment. We expect earthquakes to occur at a frequency of 10^{-4} but ships travel into areas where there may be hazards such as volcanoes, which could be pretty disastrous to a nuclear shipment, particularly in regard to the fact that it would take away a lot of the possibility for recovery measures to deal with the situation. This point occurred to me whilst some of the figures were being mentioned. We put a lot of effort into assessing the effects of seismic events and it seems that these were of the same order of frequency as the accident events that the speaker was referring to.

DATABASE RECORDING OF REACTOR AND FUEL ELEMENT
INFORMATION FOR PACKAGING AND TRANSPORT

D. Bibby

Nuclear Transport Limited
Risley
Warrington
Cheshire

INTRODUCTION

Nuclear Transport Limited (NTL) has been transporting irradiated
uranium oxide fuel elements from Pressurised Water Reactors (PWR) and
Boiling Water Reactors (BWR) in Europe to the reprocessing plants of
British Nuclear Fuels plc (BNFL) at Sellafield in England and COGEMA at
Cap La Hague in France since 1972.

NTL has transported, using its own fleet of flasks and associated
equipment to date, a total of 6750 PWR and 8500 BWR irradiated fuel
elements resulting in a total irradiated uranium weight of 3820 tonnes U
from 34 European reactors.

In the latter part of the 1970's the total weight of fuel was in the
order of 450 tonnes U per annum, equivalent to a total number of
shipments per annum of 150.

With the development of both mini and micro computing systems
becoming more readily available in this period, NTL decided to
investigate computerising its operations in order to absorb and utilize
the large volumes of data required into a computerised management control
system.

During the period 1980 to 1982, NTL installed a mini computer and
developed a bespoke software database system covering interactively all
aspects of the organisation of transport, for example, routes, equipment
specifications, approvals, maintenance, spares etc.

Following the successful implementation of the system at NTL Risley,
peripheral computer systems were installed in the NTL offices in Germany
and France.

Although the system developed covers all aspects of the business
this paper discusses the reactor fuel and transport licensing aspects of
the system.

The purpose of the software was to establish a method of analysing

reactor operating data and individual fuel element information to meet
transport commitments, to provide assurances to the consignors,
consignees and the reprocessors that transport licensing could be
programmed well in advance of transports, and to demonstrate the adequacy
of the packages to meet reactor power generation trends. The fuel
element data coupled with the major approval parameters is intended to
provide assurances that the fuel elements identified for transport are
within the scope of the approvals, to provide statistical analysis of the
shipments, to determine the heat content and activity, to record and
analyse the burn-up, enrichment and to provide an easily retrievable
information records system.

In 1984, COGEMA, after evaluating the system, installed a computer
link to NTL Risley to allow the input of fuel element data both from
French nuclear reactors and from their Japanese contracts.

To date approximately 26500 individual fuel element information
details are contained on the database together with the majority of the
European reactor site data.

REACTOR FUEL DATE

To assist in the establishment of this file on the data base, NTL
prepared a 45 point questionnaire which was distributed to all the
contracted reactor sites in Europe. The questionnaire is split into 4
areas:-

-   Reactor operating data, identifying cycle data, burn-ups, in-core
    parameters, etc.;
-   Fuel element data, array, enrichment, weight of uranium, etc.;
-   Fuel pin data, materials, dimensions, etc.;

and a final item which covers future trends.

The system was designed to accommodate up to 25 variations for each
reactor site, since some sites have up to 20 different fuel types which
can be operated in their reactors. The system also allows for the direct
comparison of reactor data for easy analysis of individual parameters and
quickly indicate to the transporters the capability of new fuel types in
known flasks.

All this data allows the flask designer to analyse the shielding,
criticality activity release, and heat transfer aspects of the flasks,
well in advance of the transports, and provides assurances that the
flasks will meet the International and National Regulations.

During the last three years, following trends indicated on the
original questionnaire, the database has been extended to accommodate the
recording of further data to identify the requirements by reactors of the
use of mixed oxide fuel and its subsequent transport.

The extension of the database to include the mixed oxide fuel
element data proved beneficial during the licensing and transport of MOX
fuel carried out by NTL in 1987.

The introduction of the data and the analysis of the data provided
validation of the analytical studies against transport measurement.

The reactor information provides sufficient data to allow the
preparation and submission to Competent Authorities of design safety

reports with a confidence that the majority of the significant items which could affect approval have been analysed. This has provided NTL with a working method of forecasting future programmes, forewarning the licensing bodies well in advance of the commitments to reactor and reprocessor contracts. It also enables NTL to identify the items of equipment such as fuel element support frames and multi-element bottles which would not be capable of meeting the fuel demands.

This procedure assists the flask designer to design new equipment, and to license and manufacture the equipment in time to meet the transport requirements.

Obviously the reactor information can only give preliminary indications, and it is only on receipt of the individual fuel element data identified for a transport campaign that the adequacy of the package can be determined.

FUEL ELEMENT DATA

To assist in this final checking and ensure that all requirements for safe transport are met, a fuel element file was developed on the data base. The individual fuel element data are linked to the equipment files to ensure that the fuel is not loaded into the wrong equipment.

It is also interactive with the reactor data files in order to check the existence of the fuel type and to check that a fuel element has not been nominated and shipped in a previous campaign.

The fuel element file requires the main approval parameters to be input so that on input of the fuel element data, checks are made on the parameters against the licensing controls.

NTL has developed a method of quickly evaluating the decay heat generated by the fuel elements and the total activity content.

The evaluation developments for the decay heat are based on FISPIN, ORIGEN and an NRC method and provide results which are sufficiently accurate to show adequacy of packaging and to give an assurance against the data received from the reactor site.

When fuel elements exceed the licensed parameters, indicators are set which give the transporter the opportunity of notifying the reactors and reprocessors of non-acceptability of fuel elements, of preparing alternative means to accommodate the fuel element and, if necessary, of selectively positioning the fuel elements in the flask.

The individual fuel element data are then given to the technical assistants who are provided by NTL at the reactor sites with an approved loading plan. The technical assistant has to identify each fuel element by number and position in the flask in the approved position.

Non-identified fuel elements cannot be loaded and a mechanism exists such that the base office is notified and corrective action or identification of another fuel element acceptable to the licensing parameter can easily be obtained via the computer system and agreement with the reactor operators.

On completion of the flask loading, confirmation of the loading is input to the system which activates the process of invoicing.

The data on each fuel element loading are held on the system to provide statistical information, enable records of the capabilities of the flasks to be monitored and give historical records for further design studies. The recording system also enables checks on the validity of the calculational methods used in the prediction of dose rates, etc.

APPROVALS

NTL transports to COGEMA in France and BNFL in England are from many different countries in Europe.

NTL operates a fleet of 18 flasks together with a further nine flasks belonging to COGEMA. With many different countries to transport from and through this means that numerous validations and transport permits, together with individual licences for national requirements need to be obtained.

To assist in the planning and application of the approvals an approvals file was developed, coupled with the fuel element and reactor files.

Whenever any equipment is being maintained or modified, or any drawings are changed, indicators are set on the approvals file to indicate action on the approval.

The file contains, from the transport scheduling database file, the countries from which approvals have to be obtained and an indicator of the time scales required.

The approvals file currently holds approximately 30 package approvals, together with all their associated validations, transport permits and national licences. To complement this file, note pad screens have been incorporated so that the status of the Design Safety Reports and supplements with basic parameters can be readily reviewed.

Quality Assurance and Quality Control procedures are currently being revised for the software and a rigid security access system input to the files to meet the current requirements. New procedures are developed for any modifications to the database files.

CONCLUSIONS

It is planned for the future to provide further assurances to the consignors, consignees and licensing authorities which will include assessments of the dose rates and temperatures for each fuel loading without having to use large mainframe computing facilities. The system will also provide a personnel health physics recording system coupled to the flask movements.

RADIOACTIVE MATERIAL TRANSPORTATION IMPEDIMENTS

J. Mangusi

Transnuclear Inc
2 Skyline Drive
Hawthorne
NY 10532
U.S.A.

ABSTRACT

Impediments which affect the smooth movement of radioactive material shipments occur from time to time and take many forms. This paper gives several examples which were encountered by Transnuclear during the performance of hundreds of transports.

New legislation and regulations, as one might expect, contribute significantly to confusing one or more aspects of a transport as shown by the legislation described below.

On 2nd October 1986, the Comprehensive Anti-Apartheid Act of 1986 was enacted. This Act prohibited the import into the Unites States of uranium ore and uranium oxide produced or manufactured in South Africa.

Later that month President Reagan signed the Consolidated Omnibus Budget Reconciliation Act, which imposed user fees on many customs entries and warehouse withdrawals, effective 1st December 1986. Then on 17th November 1986, President Reagan signed the Water Resources Development Act providing for the collection of a fee on imports and exports for the maintenance of ports and harbors. The rules and regulations implementing these three Acts generated many questions and much confusion in the industry.

In addition to impediments caused by new laws and regulations, different interpretations of existing regulations also serve to delay or complicate shipments. Examples of this category are also given in the paper.

The impediments to be discussed on Safeguard Category I and II shipments include the US requirement that only one Category II shipment be in progress at one time within US territorial limits, and the lack of an approved commerical transporter for Category I shipments.

The paper concludes with a short section suggesting that the purchasers of transportation services should not use price as the only criterion for selection.

INTRODUCTION

Transnuclear, Inc., in performing hundreds of various types of radioactive material shipments each year, frequently encounters impediments which affect smooth movement of the shipments. These impediments may arise directly from new legislation or regulations, indirectly from various interpretations of these new regulations or of existing regulations, from a reluctance of organizations or personnel to handle this class of cargo, and many other reasons which may or may not be valid. Regardless of the origin of the impediment, the results are delayed or cancelled shipments and significant extra expenses which are recoverable only in a few cases.

At the time of writing (December 1987), there were several current items which illustrate the types of actions which could constitute impediments to smooth transport.

CUSTOMS PROBLEMS

For the past 9-10 years, Transnuclear has been importing irradiated material test reactor fuel from Europe for delivery to the US Department of Energy facilities in South Carolina and Idaho.

This spent fuel is returned to the US under international agreements and has been returned duty free as waste. In April 1987, Transnuclear received notice that the US Customs service intended to classify this material as "Steam and other vapor generating boilers and parts thereof" and charge duty at a rate of 6.5%.

An initial telephone conversation with the author of the notification indicated that there was some misunderstanding on the part of US customs. It was suggested that if we would quickly respond with a letter acknowledging receipt of the notice and stating an intention to provide details at a later date further action by Customs would be delayed.

This was done and later a letter was sent stating that the newly proposed classification of this material was not applicable and furthermore that the new Harmonized System of classification, expected to be approved in early 1988 had a specific entry entitled "Spent Fuel" with a corresponding absence of duty. Therefore, it is obvious that the intention of the US Government is not to collect duty on irradiated fuel.

Shortly after sending this letter, another notice was received from a different Customs official stating that a decision had been made on this issue and duty wil be charged at 6.5% under the category of "Steam and other vapor generating boilers and parts thereof".

Completely incredulous at this decision, we telephoned the original Customs agent only to find that he had retired. After being passed to a new person, we learned that, like the first individual, she was only temporarily assigned to this group to help reduce the terrible backlog.

We asked for the basis for the decision and the reaction to our letter. It appears that during the transition of personnel, our letter had been misplaced and the decision was made without reference to our letter. We offered to immediately transmit a copy of our letter to clear this matter up but we were told that since a formal decision had been made, our only recourse was to wait until final liquidation is completed by Customs for the shipments in question and then file a formal protest.

We so instructed our custom house broker who handled these shipments for us.

The next correspondence received from Customs consisted of several invoices requiring payment by a certain date or interest would be accrued until paid. Customs requested the payment and promised a refund if the protest was successful. Our client has chosen, and we agree completely, to turn this matter over to a law firm specializing in customs matters. As of December 1987 no schedule had been established for resolution of the issue. We, and our affiliated company in France, performed these shipments in good faith and have not been compensated for the time spent on attempting to resolve what appears to be a mistake on the part of US Customs, caused in part by a confused and over worked staff. Needless to say the lawyers who are now involved will be compensated for their time.

CATEGORY II SHIPMENTS

An area in the United States Nuclear Regulatory Commission (USNRC) regulations which has placed a significant obstacle to the smooth performance of Safeguard Category II shipments is the requirement that only one such shipment can be in progress at any one time within the US territorial limits. The purpose of this rule is to prevent more than one Category II shipment from being accidentally located at the same place at the same time and which could result in a Category I quantity of material. As the regulations for Category I shipments are much more stringent than those for Category II, the combined quantity would be in violation of the applicable regulations.

This requirement, therefore, causes each shipper to reserve a time slot with the USNRC during which each shipment is to take place. Many times a project requires many Category II shipments to be completed and it is not uncommon for a shipper to reserve a period of several months to cover his requirements. Under these circumstances, it is not inconceivable to envisage a large part of an entire year to be blocked out. Accordingly, if a particular shipment is delayed for any reason, the cargo may not be able to move for several months unless a swap of time slots can be arranged with another shipper.

I believe this rule, while based on good reasoning, is not practical and that there should be some way to relax it while still offering reasonable assurances that its objective will be satisfied. As it is now, the number of Category II shipments is severely limited by available time slots.

CATEGORY I SHIPMENTS

The previous discussion mentioned Category I shipments which bring to mind another rather severe impediment to transportation in the U.S.

Prior to 1974 Category I shipments of highly enriched uranium and plutonium were handled only slightly differently from other radioactive material shipments and, therefore, costs were not significantly different. Beginning in 1974, safeguard requirements began to stiffen and, for the next four to five years, requirements were added regularly until these types of shipments became a complicated endeavor. Transnuclear, in cooperation with a commercial carrier, was able to satisfy changes until 1986 when the commercial carrier decided to leave this part of the business because the costs of maintaining the required equipment and personnel did not justify the anticipated revenue from the relatively few shipments per year.

In order to maintain the flow of material, Transnuclear requested the United States Department of Energy (USDoE) to utilize its Safe Secure Transport (SST) systems for commercial shipments. The DoE systems which is used primarily for defense related transports meets all NRC regulations for Category I shipments. After more than one year of effort, the DoE reluctantly agreed to utilize their equipment and personnel for these shipments and one shipment was completed.

Thinking we had solved the problem we began planning for a second shipment only to find out that there is strong opposition in various parts of the government for the DoE to continue providing the service and recently the DoE has stated that they will provide the service only until the end of fiscal 1988 or until a commercial carrier can again be located.

While there is a law that US Government equiment and personnel should not be utilized for commercial ventures, the intent is to make sure the government does not take work from the private sector. In this case, no one in the private sector really wants the job and, to the best of my knowledge, no one has complained that the USDoE is performing the shipments. I remember hearing many more complaints when DoE eliminated use of private carriers for government special nuclear material shipments and began performing the transports with the newly created DoE SST systems.

As part of our attempt to keep the SST program available for the few commercial Highly Enriched Uranium (HEU) shipments each year, we alerted the DoE that if the only alternative available was a very expensive commercial carrier, overseas customers might go elsewhere for HEU supply.

LEGISLATION AND REGULATIONS

New legislation and regulations, as one might expect, contribute significantly to confusing one or more aspects of a transport.

On 2nd October 1986, the Comprehensive Anti-Apartheid Act of 1986 was enacted. This Act prohibited the import into the United States of uranium ore and uranium oxide produced or manufactured in South Africa. Later that month, President Reagan signed the Consolidated Omnibus Budget Reconciliation Act, which imposes user fees on many customs entries and warehouse withdrawals, effective 1st December 1986. Then 17th November 1986, President Reagan signed the Water Resources Development Act providing for the collection of a fee on imports for the maintenance of ports and harbors. The rules and regulations implementing these three Acts generated many questions and much confusion in the industry.

While not seemingly related, the implementing regulations for the Anti-Apartheid Act and the collection of user fees under the Budget Reconciliation Act produced a potential conflict in the administering of temporary imports into the United States.

User fees, which effectively are taxes on imports, began on the 1st December 1986 at a rate of 0.22% of the cargo value and were reduced to 0.17% on 1st October 1987 to remain in effect for the following two years. Except for a few exempted items, these fees were to be collected on all consumption entries and warehouse withdrawals.

One exempted area is Temporary Imports under Bond (TIB) wherein material is imported to the US for minor processing or testing and

exported. Historically, uranium oxide and hexafluoride were not imported
under TIB conditions because no import duty was involved and there was no
incentive to get involved with the TIB requirements. However, with the
imposition of the user fees, there now was incentive to investigate
temporary importation to avoid the user fees.

Initial conversations with the US Customs officials working with the
user fee task force revealed that uranium imported for conversion or
enrichment should not be entered under a TIB because excessive processing
was involved.

However, in March 1987 the US Treasury published an interim rule to
amend the regulations which were enacted to implement the Anti-Apartheid
Act which prohibited the importation of South African uranium ore and
oxide. This interim rule was issued pending clarification of the
congressional intent of the Anti-Apartheid Act. The interim rule
permitted temporary importation of ore and uranium oxide from 10th March
1987 through 30th June 1987 unless extended.

Thus the condition was created wherein the interim rule permitted
temporary importation of uranium ore and oxide from South Africa while US
Customs personnel involved with the user fee task force were stating that
this material should not be imported under TIB requirements. The interim
rule was not extended and, therefore, after 30th June 1987, the temporary
importation under the interim rule was discontinued.

As a result of the user fees, however, much attention has been
generated in the area of temporary importation of uranium oxide and
uranium hexafluoride. It is our understanding that some U308 may have
been imported under temporary rules. The general conditions for
temporary imports are that the merchandise may be admitted into the
United States only on the condition that if any processing of such
merchandise results in an article manufactured or produced in the United
States:

(i) a complete accounting will be made to the Customs Services for all
     articles, waste, and irrecoverable losses resulting from such
     processing, and

(ii) all articles and valuable wastes resulting from such processing will
     be exported within the bonded period, except that in lieu of the
     exportation or destruction of valuable wastes, duties may be
     tendered on such wastes at rates of duties in effect for such wastes
     at the time of importation.

Normally, Transnuclear is the importer of record for shipments
under its control and since almost all uranium shipments enter the US
duty-free, the potential liability to US Customs is small. However, in
the case of temporary importations, the importer of record is the
responsible party to US Customs for any violations of the temporary
importation regulations. If material, imported under a TIB to avoid
payment of the user fee, were not exported in accordance with the rules,
Transnuclear would be liable to US Customs for applicable duties fees and
penalties. We believe the owner of the material should benefit from the
advantages of temporary importation if US Customs agrees, but not at the
expense of the importer of record.

In many cases, the importer of record, under routine circumstances,
has no direct relationship with the owner of the material, but is working
for a converter or fuel fabricator.

A contract with these parties is not sufficient since the owner of the material might, after the material is imported, decide to swap it for other material as is frequently done. In this case, the importer of record would not be able to complete the export as required by US Customs.

INTERNATIONAL REGULATIONS

Interpretations of regulations, or slightly differing regulations between countries, or between national regulations and international regulation also pose impediments to transportation.

Transnuclear, for example, was involved in planning a series of shipments of tritiated water from Europe to the US.

The US regulations stated that the concentration limit could not exceed 5 Ci/liter in order to be considered low Specific Activity Material. The 1973 IAEA regulations permitted a concentration of 10 Ci/liter and the 1985 IAEA regulations permitted a concentration of 20 Ci/liter.

Initially the United States Department of Transport (USDoT) did not want to permit the shipment to take place. After several discussions and an exchange of letters the shipment was permitted to proceed under two sections of the DoT regulations. One section permitted strong tight industrial packages to be used, if the shipment was conducted under Exclusive Use provisions. The second section permitted shipments which are in accordance with the IAEA regulations to be imported into the US and transported to destination. The real question which remains is that, given all the international meetings which are held, why should there be such a big difference in the allowable concentration of tritiated water to be shipped and classified as Low Specific Activity (LSA).

A second example in this area is in the shipping of empty UF6 cylinders. Normally, UF6 cylinders when effectively emptied still contain some material referred to as "heel". Since material remains in the cylinder, these cylinders are usually shipped with the same shipping name as when loaded but with an added explanation that the cylinder is effctively emptied of its contents. Motor carriers and ocean carriers seem to understand the situation and different rates exist for the shipment of heel cylinders and loaded cylinders.

Transnuclear however, ran into a slightly different situation in trying to ship cylinders that had been cleaned, washed and dried and thoroughly emptied of contents. Our customer had arranged for these cylinders to be shipped via ocean carrier as empty cylinders. To strictly meet the definition of empty cylinders the internal contamination must be measured and must fall within a specified limit. As is customary in the industry, the cleaning facility did not perform this procedure and, therefore, would not ship as empty and also claimed that the shipping name should be the same as for a full cylinder. If this were the case, our customer would have had to pay a much higher ocean freight rate.

After much discussion and many telex exchanges, it was decided to ship the cylinders as "Radioactive Material, nos," since all parties finally agreed that there was probably some very minor contamination in the pores of the internal walls of the cylinders but that this contamination did not warrant shipping the cylinder as a full cylinder or as a cylinder containing a heel.

PURCHASE OF TRANSPORTATION SERVICES

One last exmple I would like to mention is that of the purchaser of the transportation services being aware of how the transport is to be handled and not to be concerned only with the price being paid for the service. The provision of transportation services is a very competitive area and sometimes I fear the price is the only criterion being used to select transportation co-ordination services.

The nuclear industry is viewed, by those not involved in it, with a skeptical eye at best. For transports to be performed in anything but the best and safest manner can only bring more criticism.

For example, as most of us are aware, the insurance industry has increased the premiums for insurance coverage in many areas, not the least being the motor carrier industry. An offshoot of this has been the requirement that for some motor carriers to continue their insurance coverage, they must no longer carry uranium concentrates.

Several carriers which in the past accepted this freight no longer do so. The two major nuclear carriers in the US will accept U308 but at rather high rates and, in one case, only if the U308 is shipped in the company's special trailers. What happens then is, that during the bidding process, some transport agents (transporters) will approach motor carriers who have not carried the material previously and request a quotation. The carrier, not fully understanding the commodity and the fact that his insurance policy may not allow him to carry it, accepts the freight at probably a lower rate than would a carrier who has experience with the cargo. The transporter who approached this inexperienced carrier thus has a good chance to win the bid and perform the transport. The owner of the material, who may be several steps away from the actual transportation arrangements, is completely unaware of the situation.

We have experienced this in investigating a motor carrier who has supposedly carried U308 for someone else. When pushed to prove that his insurance coverage does not prevent him from carrying the cargo, we are met with either:

(1) no response or
(2) a response that upon checking with his insurance broker, the carrier
    will no longer accept the freight.

Nevertheless, the transporter who hired this carrier the first time won a job on that basis and those who were trying to meet all requirements and intents of the regulations, lost.

If the purchaser of transportation services does not ensure that things are being handled properly, but rather looks only at price, this practice will continue.

A somewhat similar condition could exist with regard to railroads. Some railroads have chosen not to carry U308 because they feel, in an accident, the drums will open and cause a contamination problem. One does not see a similar situation with UF6 because the packaging is stronger and rail accidents involving UF6 have not resulted in any contamination.

We believe some rail shipments have been performed under FAK (freight all kinds) rates, unknowingly by railroads who did not want to carry the U308. Some railroads have indicated to us that it is possible that some U308 could have been shipped this way and not been detected.

If a shipper or transporter has obtained a job in this manner because the
FAK rate was less expensive than the alternative truck rate and the
purchaser of the transportation services was not aware of or did not care
how the transport was performed, we again have a situation in which
material was transported incorrectly.

If an accident should occur or some other incident occur during a
transport which is being performed in other than the correct manner, the
result will probably be an over reaction by regulatory authorities, which
in the end will result in increased costs and add to the complexities of
performing the transports.

I urge, therefore, that the purchasers of transportation services
make every effort to determine that their material is being transported
in the correct manner and that price not be the only criterion used in
selecting transportation services.

APPLICATION OF RISK ASSESSMENT METHODS TO THE TRANSPORT
BY RAIL OF RADIOACTIVE MATERIALS ON THE SELLAFIELD SITE

J.O.Oyinloye[1], R.J. Williams[2], J. Fields[2]

[1]Electrowatt Engineering
Services (UK) Limited
Birchwood
Warrington WA3 7BH

[2]British Nuclear Fuels plc
Risley
Warrington
Cheshire   WA3 6AS

INTRODUCTION

The reprocessing of irradiated nuclear fuel arising from UK and
overseas nuclear power stations is carried out at British Nuclear Fuels
plc (BNFL) Sellafield works in Cumbria.

Radioactive materials associated with reprocessing operations are
transported between the various plants on the Sellafield site by methods
appropriate to the nature of the radioactive materials. One such method
is rail transport which has been established as a safe and reliable means
of moving active materials. The materials are usually transported in
containers which are housed in shielded flasks. Each container/flask
assembly is transported on purpose-built vehicles (known as 'flatrols')
which are moved by a locomotive.

The safety of off-site transport of radioactive materials within the
UK is justified on the basis of the structural integrity of the transport
flask containment as defined, for example, by the IAEA regulations which
are given legal effect in this country by the Radioactive Substances
(Carriage by Road) (GB) Regulations. For the transport of radioactive
materials on the Sellafield site, however, the justification of safety is
demonstrated by compliance with both the guidelines relating to normal or
routine operations on site and the appropriate accident risk criteria.

The paper describes the methodology for assessing the risks due to
radiological hazards arising during the transport of radioactive
materials, and the application of the methodology to the transport of
radioactive waste materials between two plants on the Sellafield site is
discussed.

SAFETY REQUIREMENTS FOR THE ON-SITE MOVEMENT OF RADIOACTIVE MATERIALS

Under the Nuclear Installations Act 1965, all operations on the
Sellafield site are carried out by BNFL under a licence issued by the

Nuclear Installations Inspectorate of the Health and Safety Executive. The conditions of the licence require BNFL to demonstrate the safety of the operations, including the on-site movement of radioactive materials.

As it is impossible to provide absolute safety, the object is to show that plants and operations will be adequately safe. The concept of adequate safety implies that there is an 'acceptable' level of risk which is usually derived by comparison with the complete spectrum of risks to which people are exposed. The risk criteria derived by BNFL (Ball and Curtis, 1983) are based on risks considered to be acceptable by the International Commission for Radiological Protection (ICRP, 1977) and by comparison with safer parts of the British chemical industry. Furthermore, it is BNFL's policy not only to ensure that all relevant safety standards are complied with but also that risks to employees and the public are kept as low as reasonably practicable (ALARP).

Separate risk citeria are established for the general public and radiation workers. The risk citeria for the general public separately address individual risks (which ar applicable to individual members of the public) and societal risks. For example, the risk or death to a member of the public from the Sellafield site as a whole is deemed likely to be acceptable if it does not exceed 10^{-6} per year. In practice, this risk would arise from the uptake of radiation emitted from the site and which could lead to cancer though it should be appreciated that annual target dose is much less than that due to natural, background radiation. For radiation workers, the basic criterion is that the fatality risk from radiological incidents should not exceed 10^{-5} per year.

RISK ASSESSMENT METHODOLOGY

The radiological risk associated with the rail transport of active materials between any two plants is the summation of the individual risks from all the accident/initiating events which have the potential to result in a radiological hazard. In general each accident/initiating event can have various different consequences, the likelihood and severity of which depend on what features are available either to prevent the accident or to mitigate its consequences.

The most convenient way of analysing the various potential accident sequences is by using event trees, especially where the consequences of the accident sequences are not known initially. Each accident sequence results in a category of activity release whose frequency is calculated by an event tree as a function of initiating event frequency and of the branching event probabilities. (In event tree terminology, the accident-mitigating features are variously referred to as 'top events' or 'engineered safety features'; the term 'branching event' is preferred here because it is more graphic and avoids the confusion with 'top events' as used in fault tree terminology. Such branching events may take the form of a physical event, such as whether a flatrol overturns following a flatrol derailment, or of an engineered safeguard). The frequency of each category of activity release can then be combined with the radiological consequences of the release to produce the risk associated with the release. The risk associated with an initiating event is the sum of the risk over all the release categories in the event. Finally, the risk associated with a given transport system is obtained as the sum of the individual risks from all the possible initiating events resulting from operating the transport system. The assessed risk is then compared with the accident risk criteria described in the previous section.

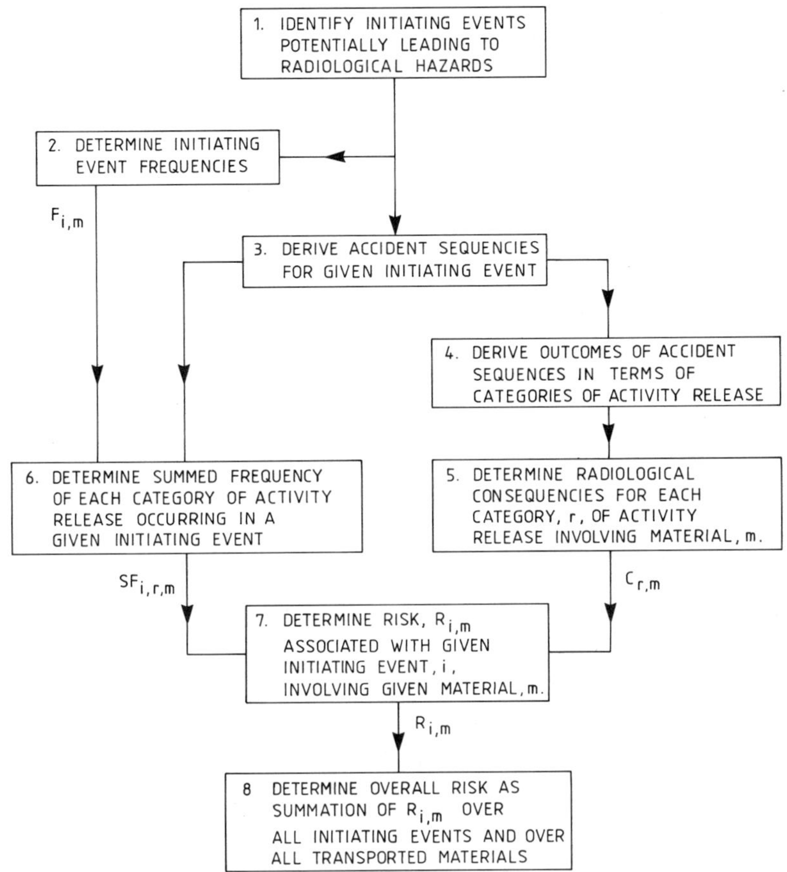

Fig. 1. Stages in the risk assessment methodology
(Symbols are defined in the text)

Thus, the methodology described in this paper for assessing the safety of a rail system for transporting various active materials between two plants on the Sellafield site comprises the stages or elements shown in Figure 1.

INTERMEDIATE LEVEL WASTE

Intermediate level waste produced by future reprocessing operations will be transported by rail across the Sellafield site for encapsulation prior to long-term storage. The bulk of this intermediate level waste will be generated in the initial shear-leach stage where fuel assemblies are to be chopped into 2-10 cm lengths prior to being passed into a dissolver containing nitric acid. This allows separation of the insoluble stainless steel or Zircaloy fuel cladding from the soluble uranium oxide fuel and fission products.

Active Materials Transported

For the sake of brevity, only one of the waste streams to be transported is described in detail in this paper. The hulls' waste stream has been chosen because it adequately covers the spectrum of potential hazards associated with all the active waste streams transported.

The hulls' waste stream will comprise one dissolver batch of hulls and coarse fines. Hulls are the non-soluble sheared lengths of fuel pins which remain after the dissolution process in the dissolver. Coarse fines are produced in the same shearing and dissolution processes as the hulls but consist of finer materials typically up to 2mm in size. Though washed prior to removal from the dissolver the hulls will be contaminated with active liquor, this will also be discharged along with the hulls and coarse fines. The container will be topped up with acidic liquor to cover both the hulls and the coarse fines. The potential hazards associated with the hulls waste stream are:

-   Its potential as a source of activity release to the environment.
-   Its potential to generate flammable hydrogen/air mixtures due to radiolysis.
-   Its potential as a direct radiation source.
-   Its potential as a heat generation source which, in the unlikely event that liquor cover is lost, might overheat the solids in the container.

Flask/Container Assembly

The waste materials will be held in lidded containers with a nominal capcity of $1m^3$. These containers will be filled in heavily-shielded plant areas and will pass through the engineered flasking-out ports into the shielded transport flasks. These flasks are top-loading and will comprise a lid and a body which are bolted together. The flask lid is made out of forged stainless steel, whilst the flask body consist of stainles-steel inner and outer walls, the annulus between which is lead-filled. Both flask and container are manufactured to exacting specifications. Each flask/container assembly weighs about 41 tonne and will accommodate a payload of about 0.5 tonne. A high standard of containment is achieved at the interface between the flask lid and body by two 'O' ring seals with an interspace test point between them.

Flatrol/Locomotive Assembly

The container/flask assemblies will be transported on purpose-built rail vehicles, known as flatrols, on a one-flask-per flatrol basis. In common with flatrols which are already in use on the Sellafield site, the flatrols exhibit such safety design features as:

-   low centre of gravity and hence inherent stability;
-   shock absorbing buffers to minimise the consequences of flatrol impacts.

The flatrols are to be moved, usually in pairs, by a locomotive. The locomotives are automatically governed to a maximum speed of 8 mph.

Rail Network

The design and construction of rail track follows established practice based on experience gained from previous work done on the Sellafield site rail network. The rail link is part of a larger Sellafield rail network and will take advantage of current developments in track design and overall control of the site railway network. Design safeguards provided on the rail network for preventing accidents will include the following:

-   The control of individual train movements using a system of Radio Electric Token Block (R.E.T.B) signalling, for minimising the occurrence of rail/rail collisions.

-   The provision of protection at level crossings to minimise the
    occurrence of road/rail collisions.
-   The installation of trap points in order to intercept runaway
    vehicles.
-   A limitation of the normal operating speed for the site railway
    system to 5 mph.

APPLICATION OF RISK ASSESSMENT METHODOLOGY TO RAIL TRANSPORT

<u>Identification of the radiologically significant initiating events</u>

Since the flask is sealed, the initiating events which can lead to a
significant radiological hazard are those which result in the loss of
containment of the flask and its container. In order to determine these
initiating events, the primary accident events which could, in principle,
result in such a failure mode of the container and flask are first
identified. For this identification, consideration was given to:

i)   The potential hazards associated with the physical and chemical
      properties of the waste material eing transported e.g. hydrogen
      ignition.
ii)   The potential hazards associated with the basic mechanical integrity
      of the flask and its container, e.g. weld failures, loss of seal.
iii) The potential hazards associated with motion, e.g. flatrol
      derailment or flatrol impact with or by an object.

The potential causes of these primary accident events are then
analysed using a 'top-down' fault tree approach. The result is shown in
Figure 2. Clearly not all of the primary accident events identified as
potential threats to flask integrity are actually threats and of the set
initially identified, only the five numbered events in Figure 2 warrant
detailed consideration. These are:

1.   Flatrol Derailment
2.   Flatrol Impact with/by an Object
3.   Container Weld Failure
4.   Flask Despatched contains Enhanced (but not flammable) Hydrogen
     Concentration.
5.   Excessive Transit Time (leading to increased hydrogen concentration)

The justification for excluding the other potential causes
identified in Figure 2 is discussed below.

It is now necessary to develop each of these five primary events
into an event tree in order to quantify the frequency and radiological
consequences of the potential activity releases. In fact, because the
accident sequences and consequences of a flatrol derailment or a flatrol
impact with or by an object in the vicinity of the Calder River bridge
are different from when the events occur away from the bridge, it is
convenient to further sub-divide initiating events 1 and 2 to give the
following four separate initiating events.

1a   Flatrol derailment occurring in the vicinity of the Calder Bridge
     Figure 3 shows a typical event tree.
1b   Flatrol derailment occurring away from the vicinity of the Calder
     Bridge
2a   Flatrol impact with/by an object in the vicinity of the Calder
     Bridge
2b   Flatrol Impact with/by an object away from the vicinity of the
     Calder Bridge.

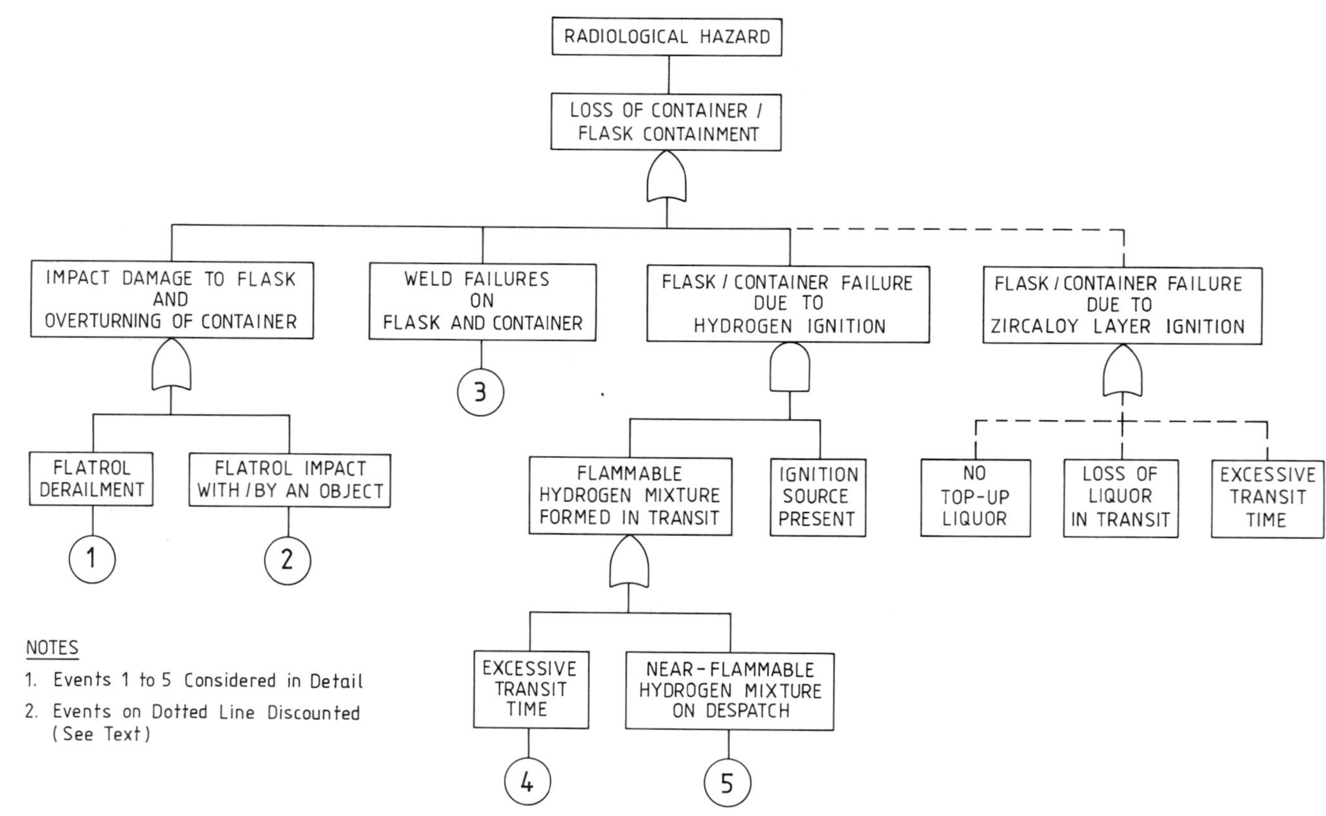

Fig. 2. Fault tree for identifying initiating events potentially leading to a radiological hazard

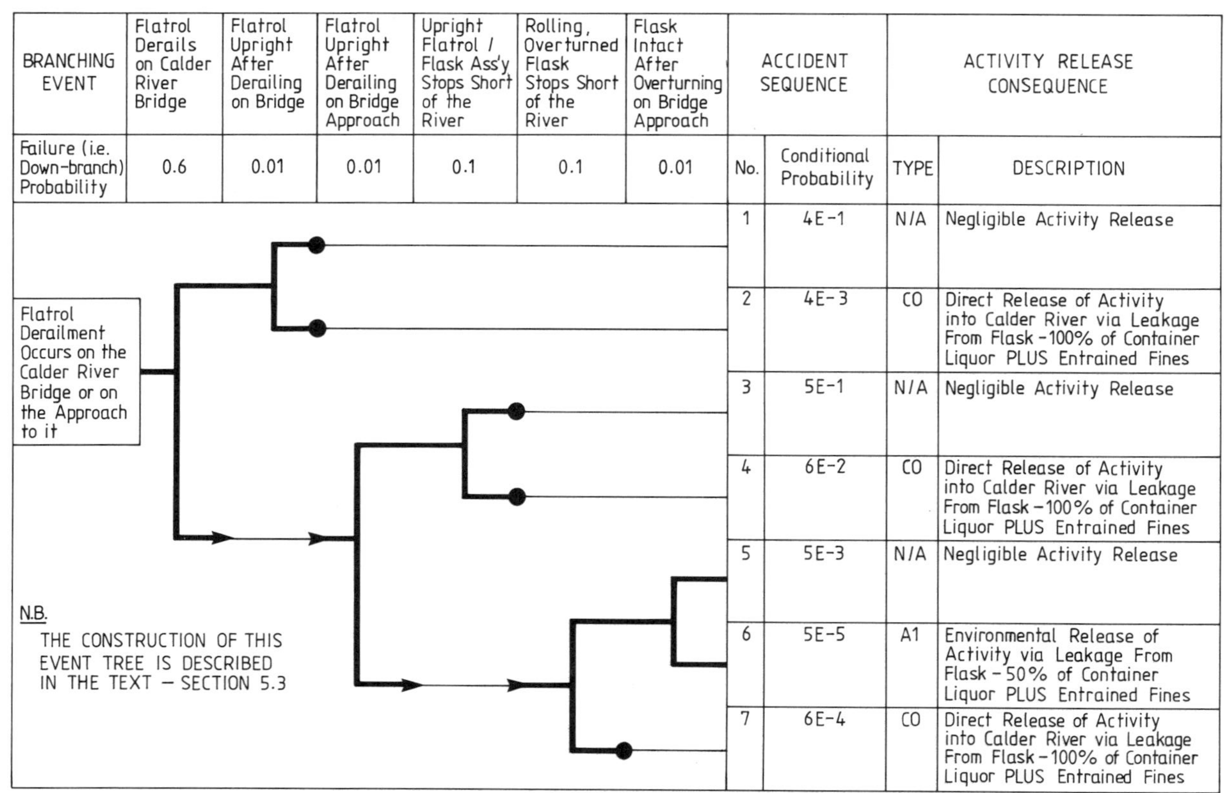

Fig. 3. Event tree for initiating event 1a: Flatrol derailment in the vicinity of the Calder River Bridge.

<u>Initiating Events Discounted as not having the Potential to Result in a
Radiological Hazard</u>

The initial survey of possible primary accident events had identified the combustion of Zircaloy debris from the fuel clad as a potential hazard. Zirconium is a well known reactive material and will burn, if powdered, in air. In practice, as explained in the following paragraph, conditions leading to the ignition of Zircaloy can not be realised in the container or flask assembly. Thus the initiating events which had been identified as potential causes of zirconium ignition could be discounted.

<u>No Top-up Liquor</u>

The two variations of this event which need to be addressed are the cases in which:

i) a batch of hulls alone (with only occluded fines) may be despatched with no top-up liquor, or
ii) a batch of hulls and coarse fines may be despatched with no top-up liquor.

In both cases, the threat of an ignition is from the Zircaloy fines only since, in experiments, Zircaloy hulls have been heated to incandescence in air without burning. Thus the second case represents the more onerous fault condition, since the quantity of zircaloy fines carried over with the hulls alone is minimal. For this second case, it has been shown that 'dry' layers of Zircaloy fines which may accumulate will remain well below the ignition temperature under the worst possible circumstances.

<u>Loss of Liquor in Transit</u>

This may be caused by a mechanical integrity failure of both the flask <u>and</u> container or by impact damage resulting from a flatrol derailment or a flatrol impact with or by an object.

Clearly, this represents a less onerous fault condition than the second of the two cases just considered, in that the heat-generating solids in the container become exposed more gradually. The temperature of any Zircaloy fines layer which may form will therefore again remain well below the ignition temperature.

<u>Excessive Transit Time (leading to Overheating)</u>

It has been shown that a flask and container assembly loaded with a maximum inventory of hulls waste will reach steady-state temperatures well below the liquor boiling-point. Hence, whatever the transit time no hazard arises.

<u>Frequencies of the Radiologically Significant Initiating Events</u>

The frequency, $F_{i,m}$ of a given initiating event, i, involving a flask of a given material, m, is given by:

$$F_{i,m} = IEP_i . NC_m \qquad (1)$$

where:

$NC_m(y^{-1})$ is the number of journeys involving the active material, m, in any given year;

IEP_i is the probability of occurrence of initiating event, i, during any journey involving a flask of any given active material.

The number of journeys, NC_m, involving each active material transported on the rail link was obtained from the plant description documents.

The derivation of the probability, IEP_i for each of the initiating events previously listed is briefly described below.

The probabilities for initiating events 1a, 1b, 2a and 2b were based on historical data taken from the several thousand container-journeys on the Sellafield rail network with an allowance being made in recognition of the greater reliability available from modern rolling stock, network control etc. It is estimated that at most 10% of any derailments or impacts may occur in the vicinity of the Calder bridge, i.e. may occur sufficiently near the bridge for a flask to enter the Calder River. The figure is pessimistic because:

-   the time spent in the vicinity of the bridge per journey does not exceed 10% of the total transit time per journey;
-   the rail tracks on the bridge are provided with side-rails;
-   the approach to the Calder is flanked by a concrete embankment which would tend to prevent a flask from rolling far.

The probability for container failures (initiating event 3) was based on the failure rate of the welds because the containers are of welded construction with no penetrations. It was pessimistically assumed that any weld failure would occur during a journey on the rail link.

Initiating events 4 and 5 could only occur as a result of operational errors of omission or commission. The probabilities for these initiating events were estimated using fault tree analysis which included basic human errors.

Derivation of the Accident Sequences per Initiating Event

The accident sequences following an initiating event are determined by the outcome of certain key factors called branching events which may be the occurrence or non-occurrence of some physical event, or the success or failure of a piece of equipment in performing its safety function. Thus, in determining the branching events for a given initiating event, due consideration needs to be given to:

-   the features and safeguards incorporated in the design to prevent the occurrence of fault conditions or to mitigate their consequences;
-   the physical environment in which the fault conditions occur;
-   the damage consequences of the fault conditions;
-   the particular hazards associated with, and the physical and chemical characteristics of the transported material;
-   the duration of the fault conditions i.e. how quickly the fault conditions can be rectified.

Once the branching events relevant to an initiating event have been determined, the accident sequences in the event can be easily derived and the event tree constructed by applying a simple TRUE/FALSE (yes/no) logic to each relevant branching event in turn. By convention, the event tree branches upwards if the proposition made by the branching event is true, and branches downwards if false.

It must be stressed that not all the branching events in an event tree are relevant to each accident sequence. For example, in the event tree shown in Figure 3, the last four branching events (which are relevant to derailments occurring on the approach to the bridge) clearly do not apply to accident sequences 1 and 2 which pertain to derailments occurring on the bridge. Thus, for accident sequences 1 and 2, the event tree is 'straight-lined' (i.e. no branching occurs) under these four branching events.

The branching events and accident sequences derived for one of the initiation events, previously identified in the section Indentification of the Radiologically Significant Initiating Events, are shown on the event tree in Figure 3.

<u>Derivation of the Category of Activity Release Associated with the Outcome of each Accident Sequence</u>

Since the activity release associated with the outcome of each accident sequence is of interest, each outcome must first be described qualitiatively in terms of the location, orientation and integrity of the flask at the end of the given accident sequence. For example, in the event tree shown in Figure 3, the outcomes of the accident sequences are as follows:

- at the end of postulated accident sequences 1 and 3, the flask would not be in the river, would remain upright and would not be breached;
- at the end of postulated accident sequences 2, 4 and 7, the flask would be in the river, would be unlikely to be upright and is presumed to be breached;
- at the end of postulated accident sequence 5, the flask woud not be in the river, would be overturned but would remain intact;
- at the end of postulated accident sequence 6, the flask would not be in the river, but would be both overturned and breached.

After similarly describing the outcomes of the accident sequences from all the initiating events identified, it was found that the resulting activity releases could be divided into the following broad categories:

Type A: Release   of   acitvity   into   the   environment   via flask/container leakage.

Type B: Release of activity into the environment via hydrogen ignition in the flask.

Type C: Direct release of activity into the Calder River via leakage from the flask.

Within each of these primary categories of activity release, there is, in principle, a continuous distribution of sub-categories, each of which corresponds to the environmental release of a particular fraction of the total activity in a given flask. In order to keep the analysis within manageable proportions, the primary release types were further sub-divided so that the activity release associated with each postulated accident sequence could be uniquely defined by:

- its primary release category, i.e. A, B or C as described above, <u>and</u>
- the proportion of the container liquor volume which would be released from the flask and container in the given accident. (Since the flask is designed to retain bulk solids, only materials which are entrained or dissolved could be released).

270

Table 1. Schedules of Activity Release Types.

Release Type and Sub-type	Percentage of Container Liquor Involved
A: Environmental Release of Activity via Leakage from Flask	
A1	50
A2	10
A3	2
B: Environmental Release of Activity via Hydrogen Ignition	
B2	10
B3	2
C: Direct Release of Activity into Calder River via leakage from Flask	
C0	100

The resulting schedule of activity released categories is shown in Table 1, along with the respective proportions of active liquor involved in the release.

Quantification of the Radiological Consequences of Each Category of Activity Release

Each category of activity release may result in several radiological consequences, depending on the pathways available for dose commitment. In order to assess the risks associated with the rail transport system and to compare these risks with various parts of the risk criteria, it was necessary to calculate:

- the dose commitment a member of the general public would incur from airborne active material via inhalation;
- the dose commitment a member of the general public would incur from airborne active material via 'all pathways';
- the dose commitment a member of the general public would incur from active material discharged into the sea;
- the dose commitment a member of the general public would incur via direct radiation from active material released into the environment;
- the dose commitment site personnel would incur via direct radiation from active material released into the environment.

Quantification of the potential radiological consequences took into account the following factors:

- the inventory and activity of the radionuclides present in each container of active materials;
- the dose equivalent, (i.e. the dose per unit activity) for each radionuclide;
- the fractional release of active material via entrainment of the container liquor and via aeriel entrainment;
- the possible duration of activity release and exposure;

-   (for aerial releases) the dispersion characteristics of the release.

<u>Quantification of the Frequency of Each Category of Activity Release in a Given Initiating Event</u>

The summed frequency:

$$SF_{i,r,m}$$

of the activity release category, r, from the initiating event, i, involving the active material, m is given by:

$$SF_{i,r,m} = F_{i,m} \cdot SP_{i,r} \qquad (2)$$

where:

$F_{i,m}$ (y^{-1}) is determined in the manner described in Frequencies of the Radiologically Significant Initiating Events and is the frequency of the initiating event, i, involving the active matrial, m;

$SP_{i,r}$ is the summed probability of all accident sequences, from the initiating event, i, whose consequence is the activity release category, r.

Now, the probability of any given accident sequence is the product of the success (or failure) probabilities of all the branching events which define the particular accident sequence. (The failure probability, p, of a given branching event is the probability that the proposition made by the branching event is false; the corresponding success probability is given by (1-p) and is the probability that the proposition is true). By convention, branching event failure probabilities, rather than success probabilities, are presented on event trees.

For the risk assessment, failure probabilities for the branching events were either estimated from historical and empirical data or were based on pessimistic assumptions regarding the stability and dynamic behaviour of the flatrol/flask assembly following a derailment/impact accident and on pessimistic assumptions regarding the impact capability of the flask and container.

<u>Quantification of the Risks</u>

In terms of the required method of calculation, the risks to be assessed may be classified into two groups:-

i)   risks which are to be quantified in terms of the products of the frequency and consequence of each accident sequence, and

ii) risks which are to be evaluated in terms of the summed frequency of all accident sequences whose individual radiological consequences exceed a specified value.

The methods for calculating these two groups of risks are now described.

<u>Risks Quantified in terms of the Product of Accident Frequencies and Consequences</u>

For risks in this group, the radiological risk, $R_{i,m}$ associated with any given initiating event, i, which involves any given active material, m is the summated risk over all the activity release categories in the initiating event, viz:

$$R_{i,m} = \Sigma r \; SF_{i,r,m} \; C_{r,m} \cdot M(C_{r,m}) \qquad (3)$$

where:

$SF_{i,r,m}(y^{-1})$ is the summed frequency of the activity release category, r, in the initiating event, i, involving the active material, m;

$C_{r,m}(Sv)$ is the radiological (dose) consequence of the activity release, r, involving the active material, m;

$M(C_{r,m})$ in (Sv^{-1}), is the risk of death to an individual as a consequence of a dose uptake of $C_{r,m}(Sv)$.

The determination of:

$$C_{r,m} \text{ and } SF_{i,r,m}$$

has been described in the sections on Quantification of the Radiological Consequences of Each Category of Activity Release, and Quantification of the Frequency of Each Category of Activity Release in a Given Initiating Event respectively.

The risk, R, is then obtained as the summation of $R_{i,m}$ over all the initiating events and over all the active materials that are transported, viz:

$$R = \Sigma_i \Sigma_m R_{i,m} \tag{4}$$

Risks Evaluated in Terms of the Frequency of Exceeding a Specified Consequence Value

For risks in this group, the summed frequency:

$$f^*_{i,m,}$$

of all activity releases (in the initiating event, i, involving the active material, m) whose individual radiological consequence exceeds a specified value is given by:

$$f^*_{i,m} = \Sigma_r SF^*_{i,r,m} \tag{5}$$

where:

$SF^*_{i,r,m}$ is the summed frequency of all active releases of category, r, (in the initiating event, i, involving the active material, m) whose radiological consequence exceeds a specified value. This summed frequency is calculated in the manner described in section Derivation of the Accident Sequences per Initiating Event.

The summed frequency, F^*, of all accident events whose individual radiological consequence exceeds a specified level is then given as the summation of $f^*_{i,m}$ over all the initiating events and over all the active materials that are transported, viz:

$$F^* = \Sigma_i \Sigma_m f^*_{i,m} \tag{6}$$

Data Processing and Presentation

A total of seven waste streams, each with different chemical and physical compositions, radionuclide inventories etc. are to be transported by rail. The risk associated with transporting each waste stream has been assessed in the manner described in the foregoing sections for the hulls waste stream. Thus the overall risk assessment involved:

- 7 different types of active materials transported at different rates per year;
- 7 different initiating events potentially involving each of the seven types of active materials;
- 7 different categories of activity release involving each of the seven types of active materials, and
- 5 different dose consequence calculations for each category of activity release involving each active material.

The large amount of data processing required was facilitated by the use of computer programmes. Nevertheless, the method of calculation and the format of data presentation needed to be efficient and compact in order to reduce the number of calculations and the volume of input and output data. The format of data presentation adopted for each initiating event comprises:

i) A generic event tree (e.g. Figure 3) which presents those input and output data items (such as the branching events, their description and failure probabilities, the accident sequences with their associated probability and activity release categories) that are independent of which active material is involved in the event.
ii) Up to seven tables, one table for each active material involved in the initiating event. These tables present data items (such as initiating event frequencies, accident sequence frequencies, radiological consequences, etc) which are dependent on the active material being transported.

Finally, the summations of:

$$R_{j,m} \text{ and } f^*_{i,m}$$

in equations 4 and 6 are carried out separately over all the initiating events and over all the active materials so that the percentage contribution from each initiating event and active material to the overall risks is calculated. This information about the significatnt contributors to the overall risks would enable the transport system to be reviewed where necessary.

Comparison of the Assessed Risks with the Accident Risk Criteria

The risks associated with the rail transport of intermediate level waste have been assessed using the above methodology and shown to be an acceptably small percentage of the risk targets discussed in the section on Safety REquirements for the On-site Movement of RAdioactive Materials. For example, the assessed risk to a member of the general public is found to be 0.3% of the target. Similarly it has been shown that the assessed risk to site personnel is about 0.2% of the appropriate target.

CONCLUSION(S)

This paper illustrates the use of probabilistic risk assessment methods in the safety justification of nuclear reprocessing plants and operations on the Sellafield site.

The basic risk assessment methodology is clearly applicable to plants and operations in other industries. For other applications certain elements of the methodology, as described and applied in this paper, may require different approaches. Two such elements of the methodology are discussed below.

In the section Identification of the Radiologically Significant Initiating Events , initiating events were derived using a 'top down' fault tree approach since one particular hazard consequence, namely, radioactivity release, was of interest. Where the undesirable or hazard consquences are not obvious, a 'bottom up' approach, such as Failure Modes and Effects Analysis (FMEA) or Hazard and Operability Studies (HAZOPS) may be used.

Also, in the section on Quantification of the Frequency of Each Category of Activity Release in a Given Initiating Event, accident sequence probabilities were calculated simply as products of the success or failure probabilities of the branching events. This is because, for the particular application described, the branching events are unrelated, i.e. do not have common causes of failure. Where the branching events, for example, are in the form of items of safety equipment which have a common power supply, accident sequence probabilities have to be calculated by deriving minimal cutsets using Boolean reduction.

REFERENCES

Ball, P. W. and Curtis, L., 1983, Safety Criteria for Nuclear Chemical Plants in Paper to Symposium of Safety and Reliability Society, Southport, England, September 1983.

ICRP Publication 26, 1977, Annals of the ICRP: Recommendations of the International Commission on Radiological Protection, Publisher, Pergamon Press.

SAFE TRANSPORT OF RADIOACTIVE MATERIAL—
A TRAINING EXPERIENCE

C.R.Chapman

CEGB Nuclear Power Training Centre
Oldbury-on-Severn
Thornbury
Bristol BS12 1RQ

When one considers the nuclear industry and the title of this
international conference "Transportation for the Nuclear Industry" it is
particularly the transport of nuclear fuel which is foremost in the minds
of most interested parties. This is understandable as fuel transport can
be considered the industry's umbilical cord - bringing in the vital feed
stock and removing the necessary waste products. Such movement, which is
imperative to maintain the nuclear industry operational, can have many
vested interests. It can, for example, be an embarrassment, it can be
resouce-intensive, it can be time-consuming, it can also be turned to
commercial advantage; but regardless of motive the very nature of the
cargo - radioactive material - requires the consignee and the carrier to
be diligent. Diligent for the sake of the employees of consignee and
carrier alike, for the sake of the public, and for the sake of the
environment. And whilst such diligence applies to nuclear fuel it is
germane to all radioactive material when moved from one location to
another.

Many of the measures necessary to safeguard the public and the
environment during any transportation for the nuclear industry are
obvious, even to the layman; however, others are more subtle, even for
the expert and some formal training is essential to provide an
understanding of the vicissitudes of this specialised transportation.

Whether the nuclear material be transported by road, rail, sea or
air, there are national and international recommendations and conventions
supported by legislation in many areas.

Training is necessary to comprehend the statutory and advisory
nature of these requirements. The International Atomic Energy Agency has
published a number of detailed documents to guide those involved with the
transport of radioactive material in safe practices. Training is
required to derive the maximum benefit from this documentation.

There is the philosophy of radiation protection and the 'as low as
reasonably achievable, economic and social factors being taken into
account' (ALARA) principle which must be fully understood. There are the
various categories of radioactive material that have been identified
internationally - those with low specific activity and those surface
contaminated objects, for example. The various types of package that

have been accepted internationally in which radioactive material may be transported and the concept of activity limits must be constantly in the thoughts of those who have any responsibility regarding transportation for the nuclear industry. The design philosophy for the type of package and their quality assurance and the methods of insurance and inspection are important facets that should be comprehended. The nature of the tests conducted to check the integrity of the containment and the shielding, and the regulatory requirements for such tests should be understood. Training can provide the means by which the required understanding and comprehension can be grasped.

One must consider the physical processes of preparing various categories of nuclear materials for transportation: the packaging of the material and its encapsulation or containment; the packaging itself has to be chosen - within the nuclear power industry, for example, a box for new fuel, a drum for enriched fuel or a flask for irradiated fuel; labelling the package and the appropriate placarding for it must be carefully identified whether the material be transported by road, rail, sea or air; the arranging of insurance and the completion of the paperwork demand that the practitioner is fully aware of his responsibilities, as well as having the dexterity to undertake these tasks. Training is the only sure way to achieve the successful completion of all the processes and tasks identified above.

Besides the preparation of radioactive material for transport and its safe packaging to comply with national and international regulations, there is its actual movement from one location to another - the subject of this conference - Transportation for the Nuclear Industry! The responsibilities of consignor and carrier for the control of material in transit, the idea of a Transport Index, the stowage of packages and what to do with those that are undelivered; all are practicalities that the nuclear industry has a moral as well as a statutory obligation to address. Training is essential, not only to meet the requirements of the control of radioactive material in transit, but to maintain the credibility of the nuclear industry.

Then there are comprehensive schemes to cope with people, plant and situations in the event of incidents. The general accident provisions - contingency plans such as the National Arrangements for Incidents involving Radioactivity (NAIR) and the Irradiated Fuel Transport Flask Emergency Plan are applied in this country, but similar provisions should exist in every country where the transport of radioactive material is envisaged. Training is imperative if the safety of the public and the environment is to be assured at all times in such circumstances.

Training is thus a key component in safe transportation for the nuclear industry. The remainder of this paper offers a method by which suitable training may be provided. It is based on experiences of organising and managing an international course in England during the autumn of 1987 for the International Atomic Energy Agency (IAEA) on Safe Transport of Radioactive Material. The programme for the course is outlined in the Appendix. However, such experiences are usefully shared here as they will enable others contemplating the provision of training in this topic to gain from the lessons learned.

A three week course was hosted by the Central Electricity Generating Board (CEGB) at its Nuclear Power Training Centre on behalf of the British Government under the IAEA Interregional Training Programme; the Agency offers a scheme to provide technical support and information for developing nuclear nations which are member states of this Agency of the United Nations.

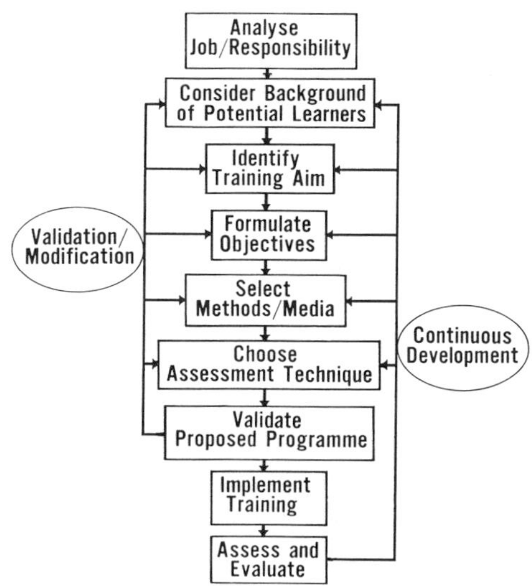

Fig. 1. Systems approach to training.

As the CEGB had been invited by the IAEA to take responsibility for the organisation of the course, a Systems Approach to Training (Figure 1) was followed, in keeping with the Board's nuclear training policy. A small panel was established under the direction of a training manager from the CEGB Nuclear Power Training Centre comprising the following individuals:

- a technical expert from the Agency (whose normal duties included specific responsibility in Vienna for the international guidelines, safe practices and technical documentation regarding the safe transport of radioactive material);
- a technical adviser from the Department of Transport (DTp) (whose normal duties included responsibility in the UK for giving approval for transportation of radioactive material);
- a Tutor from the CEGB Nuclear Power Training Centre (whose normal duties included lecturing and writing documentation on health physics, radiological protection, radioactivity and safe practices required to handle and transport radioactive substances);
- a Senior Tutor from the CEGB (whose normal duties included responsibility for all nuclear specialist technology courses at its Nuclear Power Training Centre).

The latter panel member acted as Course Manager throughout the design, development and implementation stages of the course, having had the additional experience of working in developing nuclear nations and of organising and managing other international courses for the CEGB in conjunction with IAEA. This panel, in designing the course, also drew on expertise from elsewhere besides IAEA, DTp and CEGB.

In adhering to the Systems Approach to Training the panel decided that the IAEA documentation (the Safety Standards, Safety Guides and Technical Documents) should for the framework for the participants' ultimate knowledge. It was agreed that a thorough familiarisation of the published material from Vienna should be the goal for all the participants, coupled with practical experience of dealing with radioactive material to enable them to handle it safely; this would enable them to transport (or be responsible for the movement of) such material safely, not only within their own national boundaries but also across international frontiers by road, rail, sea and air. This stage was regarded as an analysis of the jobs or responsibilities of the participants.

The next stage in the Systems Approach to Training is to examine the background of the potential learners. In practice this was achieved by the training experts of IAEA and CEGB jointly determining the minimum academic and technical backgrounds that the participants should have, publishing these standards and selecting the participants accordingly.

Having thus fixed the ultimate goals in the first stage and the initial backgrounds in the second stage, the difference between these two identified the training to be provided. At this stage the panel had to make some pragmatic decisions rather than act as idealistic training course designers. Certain aspects of the deficiencies between the goals for the participants and their backgrounds could not realistically be provided by attending a course - whatever its duration - at the Nuclear Power Training Centre. It was decided, therefore, that the aims of the couse should be to provide a thorough familiarisation of the IAEA Safety Series relating to the Safe Transport of Radioactive Material, viz 6, 7, 37 and 80, and an understanding of the reasons for the regulations embodied in these. It was further decided to address the associated skills required to implement the regulations, illustratively only, as hands-on training would have been impracticable. After the panel had focused on the general aims of the course the next stage was to discuss how they should be met by deciding the macro training objectives - the framework by which the aims could be met.

Thus a programme was designed, elaborated further below, during which stage the format of the course was chosen. There were to be formal presentations, panel discussions, participative exercises, demonstrations, and visits to relevant locations around the United Kingdom. The panel was mindful that the Agency and the CEGB wanted some indications, even assurance, that the course and the training had been effective. Thus it was felt desirable to have a means of evaluating the course and of assessing the participants - not necessarily formally but certainly in a structured way. Another consideration that the panel recognized was that the course was to be conducted in English, probably not (and in the event definitely not) the mother tongue of any participant. Time, therefore, had to be allocated within the training programme for clarification of terminology to cope with language problems as well as those of a technical nature. Another dimension, which the training experts of the panel drew to the attention of the technical experts, was that training course design can be, in fact should be, quite different from technical conference design. Course sessions, of necessity, must be structured quite differently from conference papers; for example, allowance was made for dialogue between speaker and participant within each presentation; opportunities were provided to revisit issues requiring elaboration; tutorial assistance was made available when needed.

It was only after the panel had produced a draft programme, identified potential speakers and had discussed the aims and objectives of each session with them that it became apparent that the course would take three weeks. At this stage a member of the administration group at the Nuclear Power Training Centre joined the panel to deal with the very demanding 'behind the scenes' backup essential for the smooth running of a course - particularly one of an international nature. It was agreed, for example, that every session would benefit from handouts and that these should all be provided for the participants at the commencement of the course. Speakers had to be encouraged to prepare draft texts for typing well in advance of the course to enable the CEGB Tutor to check them for uniformity - and several speakers required a great deal of encouragement. There was no logical reason for the reticence of speakers to write handouts except departure from the traditional "last minute" committing of thoughts - however logical - to paper. In practice the availability of macro-objectives proffered viable headings around which the handouts could be written.

To follow the Systems Approach to Training rigidly, the organisers should have staged a pilot course with surrogate participants to validate the proposed programme. Improvement or modifications could then have been incorporated to any part of the design stage - choice of objectives, choice of format, method of evaluation etc. In the event it was impracticable to mount a pilot course and the validation/modification became a paperwork exercise. Nevertheless this step was undertaken to ensure that "it would be all right on the night."

The procedures outlined above provided the method by which this course was designed and in the event its implementation was a great success, with compliments being poured over organisers and speakers alike. In fact gratitude is still being expressed by the participants from thousands of miles distant and after several months have elapsed. But what did we teach them while they were with us at the Nuclear Power Training Centre? What was the content of the training course?

The course, as already intimated, was built around the IAEA Safety Series 6, 7, 37 and 80 and the supporting Technical Documents. However, when the organising panel first met there were two editions of Safety Series 6 in circulation, Safety Series 7 was being rewritten and several Technical Documents were half-written. It was decided to refer to the 1985 Edition of Safety Series No. 6 throughout the course and to provide every participant with copies of the Safety Series. As a number of the participants were working in Latin America it was convenient to give them copies of Safety Series No. 6 in the Spanish language. Other participants were more comfortable reading French than English and so we gave them Safety Series No. 6 published in French. Unfortunately (for the participants) the Technical Documents were available only in English, nevertheless these were issued as part of the course decumentation. As a principle, the paragraph headings in the booklets were used as the substance of a teaching session or series of sessions.

Early during the planning stage of the course it became evident that, with the quite diverse considerations for the safe transport of radioactive material, some coherence had to be introduced. Accordingly six examples of radioactive material were identified as follows:

a) uranium ore and concentrates, uranium hexafluoride and new fuel
b) low level radioactive wastes
c) spent fuel
d) radioisotopes - such as technetium generators
e) sources used for non-destructive testing
f) excepted materials.

During the course reference was made, wherever sensible, to these six examples.

For some considerations (categories of radioactive materials) the six examples could be grouped together in one way, but for other considerations (type of package, or activity limits, or administrative procedures, or insurance) the six examples could be grouped differently. Notwithstanding, these examples provided a useful and recognisable theme throughout the course.

It was further decided that the first part of the training should concentrate on the IAEA documentation with the lecturers highlighting particular aspects and clarifying any issues in the documentation that the participants might wish to raise. It was, therefore, essential that potential speakers should be thoroughly conversant with the information and, where possible, have been involved in the original preparation of such IAEA material. The IAEA technical expert in Vienna, and indeed his predecessor, were accordingly invited to make presentations early in the course. Having planned to set the scene by elaborating on the content of the Safety Series, the organising panel felt that subsequent presentations should reflect experiences of applying the guidance in the practical world.

As the participants were from developing nuclear nations, or Member States of the Agency where the transport of radioactive material was in its infancy, it was appropriate to include lecturers with wide expertise, encompassing the packaging, transportation and licensing or regulation of such. Thus speakers came not only from the United Kingdom, countries with a major nuclear interest such as France and Germany, and countries with vast distances to consider such as the United States of America, but also those with experience in a country having a relatively small nuclear involvement such as Finland. The participants could, therefore, benefit not only by listening to practitioners expounding the virtues and merits of the international requirements but by hearing at first hand the benefits, and any difficulties, of their practical application in a variety of countries. However such an information exchange was not merely one way, and the added bonus of staging a course such as this, was for the national and international experts to learn of problems the participants had discovered in interpreting and implementing the IAEA documentation. As many of these experts were chosen because of their involvement with the production and revision of IAEA material they are now admirably placed to consider the situations drawn to their attention by the participants on our course.

As the organisation for the course in this country lay primarily with CEGB, it was considered desirable that the course itself should not be flavoured too strongly with Generating Board attitudes, however commendable. Accordingly other bodies with nuclear interests were involved with presentations such as Amersham International plc, British Nuclear Fuels plc, the Atomic Energy Authority, United Kingdom Nirex Ltd, the National Radiological Protection Board, and the Nuclear Installations Inspectorate. Whilst not having a primary interest in nuclear matters the Civil Aviation Authority, and the International Maritime Organisation (which has its headquarters in London and is yet another Agency of the United Nations), are very much involved with safe practices in transporting radioactive material, as is the Department of Transport. Potential speakers from all these organisations were contacted and involved at an appropriate stage in the Systems Approach to Training. All those who agreed to make presentations were suitably briefed and exhorted to supply drafts of their handouts well in advance of the course. Assistance was made available from the CEGB Nuclear Power Training Centre in the composition and preparation of visual aids to accompany the presentations, in addition to the editing and typing of the handouts discussed above.

Other organisations in this country deeply interested in transportation for the nuclear industry were approached. Securicor and

British Airways not only provided speakers but acted as munificent hosts
to demonstrate how they each put the IAEA regulations into practice.
Croft Associates Ltd., a UK firm offering advice on specialised packaging
and transportation, also provided on authoritative speaker. At the
Structural Test Centre at Cheddar, CEGB engineers described the
facilities and services available there to carry out drop tests and
various integrity tests on Type A and Type B packages. The participants
on the course were privileged to witness a drop test on a 1/4 scale model
of a Magnox Irradiated Fuel Element Transport Flask. The Course Manager
welcomed this demonstration as being far more illuminating than seeing a
film of a typical test. He was, though, a little disappointed that CEGB
was unwilling to stage another 'Operation Smash Hit' for the benefit of
the participants on this training course(!).

To promote the heuristice element of the course, a number of
participative exercises were devised by the organising panel requiring
the participants to work in groups of four at specified points during the
course. The composition of each group was determined by the Course
Manager after two weeks had elapsed, enabling him to assess the
personality, compatibility and language fluency of each participant. The
numbers were such that three pairs of groups acted the role of
consignor/carrier whilst the other three pairs behaved as the competent
authority/regulatory body/compliance assurance team. All groups were
given imaginary conditions under which certain radioactive material had
to be transported. The roles were then changed and other conditions
considered. At the end of the third week of the course a spokesman from
each group made a presentation to the other participants of the findings
of that group. This technique not only ensured that the participants
immersed themselves in the course but the exercises were carefully chosen
to cause the participants to constantly refer to the IAEA documents; they
were asked to quote paragraph and section numbers in their presentations
which made them thoroughly familiar with the content of the IAEA
publications.

These participative exercises brought an additional advantage during
the presentations made towards the end of the course by the participants.
It enabled the Course Manager and the technical experts from Vienna and
the Department of Transport to evaluate the effectiveness of the course
and, to a certain extent, assess the understanding by the participants of
the material they had been taught. A questionnaire was also completed by
each participant to provide the organisers with further information on
the quality of the training provided. The subsequent high quality of the
presentations by the participants and the results of the questionnaires
can be interpreted only in terms of the course being successful.

Towards the end of the design stage of this course an indication was
given of the potential high esteem in which this course was likely to be
held. Mr Noramly bin Muslim, a Deputy Director General of the IAEA and
at the time Acting Director General, expressed the desire to attend
during the presentations of the participative exercises at the conclusion
of the course. In the event his presence and concluding address
demonstrated that this course on Safe Transport of Radioactive Material
had provided a valuable contribution to Transportation for the Nuclear
Industry.

APPENDIX 1

IAEA INTER-REGIONAL TRAINING COURSE on
SAFE TRANSPORT OF RADIOACTIVE MATERIAL

6 - 23 October 1987

Tuesday, 6 October

1.　Introduction to the Course
　　1.1　Welcome to Members and Introductions - V. J. Madden
　　1.2　Administrative Arrangements - C. R. Chapman
　　1.3　The Organisation and Work of the IAEA - P. Schultze-Kraft
　　1.4　Aims of the Course - R. A. Sullivan
　　1.5　Outline of Course Programme, Participative Exercises,
　　　　　Leisure Activities - C. R. Chapman

2.　The Transport Scene
　　2.1　Nature of Radioactive Materials
　　2.2　Range of activities
　　2.3　Regulations
　　2.4　Types of Packages
　　2.5　International Aspects
　　2.6　Variety of Modes of Transport - R. A. O'Sullivan

3.　Development, Purpose & Principles of IAEA Regulations - R. B. Pope
　　3.1　IAEA Safety Series 6

4.　Introduction to Theme of the Course - D. J. Blackman
　　4.1　Uranium Ores and Concentrates, Uranium Hexafluoride, New Fuel
　　4.2　Low Level Radioactive Wastes
　　4.3　Spent Fuel
　　4.4　Radioisotopes e.g. Tc Generators
　　4.5　Sources for NDT Use
　　4.6　Excepted Materials

Wednesday, 7 October

5.　Purpose and Use of IAEA Documents - R. A. O'Sullivan
　　　　　Safety Series 80
　　　　　Refer to Technical Documents

6.　Application and Implementation of Safety Series - R. B. Pope
　　　　　Safety Series 7}
　　　　　Safety Series 37}　To be distributed

7.　Radiation Protection - K. B. Shaw
　　7.1　Quantities and Units
　　7.2　Principles
　　7.3　ALARP and Dose Limits
　　7.4　Radiation Protection Surveys

8.　Visit to Oldbury Power Station - C. R. Chapman

Thursday, 8 October

9.　Categories of Radioactive Materials - E. P. Goldfinch
　　9.1　Excepted Materials
　　9.2　Low Specific Activity LSA-I, II, and III
　　9.3　Surface Contaminated Objects SCO-I and II
　　9.4　Fissile Excepted

20.5   Special Arrangements - D. J. Blackman
20.6   National Regulatory Implementation in US - A. W. Grella
20.7   National Regulatory Implementation in France - M. Grenier
20.8   National Regulatory Implementation in FDR - F. W. Collin
20.9   National Regulatory Implementation in UK - D. J. Blackman
20.10   Panel Discussion - D.J Blackman, F.W.Collin, A.W.Grella,
        M.Grenier

Thursday, 15 October

21. General Accident Provisions
     21.1   Contingency Plans (NAIR) - G.C.Roberts
     21.2   Contingency Plans (Emergency Responses) - P.F.Beaver
     21.3   Irradiated Fuel Transport Fask Emergency Plan - R.F.Pannett
     21.4   Experience in USA - A. W. Grella
     21.5   Experience in France - M. Grenier
     21.6   Experience in FDR - F. W. Collin
DISCUSSION - C. Roberts, R. F. Pannett, A. W. Grella, M. Grenier

Friday, 16 October

22. Introduction to Participative Exercises - D. J. Blackman
     (split into six groups of 4/5 participants)
     a)   Quality Assurance Consignor/Carrier
     b)   Competent Authority/Regulatory Body Compliance Assurance Team

Monday, 19 October

23. Visit to CEGB Structural Test Centre, Cheddar
24. National Competent Authority in a Small Country - H. Koponen

Tuesday, 20 October

25. Work on Participative Exercises - R.A. O'Sullivan, D. J. Blackman,
     H. Koponen

Wednesday, 21 October

26. Control of materials in transit - R. D. Cheshire
     26.1   Consignors' responsibilities
     26.2   Carriers' responsibilities
     26.3   Standards to which carriers have to operate
     26.4   Radiation and Surface contamination
     26.5   Stowage
     26.6   Undelivered packages

27. Nuclear Insurance
     27.1   Materials Damage - D. A. Williams
     27.2   Liability and International Aspects - S. H. von Gian

28. Work on Participative Exercises - R. A. O'Sullivan, D. J. Blackman,
     R. D. Cheshire

Thursday, 22 October

29. Presentations of Exercises - R.A.O'Sullivan, D.J.Blackman, L.Price

Friday, 23 October
30. Closing Address - Noramly bin Muslim
31. Assistance and Resources Available - CEGB
32. End Of Course Review - R.A.O'Sullivan, V.J.Madden, C.R.Chapman

DISCUSSION FOLLOWING SESSION 3: Papers 5 - 8

MR.D.BLACKMAN, DEPARTMENT OF TRANSPORT:
Whilst Mr.Chapman's paper is still fresh in our minds I would like to
make one or two comments. I don't think it was made totally clear in the
paper that it was in fact a course for personnel from developing nuclear
nations and what he said at the beginning is that you have got to make
sure you know exactly what it is you are training for. One of the results
of this course, as a matter of interest, is that the Portuguese delegate,
whose only prior knowledge, when she came to this course, was really the
transport of non-radioactive dangerous goods, is now in fact representing
her country at the CEC Special Working Party on the transport of
radioactive materials as well as representing Portugal at the special
meeting on the Transnuklear Mol affair. I have had requests from a number
of course participants for them to come and visit the UK and we have the
Malaysian delegate coming to the Department of Transport for most of
September. There will also be someone from Brazil and I have had informal
requests from the delegates from Saudi Arabia and Nigeria. I personally
found it one of the most rewarding three weeks of my life. I did not
attend full time, I had other work to do, but on the occasions when I
did, I found it an exceptionally rewarding experience and Mr.Chapman and
his colleagues did an excellent job in getting this course together.
Having talked informally to some of the people at the end, I gathered
that if there were to be another one in a very short time, he would
probably get the whole lot wanting to come back and do the whole thing
again.

Following a further question on Paper 3:8 by Mr.D.G.Walton, University of
Manchester, concerning the percentage of UK personnel involved in nuclear
transport who had been through an effective structured training
programme, it is regretted that, due to a technical fault, the rest of
the discussion of this half-session is unavailable.

TRANSPORT STUDIES FOR CANDIDATE
NEAR-SURFACE REPOSITORY SITES

D.Bennett, J.Fitzpatrick

UK Nirex Ltd
Curie Avenue
Didcot, Harwell
Oxon OX11 ORH

ABSTRACT

A consideration of transport issues forms an important aspect of the overall exercise to assess potential sites for the disposal of radio-active wastes. This paper outlines the main findings and conclusions of work carried out on the following transport aspects for four candidate near-surface repository sites for solid, low level radioactive wastes (LLW):
- logistics, costs and accident rates of road and rail transport
- feasibility of linking sites to the rail network
- non-nuclear environmental impact
- feasibility of sea transport to coastal repository sites
- feasibility of transporting very large items of LLW.

The work provided information on the scale and type of transport operation required for LLW disposal in the UK.

The main conclusions can be summarised as follows:

  i) the traffic volumes as a whole would not be excessive and would only become significant in the vicinity of the repository sites;

 ii) transport of LLW by road was the least expensive and most flexible mode;

iii) transport costs were dominated by operating costs;

 iv) it was feasible to link all the sites to the main rail network;

  v) for coastal sites sea transport of LLW including very large items of LLW was feasible combined with the appropriate road connections. The results will generally be applicable to the Nirex assessment of potential deep repository sites.

INTRODUCTION

UK Nirex Ltd has been set up by the UK nuclear industry to establish facilities for the disposal of solid, low level (LLW) and intermediate level (ILW) radioactive waste in accordance with the strategy of the government. In January 1986, four candidate sites were announced for detailed investigations into their suitability for near surface disposal. These were at Bradwell, Essex; Elstow, Bedfordshire; Fulbeck, Lincolnshire and South Killingholme, Humberside. On 1 May 1986 it was further announced by the Government that only LLW should be considered for near surface disposal.

Nirex carried out a wide range of studies in order to assess the suitability of these sites. These included a number of studies on transport to provide input to the selection of a preferred site, to identify preferred transport systems and to assess the impact of transport. The work was well advanced when it was announced on 1 May 1987 that LLW would be disposed of in a deep repository along with ILW and the near surface site assessments were discontinued. This paper outlines the findings and discusses the conclusions of five transport studies commissioned by Nirex covering the following topics:

  i) The logistics, costs and accident rates of road and rail transport (Hunting Engineering Ltd, 1987),
 ii) The feasibility of linking the sites to the rail network (British Rail, 1987),
iii) The non-nuclear environmental impacts of transport (JMP Consultants Ltd, 1987),
 iv) The feasibility of sea transport to Bradwell and Killingholme (Nuclear Transport Ltd/British Nuclear Fuels plc/James Fisher & Sons plc, 1987),
  v) The feasibility of transporting intact heat exchangers from the decommissioning of the Magnox reactors by sea to Bradwell and Killingholme (Burness, Corlett and Partners, 1987).

The studies concentrated on the transport of LLW; however, the transport of repository construction materials, spoil from construction and personnel were also considered.

For these studies the following background information and general assumptions were used. This information was of a preliminary nature and was subject to change as the overall repository project was developed.

LLW was considered in two broad categories; that arising from the routine operation of nuclear facilities (operational wastes) and that arising from the dismantling of facilities (decommissioning wastes). It is appropriate to consider these as separate categories since decommissioning wastes are generally bulkier in nature than operational wastes and thus require larger, heavier packages. An option for certain decommissioning wastes, such as Magnox reactor heat exchangers, is to transport and dispose of them intact, rather than to cut them up and package them.

The quantities of these two categories considered in the transport studies were:-

- operational LLW : 13,000 m^3 per year (excludes wastes arising from the reprocessing of irradiated fuel).

- decommissioning LLW : 150,000 m^3 in total (only includes wastes from Stage II decommissioning of Magnox reactors).

These figures are based on the 1985 UK Radioactive Waste Inventory (Fletcher et al, 1986).

It was assumed for the studies that the operational LLW would be packaged, where possible, in 200-litre drums otherwise in 5 m^3 waste boxes; decommissioning LLW would be either packaged in 25 m^3 waste boxes weighing up to 120 tonnes or transported as intact items. For operational LLW 68% was considered suitable for packing in drums. The 200-litre drums would be loaded sixty at a time into ISO type freight containers for transport. Thus the packages to be transported are freight containers loaded with drums, waste boxes (two sizes) or intact large items.

The LLW was assumed to be produced at some 24 sites throughout the UK and was to be transported to a near surface repository at one of the four candidate sites. The locations of the major waste producing sites and the four candidate repository sites are given in Figure 1 and a breakdown of the LLW inventory assumed by site is given in Table 1. Subject to size and weight limitations, rail and road were considered possible modes for transport to all four sites. In addition two of the sites, Bradwell and Killingholme are coastal and thus sea was considered as a possible mode of transport to these.

It should be noted that for rail transport operations an element of road transport may still be involved to transfer the packages from the producing sites to the railheads and from the railheads to the repository. Also for sea transport there will be an element of road or rail transportation to the port of loading.

The areas of work introduced above are now briefly outlined and their main conclusions presented.

Fig. 1. Location of Waste Producing and Repository Sites

Table 1. Assumed LLW Total Arising to 2030

Site of Arising	ANNUAL Operational Arisings			TOTAL Decommissioning Arisings	
	Volume (m^3)	Container loads	Waste boxes	Volume (m^3)	Waste boxes
Amersham International					
Amersham	1500	133	60	–	–
Cardiff	500	38	20	–	–
BNFL					
Calder Hall	100	5	16	–	–
Capenhurst	100	5	16	–	–
Chapelcross	100	5	16	–	–
Springfields	2700	159	320	22700	908
CEGB					
Berkeley (inc.labs)	300	17	40	7500	300
Bradwell	200	10	32	7000	280
Dungeness	400	20	64	15000	600
Hartlepool	200	10	32	4000	160
Heysham	400	20	64	8000	320
Hinkley Point	400	20	64	16000	640
Oldbury	200	10	32	11500	460
Sizewell	200	10	32	12000	480
Trawsfynydd	200	10	32	11000	440
Wylfa	200	10	32	23500	940
SSEB					
Hunterston	400	20	64	12400	496
Torness	200	10	32	4000	160
UKAEA					
Dounreay	100	5	16	–	–
Harwell Group	2800	114	576	–	–
Winfrith	800	50	80	–	–
NDS					
Calder Hall	600	50	0	–	–
Harwell Group	100	9	0	–	–
MoD					
Rosyth	100	8	4	–	–
Devonport	100	8	4	–	–
Faslane	100	8	4	–	–
TOTALS	13000	744	1652	154600	6184

Note: Annual arising data based on 1985 UK Inventory. Decommissioning
arising based on Nirex Estimates. This data has now been updated.

OPERATIONAL CHARACTERISTICS OF ROAD AND RAIL TRANSPORT

The operational characteristics of road and rail transport were studied in order to:

- identify potential routes from waste producing sites to the repository sites,
- evaluate transport logistics, equipment needs and costs,
- evaluate traffic densities,
- estimate road and rail accident rates.

Specific assumptions made in addition to the general assumption given above are as follows.

The operating life of the repository was taken as 40 years and over this period the system would transport a minimum of 90% of the available waste from each site in each year.

For road transport a tractor and trailer unit to comply with the gross vehicle weight limit of 38 tonnes and lorry payloads of either one freight container or up to two 5 m^3 waste boxes, with an average of 1.5 boxes were assumed. In addition, to study the transport of decommissioning waste by road it was assumed that the waste was size reduced to fit within 5 m^3 waste boxes.

For rail transport, a 'Block Train' system was assumed and trains travelled directly between the nearest railheads at each waste producing site and each candidate repository site. For operational LLW, trainloads comprised of either 8 or 12 four-axle wagons, dependent on the rate of LLW production at each site, and for decommissioning LLW, 4 eight-axle wagon trainloads were assumed. The waste producing sites requiring 12 wagon trains were Springfields, Winfrith, Harwell Group and Amersham. Wagon payloads were assumed to be either:

- one freight container,
- three 5 m^3 waste boxes or
- one 25 m^3 waste box.

Data on the nearest railheads and the most probable rail routes to be taken were provided by British Rail. The railheads are listed in Table 2. For road transport, routes were selected from suitable road maps. These routes were reasonably direct and maximised the use of motorways and dual carriageways.

Having established the routes a simulation model written in ECSL (Extended Control and Simulation Language) was used to evaluate the logistics of suitable transport systems and thereby determine the transport equipment requirements and costs, together with route traffic densities and weekly arrival rates of LLW at the candidate repository sites.

The model provided histograms of arrival rates which indicated an average of 3.5 trainloads per week or 47 lorry loads per week. An evaluation of costs showed that road transport offers a significant cost advantage over rail. The average cost to transport LLW by road was estimated as £110/m^3 compared to an average cost to transport by rail of £350/m^3. Total transport costs for the 40 year period had marginal differences between the four candidate sites ranging from total costs for

Table 2. Suitable Railheads and Seaports for Despatch of Radioactive
         Waste

Site of Arising	Railhead	Seaport
Amersham International		
Amersham	High Wycombe	Sharpness
Cardiff	Cardiff	Cardiff
BNFL		
Calder Hall	Sellafield	Workington
Capenhurst	Chester	Ellesmere Port
Chapelcross	Annan	Workington
Springfields	Springfields	Heysham
CEGB		
Berkeley	Berkeley Road	Sharpness
Bradwell	Southminster	Lowestoft
Dungeness	Lydd	Folkestone
Hartlepool	Hartlepool	Hartlepool
Heysham	Heysham	Heysham
Hinkley Point	Bridgwater	Sharpness
Oldbury	Berkeley Road	Sharpness
Sizewell	Leiston	Lowestoft
Trawsfynydd	Trawsfynydd	Holyhead
Wylfa	Valley	Holyhead
SSEB		
Hunterston	Fairlie	Ardrossan
Torness	Torness	Torness
UKAEA		
Dounreay	Georgemas	Scrabster
Harwell Group	Didcot	Sharpness
Winfrith	Winfrith	Poole
NDS		
Calder Hall	Sellafield	Workington
Harwell Group	Didcot	Sharpness

Note: The ports listed were assumed for study purposes only and
      do not represent a final choice.

rail from £221M for Elstow to £253M for Killingholme and for road from
£62M for Elstow to £87M for Killingholme.

Over the 40 year period costs were dominated by the operating costs
as opposed to the capital costs. In the case of rail operating costs were
over 90% of total costs whilst for road the percentage was over 85%.

In addition to the transport assessment, accident data were compiled
to provide preliminary information for a risk assessment study. Overall
it was found that relevent accident data were not easily obtainable and
general statistics had to be interpreted to give useful accident
probabilities which in turn will provide meaningful risk estimates for
both rail and road. The information that was available indicated marginal

differences in the accident rates for each mode of transport between the four sites.

RAIL FACILITY IMPROVEMENT FEASIBILITY

There are significant advantages in having a direct rail link into a repository site. The feasibility of providing a link to the four candidate sites together with the route and cost were examined. Alternative routes were considered and a preferred one selected taking into account the ease and cost of construction, operational flexibility, impact on local communities and other environmental considerations.

The design criteria applied were single line, new jointed track with a 25 tonne axle load, a maximum speed of 70 kmph, a maximum curvature of 350 m (250 m at junctions) and a maximum gradient of 1 in 20.0. For Bradwell the new spur would have been 13 km long involving both embankments and cuttings but no major bridges or tunnels. The overall cost was estimated at £8.5M and construction time at 27-33 months. For Fulbeck the new spur would have been 5km long over relatively 'flat' terrain. The overall cost was estimated at £2.6M and construction time at 15-18 months. Rail links to Elstow and Killingholme were to be very short and presented no difficulties. A Private Parliamentary Bill provides the preferred method of obtaining planning consent for a new rail link.

As an alternative to a direct rail link to the near surface facility, the requirements for improving terminals for unloading at Newark (for Fulbeck) and Southminster (for Bradwell) were examined. The improved terminals would provide facilities up to 300 metres long in the form of a reception siding, a departure siding and a run round siding, all clear of the main line. The estimated cost for such improvements at both terminals was £0.7M plus the cost of suitable gantry cranes which were estimated to cost up to £0.3M each.

NON-NUCLEAR ENVIRONMENTAL EFFECTS OF TRANSPORT

The non-nuclear environmental aspects of transport are an important aspect of the overall transport assessment. It covers the environmental effects (other than radiological) of the construction of new transport facilities and the operation of a transport system. The factors to be considered include: land take, pollution (air pollution, noise, vibration and visual intrusion), community severance, land severance, ecology, social concern plus road widths, gradients, bends, junctions, etc. The objectives of this work were to assess the environmental aspects of rail and road transport for the four candidate repository sites. This phase of the work concentrated on the impact of transport in the locality of each candidate site and considered movement of waste, construction and spoil material and people. Sea as a mode of transport was considered in less detail as part of the sea transport feasibility study at this stage.

In the case of rail transport it was assumed that the final leg to the site could either be by road from the nearest railhead or via an extension of the main line to the site. It was also assumed that some 70,000 tonnes per annum of construction material (about 90 lorry loads per week) were moved to the site during an initial 2 year period falling to 40,000 tonnes per annum after that. Spoil was assumed to remain at site for the first 2 years and after that an amount equivalent to some 80 lorry loads per week was required to be moved. Initial estimates indicate a staff of about 200, combining this with visitors to the site it was assumed that there would be some 300 two-way personnel movements daily.

The first stage of the work considered a number of different route options for each mode from the main UK trunk road or rail network and assessed their relative environmental impacts using a recognised assessment framework. The second stage compared selected options for each site and each mode for the four candidate sites.

All four candidate sites could be accessed by both road and rail, though with varying degrees of financial cost and environmental impact. In the case of road transport at Fulbeck a new access road to the site bypassing one of the local villages was identified as a means of reducing local impact. In the case of rail, it was considered that extensions to each site would minimise the environmental impact from transport movements. The rail alignments to the various sites were examined in the parallel exercise discussed earlier.

The existing high quality roadways for industrial areas around Immingham docks, and the high traffic levels of heavy goods vehicles (HGVs) means that road transport for Killingholme would have a very small environmental impact.

For rail Elstow and Killingholme both would require relatively low outlay on a new, short spur, should provoke little environmental impact and, with careful scheduling, could have been catered for by the existing network.

SEA TRANSPORT FEASIBILITY

Bradwell and Killingholme are coastal sites and therefore provide the option of sea transport of wastes. The feasibility of using sea transport to these sites has been assessed. The complete transport system for packaged LLW from sites of arising to the repository was addressed and included consideration of the transport of LLW from each waste producing site by road and/or rail via appropriate routes to a nearby seaport, loading of the transport packages into a suitable ship for transportation to either of the two sites in question and off-loading the packages at a berth at or close to the repository.

For the land transport part of the system road transport was selected since many of the waste producing sites, the loading ports and the repository sites have either limited or no rail facilities and transport to nearby railheads involves additional handling operations.

A number of ports around the coast of the United Kingdom were considered and a final list of suitable ports selected for this study on the following basis:

- proximity to waste producing site,
- reducing land journey distance,
- public and environmental issues,
- history of radioactive material traffic.

Thirteen ports were selected (see Table 2) each of which could adequately accommodate the ships indicated for the service and are equipped with the necessary facilities and labour force to ensure efficient and safe cargo operations either by roll-on/roll-off or lift-on/lift-off. In the main, major ports were not selected in favour of ports with lower sea traffic densities. Also it was considered that a smaller port had the ability to provide a dedicated service on a regular basis.

Suitable road routes between the waste producing site and their loading ports were selected bearing in mind local population and traffic densities, use of most appropriate roads and journey distances.

Barges, hovercraft and ships were all examined as potential vessels. The tow line associated with towed barges is susceptible to 'breaking', resulting in the barge becoming adrift with no means of propulsion. Therefore a self propelled vessel was considered desirable and towed barges were discounted. However a pushed barge system and barge carrying ships were also considered feasible. Hovercraft were considered unsuitable due to a number of factors including size/weight ratio to cargo capacity, high fuel consumption and noise levels.

Either a lift-on/lift-off ship with an on-board crane or a roll-on/roll-off ship with stern ramp were judged suitable. The ships considered had been designed and built for other specific purposes and it was felt prudent to consider a purpose built ship for carrying LLW packages.

A conceptual design was prepared which combined the features of a lift-on/lift-off and roll-on/roll-off type of ship. The conceptual ship which is illustrated in Figure 2 was approximately 2500 tonnes deadweight having principal dimensions of 100m x 16m x 4m draught. Such a ship could be accommodated in the selected loading ports.

Bradwell does not have a suitable port in the vicinity. Felixstowe is the nearest feasible port, being some 65 miles distant, but is not suitable for the ship design proposed. Therefore a specially constructed, dedicated berth would provide the best option.

Immingham is the nearest port to Killingholme and could accommodate the proposed ship. Alternatively a dedicated berth could be constructed at Killingholme.

The sea transport of cargoes of aggregates, cement or spoils is also feasible. Further ships and unloading facilities will be required for these materials.

Having established the technical feasibility of sea transport and having identified a suitable system albeit in preliminary and conceptual terms, a preliminary transport plan was worked up which provided an example of how the system could operate in practice.

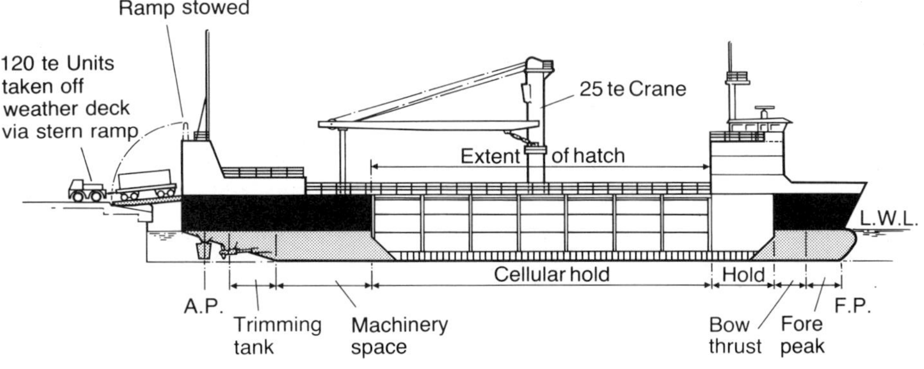

Fig. 2. Conceptual design of ship suitable for LLW transport.

The schedule devised indicates that UK arisings of LLW could be readily transported by the proposed system. This plan was also used to provide a preliminary estimate of relevant transport accident frequencies and to give a preliminary indication of the associated environmental aspects. Finally the schedule was used to draw up a business plan covering the capital and operating aspects of cost. The cost estimates indicate that the split between capital cost and operating cost is dependent on whether the ships are owned or chartered. For the transport of both operational and decommissioning LLW the cost per unit volume ranges from:

$$c. \quad £200/m^3 \text{ to } £300/m^3.$$

TRANSPORTATION OF WHOLE MAGNOX HEAT EXCHANGERS

The stage II decommissioning of Magnox reactors includes the dismantling for disposal of the heat exchangers (on stations where they are external to the biological shield). The waste management options are either to dispose of them intact filled with grout, or to cut them up and package them. The latter would probably entail significant radiation doses to the workforce due to the presence of internal contamination. The transport and disposal of intact heat exchangers therefore offers significant advantages.

Due to their large size transport of intact items over any significant distances is only practical by sea. The feasibility of transporting them from coastal Magnox stations by sea to Bradwell and Killingholme was investigated.

The largest of the Magnox heat exchangers is some 27 metres long, 6.7 metres in diameter and weighs up to 2000 tonnes when filled with grout and prepared for transport. Approximately 100 heat exchangers would arise from the Magnox reactor decommissioning programme over a period of ten to fifteen years. In the study seven power station sites were considered: Berkeley, Calder Hall, Chapelcross, Dungeness, Hinkley Point, Hunterston and Sizewell.

Potentially suitable locations for unloading berths at Bradwell and Killingholme were identified, together with the berthing and handling facilities required, preliminary operational, regulatory, environmental, radiological, safety and budget costing aspects of the transport cycle.

The approach taken for the study was to divide the total transport cycle into a number of discrete activities:

- Dismounting the heat exchangers from their installed position and loading onto heavy transporters;
- Transportation from installation point to suitable loading facility and loading onto sea transportation units;
- Sea transport to suitable berth at Bradwell or Killingholme;
- Transportation from the berth to the disposal site;
- Off-loading from transporters at repositories.

The findings of the study can be summarised as follows:

a) Proven equipment is available for the lifting of large individual loads of the order of 2000 tonnes from four heavy lift specialists using the mast lifting system. These utilise modular lattice masts with hydraulics to provide the motive jacking force. When lifted, the heat exchanger can be positioned above the transporter units and lowered from the vertical to the horizontal position using suitable tailing equipment.

b) From a consideration of five possible methods of transporting from the installation points, the use of custom designed self propelled transporters is preferred.

c) Of the seven power station sites considered all except Chapelcross are close to an open coastline and do not need to use public roads. Hence they would be suitable for the transfer of heat exchangers to sea going vessels on self propelled transporters. Chapelcross is approximately 3 miles inland and would be impractical because the size and manoeuvering characteristics of the transporters will require a substantial upgrading of any public roads.

d) While all sites are adjacent to water of restricted depth most are within easy reach of deeper water for navigation and thus with the right selection of sea transport unit can be used for loading.

e) The main ship options considered were:

-   heavy lift ship
-   roll-on, roll-off ship
-   dock lift ship
-   semi-submersible heavy lift ship
-   barge
-   purpose built vessel.

The ship selection will be dependent on many factors including safety, cost, availability, speed etc. The towed barge option offers the most cost effective system using readily available equipment. A push-tow arrangement for a barge or semi-submersible barge carrier or a purpose built roll-on, roll-off vessel would provide progressively more expensive options with lower accident rates.

f) Both Bradwell and Killingholme are considered suitable sites for the reception of incoming barge cargoes by sea. A number of feasible options for each site were considered.

g) From a brief radiological impact assessment it was concluded that there are low levels of radiological exposure to the work force and insignificant radiological risk to the public.

h) A budget estimate of the operational transport costs for each heat exchanger was found to average between £220,000 and £300,000 per unit. Berth facilities at each of the loading sites will require a capital cost of £1.7M.

At the unloading site the total civil costs for Bradwell and Killingholme were estimated as £1.85M and £2.95M respectively. These berthing facilities can be used for the handling and transport of other materials such as large boxes of decommissioning waste.

Overall this preliminary study established that the concept of transporting intact decommissioned heat exchangers was feasible, insofar as the equipment necessary for such an operation is available and proven and that in general the power station sites and the repository sites are suitable both topographically and hydrographically for a sea transport operation. The vessels could also be used for the transport of the large boxes of decommissioning waste.

DISCUSSION AND CONCLUSIONS

The work carried out in these studies identified the scale of system and operations required for transport, and provided a suitable input to an overall assessment of the sites.

With regard to road transport, it is clear that road access is a requirement for any site. Personnel and some construction materials will need to travel by road. A road transport system for LLW is inherently flexible, which would be of benefit to small waste producers. Road is the cheapest mode of transport. However, it is unsuitable for some of the decommissioning wastes.

Rail transport lacks the flexibility of road transport and, based on the budget cost estimates, is about three times as expensive. However its use does reduce the number of vehicle movements into a repository site and it does have the capability of carrying heavier waste packages than would be possible by road transport.

Sea transport is operationally inflexible due to the small number and large size of vessels. The costs are less than the indicative costs of rail transport. The system identified is one which minimised the land transport element and consequently is a complex operation involving a significant number of ports. Sea transport however does have a far greater load carrying capability than road or rail transport and the sea leg of the operation is essentially out of the public domain and thus has a low environmental impact.

The main conclusions about transport to the four candidate near surface repository sites that can be drawn from the work are the following:

1.   It would be feasible to transport to the LLW to all four sites by road and, for the two coastal sites, by sea. In addition, the transport of construction materials, spoil and personnel would not present problems.

2.   The levels of traffic (i.e. for waste, spoil and construction materials) associated with an all road transport system for a near surface repository for LLW would be about 200 lorry loads per week, of which about 40 are waste. This is comparatively low and also would have minimal environmental impact.

3.   The differences between the sites would be small in terms of overall vehicle miles (since all four sites are in Eastern England). The transport costs and the accident frequencies would therefore be similar between sites for a particular mode of transport. The associated radiological risks of transport are expected to be very low.

4.   The 1987 average cost for road transport is $£110/m^3$ compared to $£350/m^3$ for rail and between $£200/m^3$ and $£300/m^3$ for sea. Capital costs comprise 10-15% of these figures.

5.   The road routes to the site would be on motorway and 'A' roads where possible. These classes form over 90% of the total route miles in all cases.

6.   Local road access routes have been examined for each site. Their environmental impact has been assessed and routes with the minimum environmental impact have been identified.

7. There are advantages in providing a direct rail link to a repository site particularly for the transport of decommissioning wastes. This would be feasible for all the candidate sites. The links required would range in length from 0.7km (Elstow) to 13 km (Bradwell). The total cost of new track would be of the order of £0.6M per kilometre.

8. It would be feasible to establish a sea transport system to both Bradwell and Killingholme. A suitable ship design and unloading facilities in the immediate vicinity of the sites have been identified. It would be necessary to use about 13 ports in order to minimise land transport.

9. The transport of intact Magnox reactor heat exchangers by sea direct to Bradwell or Killingholme would be feasible for all reactor sites with the probable exceptions of Trawsfynydd and Chapelcross. Suitable commercially available equipment has been identified for overland and sea transport. The 1987 operational cost of transport is £220,000 – £300,000 per heat exchanger. Minimal berthing facilities would be required for loading and unloading.

10. In general terms road transport is the cheapest and most flexible system and road access would be required to any site for certain materials. Rail transport is expensive but has the heavier load capability whilst sea transport can cope with very heavy loads, may be cheaper than rail, but is the least flexible mode.

Although the studies carried out were extensive, further work was planned to:

i) assess transport risks (radiological and conventional) for all sites and modes of transport;

ii) bring the result of all the transport studies together into an assessment framework in order to identify preferred modes of transport to provide an input to site selection.

Nirex is now considering sites for the construction of one deep repository for the disposal of both ILW and LLW. Similar studies to those carried out for the candidate near-surface repository sites will be undertaken and will make use of much of the information produced from the near surface repository transport studies. Transport as a factor in the assessment and selection of sites will increase in importance if remote mainland or island sites are considered, due to the greater distances involved and the wider range of geographical locations.

ACKNOWLEDGEMENT

The work reported in this document was carried out under contract to UK Nirex Limited by Hunting Engineering Limited, British Rail, JMP Consultants, Nuclear Transport Limited/British Nuclear Fuels plc/James Fisher & Sons plc, and Burness Corlett and Partners. The authors would like to thank those contractors and our colleagues in UK Nirex Limited for their assistance in carrying out this work.

REFERENCES

British Rail, 1987, Rail Facility Improvement and Feasibility Study _in_ Nirex Report No. 40.
Burness, Corlett and Partners, 1987, Transportation of Whole Magnox Heat

Exchangers by Sea to Bradwell and Killingholme - A Preliminary Feasibility Study <u>in</u> Nirex Report No. 39.

Fletcher, A.M., Wear, F.J., Haselden, H., Shepard, J. and Tymons, B.J., 1986, The 1985 United Kingdom Radioactive Waste Inventory, <u>in</u> DoE Report No. DOE/RW/86/087

Hunting Engineering Limited, 1987, Transport Study Associated with the Disposal of LLW in Near Surface Repositories - Rail and Road, <u>in</u> Nirex Report No. 38.

JMP Consultants Limited, 1987, Transport Assessment - Road and Rail Near Surface Repository Project - Non Nuclear Environment Assessment, <u>in</u> Nirex Report No.41

Nuclear Transport Limited/British Nuclear Fuels plc/James Fisher & Sons plc, 1987, Feasibility Study for Transporting Low Level Radioactive Waste by Sea, <u>in</u> Nirex Report No. 37.

EXPERIENCE IN THE TRANSPORT OF SPENT NUCLEAR FUEL

J.E.Middleton[1], S.Blackburn[2]

[1]British Nuclear Fuels plc
& Pacific Nuclear Transport Limited
Risley
Warrington
Cheshire WA3 6AS

[2]Nuclear Transport Limited
Risley
Warrington
Cheshire WA3 6AS

INTRODUCTION

Unique experience has been gained by three Companies, BNFL, PNTL and NTL, through the transport of 8500 tonnes of Magnox and Oxide spent fuel over the past 20 years. These transports have taken place from the reactor sites in eight European countries and Japan to the reprocessing plants at Sellafield and La Hague.

The formation, composition and purpose of the Companies, stemming from their predecessors in the early 1950s, is explained. Transport routes have been developed crossing national boundaries in Europe and using road, rail and sea. For large quantities from Japan, crossing the Oceans to Europe, special-purpose ships have been developed.

The division of responsibilities between the Companies has produced specialist experience which is complementary in covering the total scope of alternative transport methods. Principal requirements to be met, the designs of spent fuel and flasks are described briefly. A summary is given of the quantities of fuel transported without significant incident, indicating the high safety standards achieved.

Detailed consideration is given to flask handling and checking methods with examples of the early difficulties encountered and their solutions. From this it is deduced that equipment and methods have reached an advanced stage of development and are consistent with the sensitivity of the industry.

Spent fuel transport is an essential part of the nuclear fuel cycle and continues to expand in quantity. A policy of continuous improvement is necessary and standards must be maintained at the high levels established, to ensure the continuing safety and success of nuclear power in the future.

British Nuclear Fuels plc (BNFL), its subsidiary company Pacific Nuclear Transport Limited (PNTL) and its associated company Nuclear

Transport Limited (NTL) have been responsible for most of the international transport of spent nuclear fuel which has taken place in the world to date. Considerable quantities have been transported safely over more than twenty years and unique experience has been accumulated.

The companies have adopted a common policy and developed similar procedures which are applied throughout the transport operations and at the reactor and reprocessing sites. This requires a comprehensive quality plan, clearance of designs and procedures through a safety committee structure and witnessing of operations by specialist engineers to ensure continued compliance at all stages.

This paper describes the services and experience of these international companies including types of fuel and container transported, the modes and logistics of the operations and arangements necessary to maintain high standards of safety and alleviate public concern.

COMPOSITION AND PURPOSE OF THE COMPANIES

BNFL and the French Company, Companie Generale des Matières Nucléaire Atomique (COGEMA) are the world's foremost nuclear fuel services companies. Spent fuel has been transported to the reprocessing plants at Sellafield in Great Britain and La Hague in France, respectively.

Since few of the power generation utilities wish to be responsible for the spent fuel after it leaves the reactor sites, the reprocessor must be able to offer to transport the fuel. The quality of the transport service is of utmost importance to the reprocessor since it takes places in the public domain and can potentially jeopardise the future of the industry.

The reprocessors were therefore anxious to ensure that they could rely on a uniformly high standard of transport which would provide economic transport prices based on optimum flask utilisation. For this reason BNFL and the French and German flask design companies, Transnucléaire and Transnuklear concluded that the interests of the European nuclear industry as a whole would best be served if they pooled their existing flasks and investment plans into one company, Nuclear Transport Limited, which would specialise in spent fuel transport.

NTL was formed in 1972 as a Company registered in Great Britain with an equal one-third shareholding by BNFL, Transnucléaire and Transnuklear, and has been responsible for extensive transport of oxide spent fuel in Europe since this date.

The United Kingdom Atomic Energy Authority (UKAEA), of which the Production Group was later to become BNFL, first began spent fuel transport in the early 1950s. Shipments from overseas to Sellafield began in 1966 with Magnox fuel from the Latina reactor in Italy and is still continuing today using a BNFL special purpose ship. The main involvement by BNFL is through PNTL which was incorporated as a British Company in 1975 specifically to provide transport services for magnox and oxide spent fuel from Japan to Europe.

BNFL are the majority shareholder in PNTL with 62.5%. COGEMA hold 12.5% of the shares and the remaining 25% are held by a consortium of Japanese power utilities and trading companies.

TRANSPORT ROUTES

     Fuel is transported to the reprocessing plants at Sellafield and La
Hague from European and Japanese reactors and the geographic locations
impose the need for flexibility in transport mode using road, rail and
sea.

     First transports of European LWR spent fuel took place in 1968.
Garigliano BWR fuel was moved by road to the port of Anzio in Italy, then
by sea to the port of Barrow in Great Britain and by rail to Sellafield.
Routes to Sellafield and La Hague have since been established from eight
countries, viz. Sweden, Holland, Germany, France, Belgium, Switzerland,
Spain and Italy. Reactor and port locations are shown in Figure 1.
Transports to La Hague are mainly by rail, sometimes crossing several
national boundaries, to the COGEMA rail terminal at Valognes. The last
stage, approximately 60 km, is by road.

Fig. 1. Reactor locations in Europe served by NTL.

Some of the transport to Sellafield has been by road to ports in Italy, Sweden and Spain, thence by chartered ship to Barrow and lastly by rail. Movements from Germany, Belgium and Switzerland have been by rail to either Zeebrugge in Belgium or Dunkirk in France, crossing the Channel on passenger or freight ferries to the British ports of Harwich or Dover, then continuing by rail. The route from the reactor in Holland has been by road to Rotterdam, by passenger or freight across the North Sea to the British port of Hull and then by rail. Road transport to Sellafield in Great Britain were ceased in 1978 due to unsuitability of access roads in the County of Cumbria.

Movements from Europe must involve the use of a ship. In the past, extensive use was made of roll-on, roll-off ferries, which also carried private cars, passengers and freight. Although no fears were expressed by passengers, for public relations reasons this practice has now ceased and freight-only ships are used. The present preferred routes for short-haul sea transport from Europe to Great Britain are rail-freight-only ferry, Dunkirk to Dover, and road-only ferry from Rotterdam to Immingham (near Hull) where the transfer from road to rail conveyance is made.

Transports from Japan consist of a short road journey from reactor to reactor port (or to a local commercial port, in one case about 10km). The long sea voyage is made using special purpose ships to the ports of Cherbourg in France and Barrow in Great Britain. In France, the flasks are taken by rail to Valognes and then by road to La Hague. In Great Britain the flasks are off-loaded onto rail wagons for transport to Sellafield. The preferred sea-route is via the Panama Canal which is the shortest distance and gives the most clement weather, but voyages have been made via the Cape of Good Hope and this remains a viable alternative. Reactor locations in Japan are shown in Figure 2.

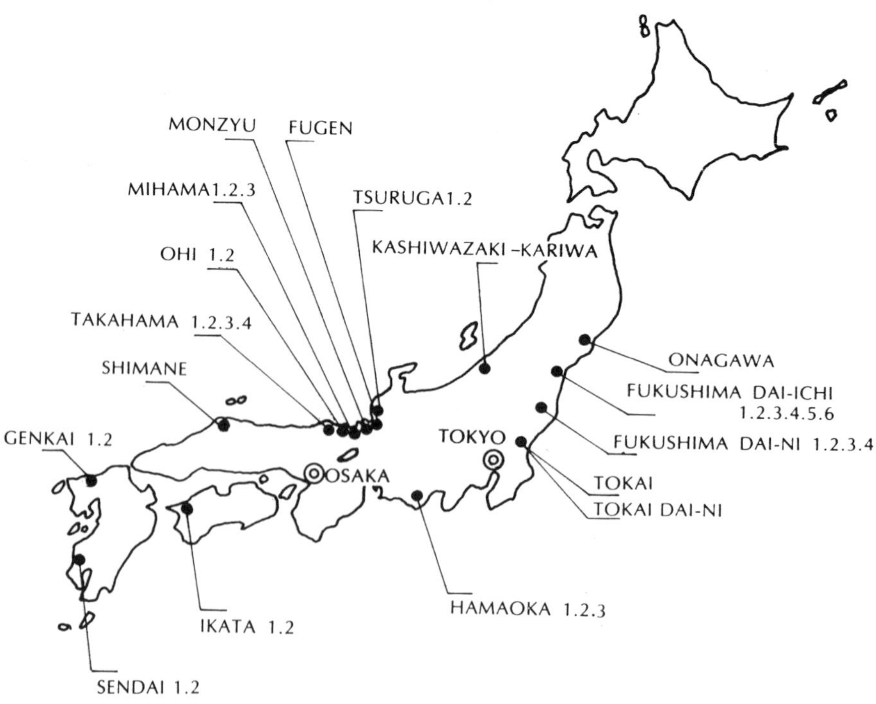

Fig. 2. Reactor locations in Japan served by PNTL.

The division of responsibilities between the companies has produced specialist experience which covers the total scope of the alternative transport systems. NTL has been dominant for transport of uranium oxide fuel within Europe, involving complex road and rail journeys across many national boundaries with the use of ferry vessels for short-haul by sea to Great Britain. BNFL and PNTL have specialised in the movement of large quantities of both oxide and magnox fuel by sea over long distances with only short road and rail journeys in Great Britain.

In Europe, spent fuel transport commenced with flasks of relatively simple design, low capacity and relatively low overall flask weights ranging from 30 to 50 tonnes, and road transport was quite common for this range of flasks. With the advancement of modern-day requirements for higher and more economical pay load, higher fuel burn-up and decay heat, the development of flasks in the range 70 to 110 tonnes has been necessary. Special purpose rail wagons are necessary which are designed to transport the heaviest of modern flasks and these have become the main mode of land transport.

The special purpose rail wagons are built to international standards and may include a locked canopy covering over the flask (Figure 3). Under normal circumstances the wagon and flask proceed along an uninterrupted journey to its destination attached to a normal freight train. It is therefore approved to travel at speeds of up to 100 kilometres per hour.

The flask contents are classified as Dangerous Goods and as such the appropriate regulations are applied for both land and sea movements of the flask. This restricts the number of flask-loaded wagons to a maximum of two per rail-freight-ferry crossing.

Prior to despatch from the reactor or reprocessing plant, each flask and rail wagon must be thoroughly checked and certified by health physics, safety and quality assurance experts. On arrival at its destination and before removal of the flask from its rail wagon, a full

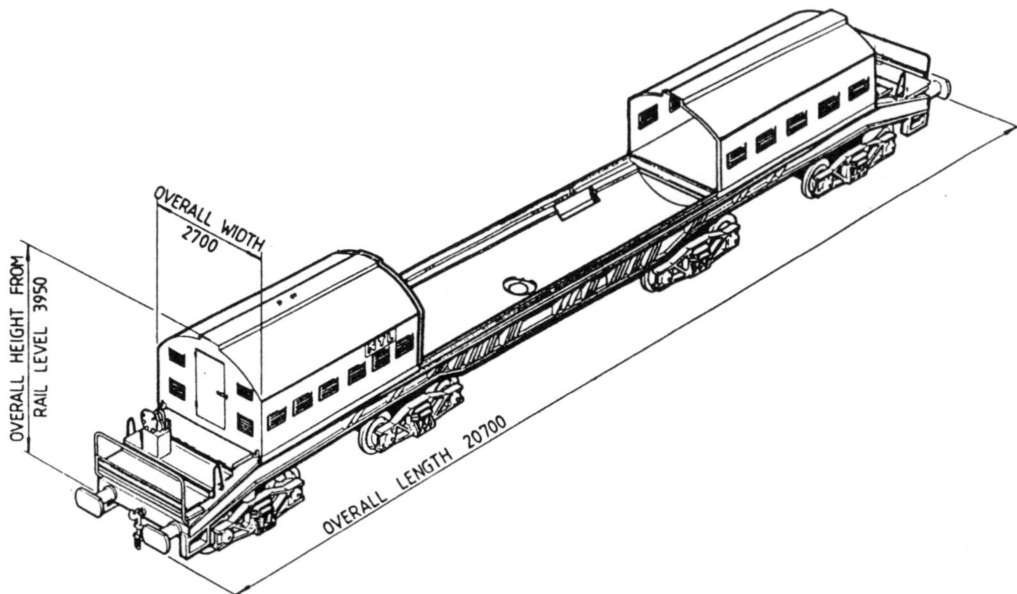

Fig. 3. Special purpose rail wagon.

radiological survey is again required. If during the journey it is necessary to perform an inter-modal transfer of the flask, eg. road to rail or vice versa, then health physics surveillance is again mandatory.

These international transport operations require an international organisation with offices or close collaborative arrangements in each of the major countries concerned. It is essential in each country to create a national image, to establish confidence and form close relationships with all concerned and particularly with those whose cooperation is vital. Many who become involved may have no practical or commercial interest in the success of the operation, eg. police, local authorities, railway officials, station masters, port authorities, border controls, customs and excise. It is detailed communications, relationships and understandings with these groups which have contributed to the safe and calm progress of international spent fuel transport.

With the signing of contracts by the Japanese utilities for large quantities of spent fuel to be transported to both Sellafield and La Hague, the development of a safe and publicly acceptable system became necessary for this purpose. There are no international regulations applicable to the design of spent fuel carrier ships and at that time there were no such ships.

BNFL/PNTL sought the advice of the Salvage Association in London, (which has a wide experience of the causes of marine disasters), on those features which could be incorporated into the design of a special ship to minimise hazards at sea. About the same time, the Japanese Ministry of Transport were independently formulating regulations to be applicable to this type of ship operating in Japanese Territorial Waters. From these deliberations a new class of ship was designed (Reference 1). The first, the Pacific Swan, made her first voyage (round the world) in 1979, followed by four sister ships, Pacific Crane, Pacific Teal, Pacific Sandpiper and Pacific Pintail.

To maintain a uniform high standard, the charter ships previously used for movements of fuel from Italy were replaced by a further special-purpose vessel, the Mediterranean Shearwater, which though smaller than the Pacific ships has all the same safety features. This vessel completed the present BNFL/PNTL combined fleet of six ships.

With the increased scale of operations came the necessity for special port facilities at Cherbourg in France and a purpose-built marine terminal at Barrow in Great Britain. The larger numbers and sizes of flasks require heavy lift cranage, rail wagons and vehicles to complete the land transports to the reprocessing plants. BNFL/PNTL/NTL operate over 200 flasks and twenty special rail wagons.

Since the ships operate on a 30,000 mile round trip and for long periods are over a thousand miles from land, it has been necessary to develop comprehensive emergency procedures and contingency plans both for flask radiological protection and for marine services. The former utilises the skills of BNFL experts based in Great Britain and similar arrangements in Japan from the expertise of the nuclear power utilities. For the latter the support of the world's leading salvor, Smit International of Rotterdam has been engaged.

The major transporters in Great Britain (BNFL, PNTL, NTL, UKAEA, the utilities CEGB and SSEB) have combined their resources with those of the local community emergency services to form the Irradiated Fuel Transport Flask Emergency Plan (IFTFEP). This provides a rapid radiological monitoring response followed by comprehensive flask handling capability.

Both sea and land emergency procedures are rehearsed at frequent intervals to ensure a high state of preparedness at all times.

FLASK DESIGN CRITERIA

Principal requirements for any spent fuel transport flask are:

<u>Radiation Shielding</u> - To reduce levels of radiation external to the flask to the very low levels necessary to meet regulatory limits and protect transport workers and the general public.

<u>Criticality</u> - The fuel support frame must be of a material and hold its cargo in such a way that in both normal operation and in accident conditions the array so formed is always sub-critical.

<u>Containment</u> - A high standard of leak-tightness must be achievable to contain gaseous and particulate radioactive substances.

<u>Heat Transfer</u> - Nuclear decay heat from the fuel elements must be transferred efficiently to the atmosphere, so that the temperatures of the fuel and flask are kept sufficiently low to avoid any damage to the fuel and permit safe handling and transport of the flask.

The flasks are designed to meet the requirements of the International Atomic Energy Agency (IAEA) Regulations, as defined in Safety Series 6, which specifies standards covering both normal and "accident" situations. It must be shown that the flask will withstand the impact of free fall from a height of 9 metres on to a hard, unyielding target, followed by a 30 minute duration totally enveloping fire at 800°C, without significant loss of leak tightness.

Design standards, detailed design features, material selection and manufacturing techniques are applied mostly to the British Standards range with some recourse to United States Standards. In addition the design must satisfy BNFL/PNTL/NTL Company Policy regarding construction, overall safety, and public acceptability.

Quality assurance in manufacture is to BS 5882 and 5750 Part 2, which requires precise definition of every item including material, manufacture and testing. It is the duty of the manufacturer to provide documentary evidence in support of his achievement of the specified standards. Before selection to tender for the work he must be able to demonstrate the ability to meet routinely the standards of workmanship demanded and an adequate management system to control the work. The QA system is extended to cover maintenance during the flask operational life when work must be certified by a BNFL inspector before the flask can be returned to service.

APPROVAL OF FLASK DESIGN

Before a flask can be used to transport fuel it must have "Package Design Approval" from the "Competent Authority" of each country through which it is to pass. Within Great Britain this is the Secretary of State for Transport whose responsibilities are exercised through the Transport Radiological Adviser in the Department of Transport. Approval is sought through the presentation of a Design Safety Report (DSR), which describes the design, proposed manufacturing methods and materials. The DSR also contains a detailed performance analysis as evidence of behaviour in normal and the regulatory test conditions.

SPENT FUEL DESIGNS

Magnox Fuel

A Magnox fuel element consists of a round section bar of uranium metal, about 2.5cm in diameter, encased in a close fitting cylindrical can made of Magnox, which is an alloy of magnesium and aluminium. The outer surface of the can has extensive integral finning to enhance the heat transfer characteristics from the element. The diameter over the fins is about 5cm and the length varies between 50 and 112 cm. Longitudinal metal or graphite struts are attached to give support in the reactor channel. The designs for Latina and Tokai-Mura reactors are illustrated in Figure 4.

LWR Oxide Fuel

Light water reactor (LWR) fuel is of two types which though fundamentally the same are different in detail to suit either the pressurised water reactor (PWR) or the boiling water reactor (BWR) concept.

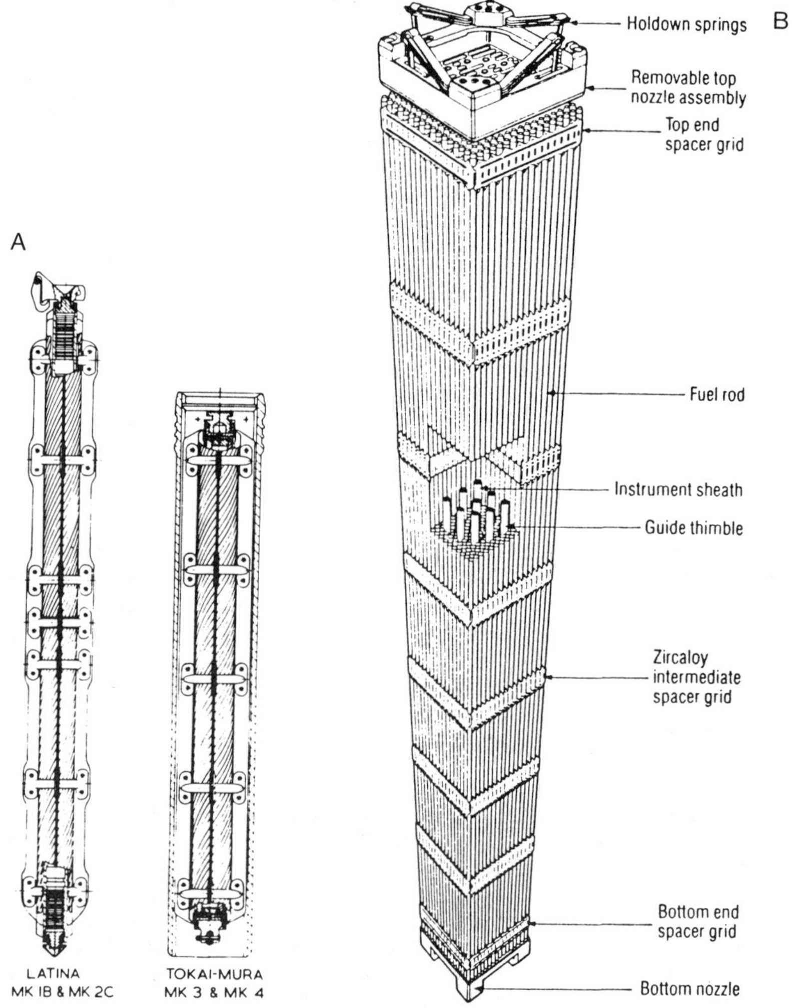

Fig. 4. Fuel element designs. (A) Magnox fuel; (B) LWR/PWR fuel.

Both types consist of a square array of fuels rods constructed from zircaloy tubing filled with uranium oxide pellets. The fuel assembly of rods is held together by top and bottom nozzles (PWR) or tieplates (BWR) with intermediate spacer grids over a total length which ranges from 3 to 5 metres. Most PWR fuel consists of a 17 x 17 array with 264 fuel rods, 24 guide thimble tubes and an instrumentation tube used for control purposes in the reactor. Rods are about 0.375 cm diameter. BWR fuel assemblies comprise 64 fuel rods held in an 8 x 8 array. A typical design is shown in Figure 4.

SPENT FUEL TRANSPORT FLASKS

Fuel and flask design criteria and the acceptance limitations of the reprocessing plants, reflecting the different policies of BNFL and COGEMA, have moulded the modern day large transport flasks into two basic types, wet flasks and dry flasks. Those used for transport to Sellafield are of the wet type and there are two designs, the Magnox flask for transport of gas reactor fuel and Excellox type flask for LWR fuel. For transport of LWR fuel to La Hague both the dry type, Transnucleaire (TNP) design and the wet "Excellox" type are used.

Magnox Flask (Wet)

This design is illustrated in Figure 5. It is used by BNFL and PNTL for transport of Latina reactor fuel from Italy and Tokai-Mura reactor fuel from Japan to Sellafield. This type is also used extensively by the British utilities, CEGB and SSEB for domestic movements of Magnox fuel in Great Britain.

It consists of a single forging, cuboid steel box of extremely heavy construction with a similarly massive lid. The fuel elements are placed horizontally into an open top steel container (skip) which is loaded into the flask. The flask is partly filled with water, sufficient to completely immerse the fuel, and assists in heat transfer to the flask body.

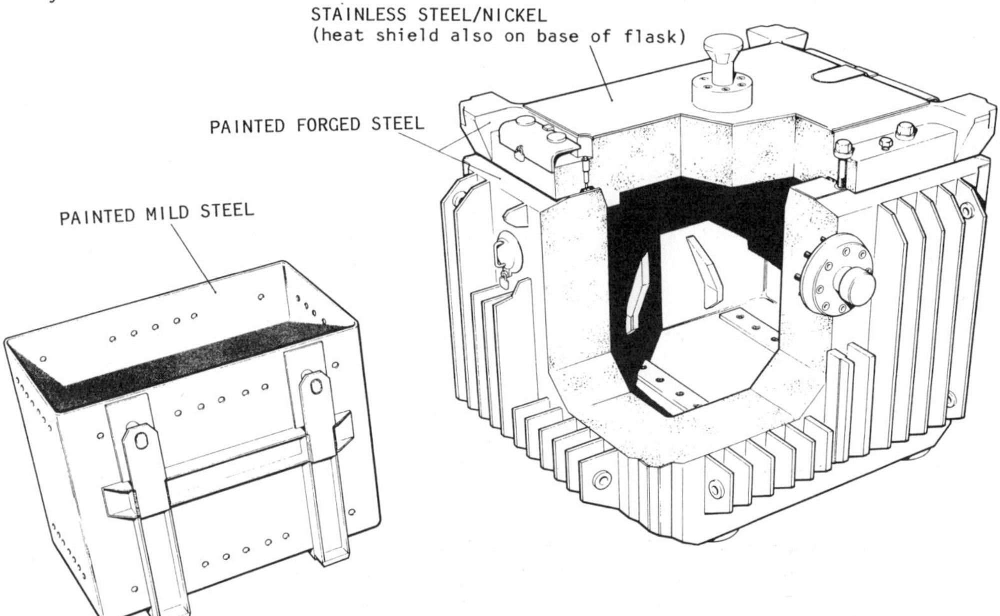

Fig. 5. Magnox flask design.

The flask walls are 36.8cm thick with vertical cooling fins welded to the outside faces to improve heat dissipation. The lid is bolted to the body by sixteen substantial bolts and is sealed by double '0' rings. The lid is covered during traqnsport by a fabricated stainless steel shock absorber which also prevents unauthorised removal of lid bolts.

The operational laden weight is about 50 tonnes. Lifting trunnions are bolted to two side faces of the body and lifting is achieved by use of a special yoke. A lifting cap is provided on the lid for lid removal. Holes in the fins at each corner of the flask are intended for anchoring the flask onto transport vehicles. Special guide fins and corners of the flask are used for location when lowering into a transport frame.

Oxide Flask: (Wet)

There are several detailed variations of this design which is used extensively for transport of LWR fuel from European and Japanese reactors to Sellafield. (Figure 6). The differences are necessary merely to transport various fuels.

The design consists of a hollow cylindrical body with circumferential external fins and welded-on base. The lid is bolted to the body and sealed with two concentric Viton '0' rings. The lid and base are protected during transport by bolted-on shock absorbers. A hollow cylindrical lead liner, encased in a steel shell, is fitted inside the flask to provide increased radiation shielding. A steel Multi-Element Bottle (MEB) fits into the cavity formed by the lead liner and contains the fuel elements which fit vertically into individual compartments. The MEB is fitted with a bolted lid. Both the MEB and the flask space are partly filled with water to give good heat transfer from the fuel to the flask body.

During sea transport and rail movements the flask is mounted on a transport and tilting frame which utilises pairs of trunnions bolted to the body. The frame also facilitates transfer of the flask from the horizontal to the vertical position for loading and unloading of fuel.

Oxide Flask (Dry)

Again there are several detailed variants to suit the different oxide fuels to be transported. This type of flask is used exclusively for movements of LWR fuel from European and Japanese reactors to La Hague (Figure 7). Though loaded with fuel under water in the reactor pond, before transport it is emptied of water and vacuum dried. Low pressure nitrogen is introduced in the cavity to prevent oxidation of the fuel assemblies at the temperatures which result from a dry heat transfer system.

The body consists of a massive forged carbon steel cylinder with a welded-on base. Internal surfaces are overlaid with stainless steel. The lid is bolted to the body and sealed with two concentric Viton '0' rings. Nickel-chromium plated copper fins are welded to the body outer surface and penetrate through a thick resin layer which forms part of the radiation shielding. The fins are to improve heat dissipation to the surrounding atmosphere.

A close-fitting basket is located inside the flask and receives the fuel elements in individual compartments. The basket is constructed of cast aluminium-silicon alloy, boron carbide and a stainless steel mesh. Both the lid and the base are protected by bolted-on shock absorbers during transport. Normally the flask is mounted on a transport frame in

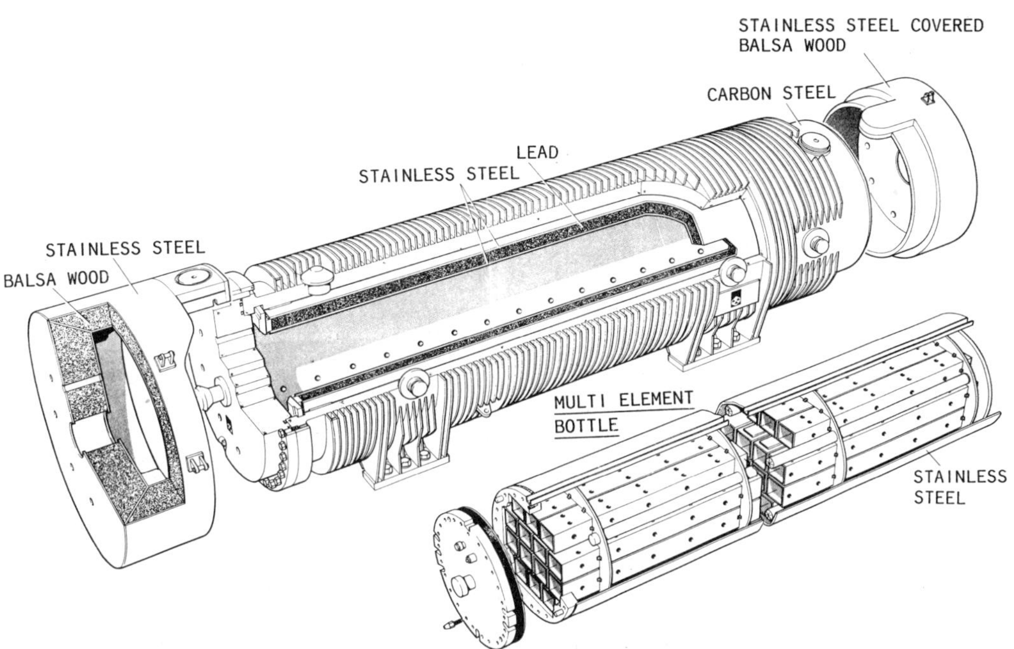

Fig. 6. Oxide fuel flask (wet) - 'Excellox' type.

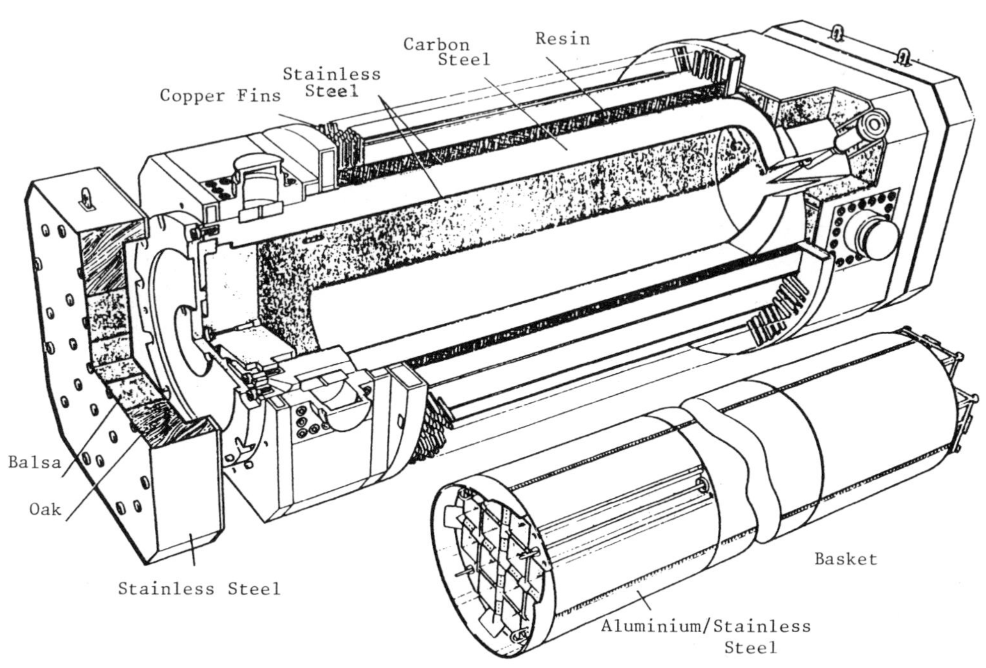

Fig. 7. Oxide fuel flask (dry) - TNp type.

the horizontal orientation using pairs of trunnions attached to the body.
The flask is tilted to the verticle attitude for fuel loading and
unloading operations.

QUANTITIES OF FUEL TRANSPORTED

The tonnage of spent fuel transported by NTL from European reactors
to Sellafield and La Hague exceeded one thousand tonnes by 1980. The
larger capacity dry flasks were then brought into service transporting
fuel to La Hague thereby increasing the annual tonnage. By the end of
1987 the total tonnage transported by NTL had increased to 4000 tonnes.
This is equivalent to 1910 transports comprisings 14,400 fuel elements.

Magnox spent fuel shipments to Sellafield from Latina in Italy began
in 1966 and from Tokai-Mura in Japan in 1969. By the end of 1987 about
2000 tonnes had been transported jointly by BNFL and PNTL, which consists
of approximately 185,000 fuel elements. Oxide fuel movements from Japan
to Great Britain began in 1973 and to France in 1979. At the end of 1987
the total transported exceeded 2500 tonnes, (about 9000 elements). The
combined tonnage so far transported by BNFL and PNTL is therefore about
4,500 tonnes.

The three transport companies, including their predecessor, The
United Kingdom Atomic Energy Authority in the early days, have therefore
safely transported over 25 years a combined grand total of about 8,500
tonnes of spent fuel of various types, using a range of flask designs
over a diversity of routes. A summary of the build-up of these quantities
is provided in Figure 8.

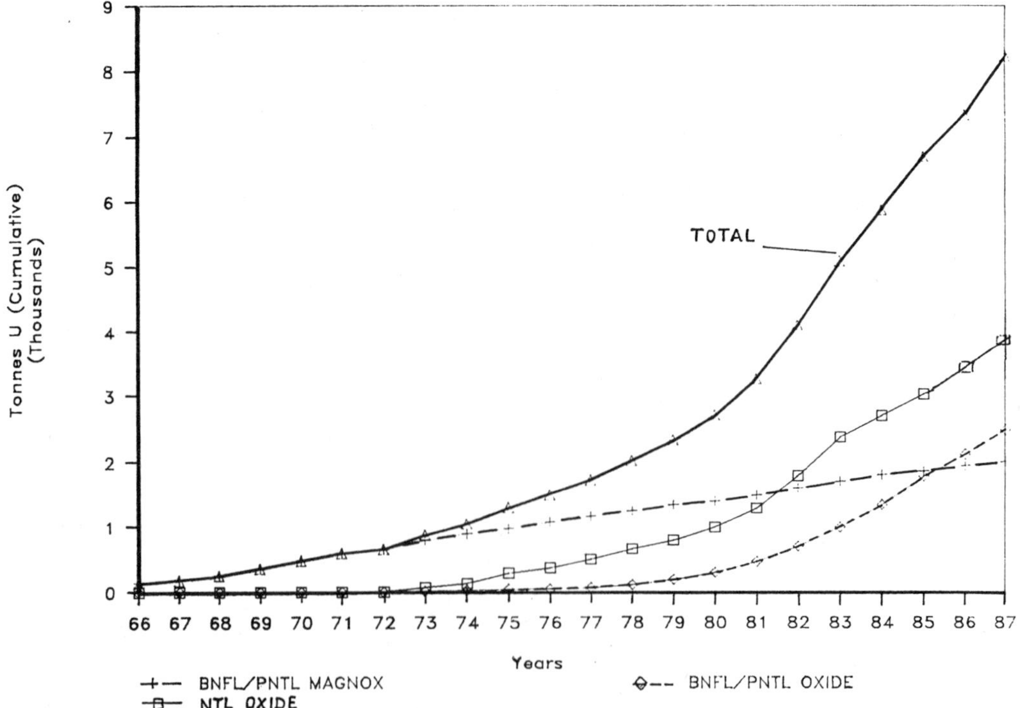

Fig. 8. Quantities of fuel transported.

FLASK HANDLING

To ensure the safe transport of spent fuel, strict attention is given to the specialised tasks of flask handling and fuel loading by employing a highly trained workforce. Each stage of the operation is independently checked for compliance with procedures specified in the quality plan. This is achieved through a combination of experienced reactor operations staff and specialist operations engineers from NTL and BNFL/PNTL who are in attendance at European and Japanese reactor sites.

The engineers advise and assist the reactor operations staff, check that there is no divergence from the set procedures, verify that fuel to be transported complies with acceptance criteria, check fuel element identities, ensure that preparations for flask despatch are acceptable and that flasks are correctly positioned and secured to road or rail vehicles or into the ship. An engineer is available throughout the transport operation and also attends at the reprocessing plant to assist with flask receipt, fuel unloading and ensure compliance with flask turnround and maintenance procedures.

On receipt at the reactor site, flask handling procedures are basically the same for both wet and dry type flasks. The first operation is the health physics radiation monitoring check of the empty flask. It is then lifted from the vehicle and usually tilted from the horizontal to the vertical before transfer into the reactor building to stand vertically in the flask preparation area. Preparations for submergence into the fuel loading pond include leak-tightness checks, if appropriate, the fitting of a metallic skirt over the finned area, removal of the flask lid (and MEB lid, if fitted), and filling of the flask (and MEB) with water. The flask is then lowered into the fuel loading pond and lifting equipment is disconnected.

The identity of the individual fuel elements to be loaded is checked and they are usually stored separately from other fuel in readiness for loading. The fuel is loaded in the correct orientation into the basket or MEB in the flask, the lid is refitted to the MEB (if applicable) and it is ullaged (a predetermined gas space is introduced by injecting nitrogen through a penetration valve). The flask lid is refitted in the pond and the lifting equipment is attached. The flask is then removed from the pond and returned to the preparation area.

External surfaces are decontaminated. In the case of a wet flask, the internal water level is adjusted to leave an "ullage" gas volume inside the flask. With a dry flask, the water is completely drained and the cavity and fuel are dried under vacuum, leaving a sub-atmospheric pressure. Special valve tool equipment is used for access to the flask cavity for draining and drying enabling the flask orifice plugs to be operated remotely. If fitted, the metallic skirt is removed from the flask.

Following radiation and contamination monitoring the flask is transferred from the reactor building to its transport vehicle. Final radiological checks are made, documentation completed and the flask is formally despatched. Spent fuel transport has become a well established operation in the nuclear fuel cycle. Flasks and their ancillary equipment have reached an advanced state of development. Many of the earlier problems associated with approvals, compatibility of flask dimensions with reactor and reprocessing site receipt facilities, contamination build-up, crud depostion from fuel, ullaging and vacuum drying have been overcome. The following examples illustrate some of the earlier difficulties and the solutions adopted:-

Fuel Quality

The nature of the cooling cycle in many BWR'S produces a depostion of "crud" (metallic oxide from the reactor circuit) upon the surfaces of the fuel rods. This crud has influences on fuel heat transfer within the reactor but can also create serious problems for the fuel transporter and reprocessor. During transport and the heating and cooling cycles to which the fuel is exposed in the flask, crud can become detached and accumulate in the bottom of the flask and the internal orifices. The main active component of the crud is cobalt producing intense gamma rays. In the case of transports to Sellafield, the problem has been resolved by enclosing the fuel in what is known as a Multi-Element Bottle and removing the bottle with its full contents on arrival at the reprocessing plant. This prevents the crud from building up in the annulus between lead liner and the steel wall. In the case of transport to the La Hague plant the reprocessor has insisted that the fuel should be cleaned by mechanical and ultrasonic methods before transport. As a result of these measures, crud no longer presents a major problem, although it caused considerable difficulties in the mid-1970's.

Failed Fuel

Failed fuel will only be accepted by reprocessing plants if it is enclosed in a capsule which prevents release of fission products into the flask and subsequently the reprocessing plant storage pond. The use of such capsules significantly reduces the normal capacity of a flask but enables failed fuel to be transported under safe conditions.

Flask Surface Contamination

External contamination of flasks has caused problems in the past due to the sweating out of contamination from both painted and stainless steel surfaces. This occurs after the surfaces have been properly decontaminated and checked. The problem can be resolved by cleaning the flask initially to one-tenth of the permissible limit so that subsequent sweating out is still within the limit. It is, however, the normal practice to protect the surface of the flask from contact with contaminated pond water by the use of metal or plastic skirts. The most practical solution is a rigid metal skirt which covers all of the finned area and is sealed to rings at the ends of the flask.

Use of Protective Skirts

In the early days flask surfaces were not protected from contamination when in the fuel loading ponds. This was mainly because the low burn-up fuel did not demand large finned external areas on the flasks for heat dissipation. Consequently this type of flask with simple surfaces could be decontaminated in a short time, using water sprays and wiping clean. With the trend towards high burn-up fuel and large capacity flasks the surfaces have become densely finned and in the case of the dry flasks an external neutron shield was necessary. Methods to prevent decontamination became important involving the development of a plastic cylinder or skirt, with the interspace between the flask and skirt being filled with clean water. This method was first used in the early 1970's, and with many improvements to the basic design, is still being employed today. The disadvantage is the time-consuming fitting operation, and the difficulties associated with maintaining a leak-tight seal to the flask.

The natural progression from a plastic skirt was to a light-weight metallic skirt which was first used in the late 1970's. The sealing of the skirt to the flask is by means of air inflatable seals which proved effective. Variations of the basic design have produced metallic skirts of different sizes ranging from the single one-piece skirt to the advanced multi-purpose six-piece skirt which is suitable for a range of flasks. The metallic skirt provides mechanical and contamination protection to the multi-finned areas of the flasks, additional radiologcal protection for operators, and an external means for cooling the loaded flask.

Flask Orifice Operation

The older flasks are equipped with valves or quick connections for the venting and draining operations. Whilst these function satisfactorily, the newer generations of flasks are fitted with plug type closures at the main flask orifices giving the following advantages:-

a)   permits remote removal of the plug from the flask orifice giving remote access to the flask cavity for the purpose of checking internal conditions, (filling, draining and vacuum drying),
b)   gives additional radiological protection to the operator,
c)   is robust, increasing the security of the flask containment particularly under abnormal conditions.

To facilitate remote operation of the orifice plug a unique piece of equipment called a Special Valve Tool was designed. The tool is a one piece unit easily handled by the operator, and it is designed to work under either vacuum or pressure conditions. This tool also forms part of the flask emergency equipment.

Cavity Drying

All dry type flasks are subjected to the process of vacuum drying to remove residual water following flask loading and draining. The consequence of not removing the residual water, which adheres and remains in pockets and on cavity surfaces, would be to create evaporation and a build-up of steam pressure during transport.

Techniques have changed over the years from blowing dry air through the flask and controlling the process by measuring the moisture content of the outgoing air, to the modern vacuum drying process. This reduces considerably the total drying time and the levels of activity discharged to the reactor active ventilation system. The principle of the vacuum drying process is to reduce the pressure of the flask cavity to the saturation pressure of water at the temperature of the cavity surfaces. The pressure will then tend to remain constant while the remaining water in the flask cavity is evaporating. With a relatively slow pumping speed, all of the water will evaporate before the pressure is reduced to a level which corresponds to the vapour pressure of ice at a temperature of zero degrees celsius. Once the flask cavity has attained the correct level of dryness, the cavity is purged with nitrogen and the pressure adjusted to sub-atmospheric in readiness for transport.

Leak Testing

Leak testing of the dry flasks was originally carried out by evacuating and isolating the flask cavity following drying. This method was not satisfactory because pressure rise in the flask could be generated by evaporation of residual water, surface outgassing, or air in-leakage. Since it was impossible to differentiate between these sources, it was necessary, conservatively, to assume that all the observed pressure rise was due to in-leakage of air. Under this system, flasks had to be much drier than the criteria specified in the flask safety report, humidity in the flask atmosphere was included in the interpretation of the rate of in-leakage. In addition the flask cavity volume was very large for test purposes, which reduced the test sensitivity and could only be compensated for by extending the duration of test.

Current flask designs have improved leak-testing arrangements incorporating primary and secondary seals with an interspace test point. The vacuum drying function is thereby separate from the leakage-testing function, allowing each orifice to be individually leak-tested. As the orifice interspace volumes are small, the test sensitivity is high and the duration of each test is only a few minutes.

Leak-testing of the wet type flasks can be achieved by employing the pressure drop method to the flask cavity. However, individual orifice interspace test points are fitted and used as the preferred testing method.

Flask Lifting Beam Operations

Following flask submission in the pond, several methods have been employed for disconnection of the lifting beam prior to fuel loading operations. Initially the side arms of the vertical beam were completely removed from the flask upper trunnions which made reconnection difficult. Modern beams are equipped with short self-standing sidearms which remain connected to the upper trunnions during pond operations. The connection and disconnection of the beam long side arms to the short side arms is done with the use of a tool operating individual drive units situated at the bottom of the long arms.

General Development

The development of handling techniques and equipment has greatly contributed to the efficiency and safety of the operations and minimised the radiation dose uptake to operators. The improved surface finish of modern flasks, taking both painted and polished stainless steel, has greatly eased the task of flask decontamination.

CONCLUSION

The transport of spent nuclear fuel is a sensitive operation since it takes place in the public domain and is subject to intense scrutiny. Substantial quantities of both Magnox and Oxide spent fuel have been transported from European countries and Japan to the reprocessing plants at La Hague and Sellafield. This has been carried out safely over a period of twenty-five years without any incident involving significant release of radioactivity.

Safety arrangements have been based on the Regulations of the International Atomic Energy Agency which are adopted into the laws of most countries. Compliance with safety requirements at all stages has been ensured by the application of comprehensive quality assurance and control.

Transport systems have been developed which encompass road/rail and sea movements, including the use of ferries and special purpose ships. Arrangements are made for trans-shipment and for frequent crossing of national boundaries and the necessary infrastructure is established. The requirement for transport of large consignments of flasks over long sea routes has led to the pioneering of special-purpose ship designs.

Equipment has reached an advanced stage of development with early difficulties overcome as they have been encountered.

The quantities transported worldwide will increase with the expansion of nuclear power and a policy of continued improvement is necessary to meet the anticipated changes in fuel specification and to safeguard the future of the nuclear industry.

REFERENCE

Spink, H.E., (1988). 'Design of a Special Ship for Spent Fuel Transport'. Conference on Transportation for the Nuclear Industry, Stratford-upon-Avon. May 1988 in these proceedings.

TRANSPORT OF SPENT FUEL FROM GARIGLIANO
POWER STATION TO AN AFR REPOSITORY

R.Bertini[1], C.D'Anna[1], A.Ceccolini[2],
G.Cuttica[2], A.Linari[3]

[1] ENEL DPT
Via Torino 6
Roma
Italy
[2] ENEL DAA
Via G.B.Martini 3
00198 Roma
Italy
[3] ENEL Centale Garigliano
Caserta
Italy

ABSTRACT

The Garigliano 160-MWe BWR station, locted in the south of Italy,
was put in a decommissioning state in March 1982.

The initial problems facing the owner of the plant – Ente Nazionale
Energia Elettrica – were due to the presence of fresh and irradiated fuel
at the site of the plant; therefore, the first initiatives were directed
towards the transport of this fuel away from the site in order to tackle
the subsequent decommissioning phases. ENEL began the various activities
related to the development of an intermediate repository where the spent
fuel would be located, the design and construction of a transport cask,
the preparation of a safety report and of the administrative procedures
with the regulatory authorities and, finally, the transport campaign with
its related health physics support.

At the same time, an Away-From-Reactor (AFR) repository had to be
designed and built; this repository was located in the north of Italy,
which implied a series of 46 return journeys of about 2,000Km each
through the peninsula.

The transport of the 64 fresh fuel assemblies destined for the sale
of the uranium content and of the 322 irradiated fuel assemblies for
intermediate storage were successfully completed by the end of 1987.

INTRODUCTION

ENEL activities in the field of irradiated nuclear fuel transport
started in 1966 with the shipments of the first spent fuel elements
unloaded from Latina power plant, in operation since 1963.

The last important transport campaign involved the shipments of 322 irradiated fuel assemblies, containing about 66t of enriched uranium, from Garigliano power plant to an Away-From-Reactor intermediate storage facility at Saluggia. This transport campaign took place from September 1985 to December 1987 and is the subject of this paper.

THE POWER PLANT

<u>Description</u>

The Garigliano Nuclear Power Station is located on a bend of the Garigliano river, at about 7Km from the sea, in the municipality of Sessa Aurunca (Caserta province), about 160Km south of Rome.

It utilized a General Electric dual-cycle boiling water reactor designed for an output of 160,000KWe. Several Italian companies participated in the construction, their technology being responsible for approximately 70% of the total cost. The plant consisted of the principal buildings for the reactor and turbine-generator and of several other buildings for the auxiliary systems and plant services. (Figure 1)

The reactor building was a steel containment sphere, of about 49m (160ft) in diameter, which contained the reactor vessel, the two secondary steam generators, the auxiliary systems for the nuclear area, the fresh fuel vault, the irradiated fuel pond and all the equipment for fuel handling.

The reactor core consisted of 208 fuel assemblies, 89 bottom-mounted cruciform control blades and structural components as the upper and lower support grid.

Fig. 1. Garigliano nuclear power station.

The main coolant entered on the bottom head of the reactor as subcooled water at 266°C, and left the vessel, at about 1.5m above the core, as a steam-water mixture at 285°C, to reach, through eight risers, the primary steam drum where the steam generated in the reactor was separated from the steam-water mixture and from here, with appropriate steam lines, it was conveyed to the secondary steam generators and to the turbine.

The second principal building contained the conventional systems (turbine-generator and feedwater), the offices, laboratories, workshops and the control-room.

The circulating water, for the main condenser, was provided by the Garigliano river at a flow of about $10m^3$/sec.

History

Plant construction began at the end of 1959 and in April 1964 the plant started its commercial activity.

When the plant was in operation, one fourth of the fuel assemblies of the core was replaced every 20 months. The fuel assemblies discharged from the core were stored in the spent fuel pool and then sent to the BNFL reprocessing plant.

In August 1978 the plant went into forced outage because of a through-wall leaking crack found in the weld region of the lower head closure of the secondary steam generator (SSG). Later, other significant indications of cracks were also found in the lower weld regions of both loops of the two SSGs. The through-wall failure did not produce any radioactive release in air or in water and no plant personnel were contaminated.

After the discovery of these cracks the Italian Nuclear Regulatory Authority, ENEA-DISP, requested inspections to be carried out, both on the failed components and on some components of the primary loop, reactor vessel, steam drum and piping.

The in-service nondestructive inspections confirmed the good state of all the examined components except the two SSGs; ENEL started studies to modify the plant with a SSG by-pass program, which would allow the plant to be operated as a single-cycle boiling reactor, with a 20% reduction in output.

In the meantime, ENEA-DISP started the total revision of the state of maintenance and operation of the plant and asked for an evaluation of the global safety margins of the plant with reference to the actual safety criteria accepted for power plants of the new generation. However, the estimated future plant life did not justify the economic effort for the amount of work required to achieve this goal.

On the basis of these considerations ENEL decided, in March 1982, to put the plant in a decommissioning state.

The Garigliano power plant, during this operating life, had sent out 12.5 billion KWh.

Decommissioning

After the Station shutdown, a permanent organization was set up, its tasks being a technical-economical evaluation of the decommissioning problems to be tackled and the related time scale.

One of the first actions was to accomplish the complete removal of fuel from the reactor to obtain another licence for zero-power plant and reduce the number of the permanent staff required by law for a nuclear reactor.

It was immediately apparent that the accomplishment of an AFR repository would allow the removal of irradiated fuel, both from the reactor and from the cooling pond.

In addition, it was thought that the fresh fuel still in store (13,316 Kg at 2,7% av.) would provide a source of revenue; in 1985 an international request to bid was issued, its main characteristic being that the cost of the disassembly operations would be paid by ENEL with part of the enriched uranium so recovered.

In this manner, no cash outflow was required; in addition, the quantity of enriched uranium recovered, instead of increasing the ENEL inventory, was traded for an equivalent amount of natural uranium.

The main decommissioning activities have been scheduled as follows:

(a) transport of the irradiated fuel assemblies out of the nuclear plant;
(b) radwaste treatment;
(c) arrangement of the reactor containment system, with its contents, in stage-1 decommissioning status;
(d) dismantling and decontamination or recycle of materials contained in buildings external to the reactor containment sphere.

According to the present plans, the station will be placed in a Storage Surveillance Status by 1992 or 1993. Further stages of decommissioning of the plant will be performed after about 30 years.

Scheduled activities are in accordance with the program of the other European plants being decommissioned, like Chinon 1, Lingen and Gundremmingden.

INTERMEDIATE STORAGE

Pending the definition of a reprocessing program, intermediate storage of the unloaded fuel presents a definite financial advantage in accordance with ENEL policy. The availability of an appropriate repository to accommodate the spent fuel both from the Garigliano and the Trino station, would meet this requirement. It was decided to design the storage capacity for an expected life of about 30 years.

ENEL, therefore, placed an order with FIAT, the owner of the AVOGADRO experimental reactor (already decommissioned), to transform the redundant pool of the plant into a temporary spent fuel assembly storage pond. This pond allowed the storage of the 322 Garigliano and 216 Trino Vercellese assemblies for a total of 135 tU.

Description

The Away-From-Reactor (AFR) facility of Saluggia constitutes an interesting example of its kind in Europe; a few years of operation will allow a reasonable comparison with dry storage facilities of similar capacity, given that optimization of an entirely new wet-type intermediate storage facility would probably lead to a much greater fuel capacity.

The location near the operating PWR plant of Trino Vercellese provides additional flexibility and delays the necessity of increasing the storage density at the plant site.

The facility, in its final configuration, meets the safety requirements peculiar to such types of plants, as demonstrated to ENEA-DISP by means of an appropriate safety analysis.

After final shutdown of the AVOGADRO 7 MW(th) reactor, only the core and a few other active components had been removed from the pool, thus before starting the transformation works, a partial decommissioning of the facility and a complete decontamination of the pool were carried out. In order to achieve the final storage facility configuration, the following structural and mechanical works had to be performed:

- Construction of the cask handling equipment and structures, i.e. cask-handling crane (60 t) unloading pit, decontamination bay.
- Construction and installation of the high-density storage racks.
- Construction and installation of the fuel handling bridge and hoist.
- Modification and updating of the existing auxiliary systems (with their instrumentation), i.e.

    (a) pool water cooling purification and makeup;
    (b) ventilation, filtration and air conditioning;
    (c) liquid radwaste storage and disposal;
    (d) radiation monitoring.

The pool and the cask/fuel handling equipment are located within the former cylindrical concrete reactor containment building (now called the storage building), with the main auxiliary systems underneath.

The auxiliary systems and equipment meet the following basic requirements:

- Maximum thermal power to be dissipated: 500 Kw (with a minimum previous decay time of six months in the reactor pools)
- Pool water bulk temperature (normal operation) $28^{\circ}C$
- Pool water radioactivity (normal operation, max): 3.7×10^7 Bq/m^3 (1×10^{-3} Ci/m^3)
- Fuel transport cask weight: 55 t.

Spent Fuel Characteristics. The spent fuel assemblies to be stored are of two different types, as follows:

	Trino V.(PWR)	Garigliano (BWR)
Enrichment (fresh fuel)	4.75% MAX	2.7% AV.
Cladding	AISI 304	Zircaloy-2
Average burn-up	30,000 MWd/MTU	27,500 MWd/MTU
Uranium weight	312 Kg	206 Kg

The racks, designed by Nuclear Services International Corporation and manufactured by FIAT, are of the high-density type using stainless steel in the form of square channels to contain each fuel assembly. No special neutron poison material (such as Boral or others) is used. The main parameters of the racks are as follows:

	PWR Fuel	BWR Fuel
Lattice Pattern (mm)	Rectangular	Square
Spacing (mm)	317 x 330	240
Channel Inside dimension (mm)	212	195
Wall thickness (mm)	5	5

While the Trino fuel assemblies, if not damaged, are transported and stored in the pool as they are, the Garigliano fuel assemblies are individually contained, both for transport and for storage, in stainless steel "bottles" to prevent "crud", which could come loose during transport, from contaminating the pool water.

The design of this container was carried out by CISE with a thorough investigation of some specific aspects upon which its good performance depended, in particular:

- criteria to be followed in the choice of the bolted cover;
- behaviour under pressure;
- choice of materials.

The behaviour under pressure was examined both from a theoretical point of view, by evaluating the elastic and stress condition with a calculation code and, through tests, by putting a container prototype under increasing pressure and measuring the total and permanent deformations. In the case of the bolted cover, two solutions were taken into account and the fabrication and pressure test of the two prototypes allowed a choice to be made. With respect to the choice of materials, the class of stainless steel was of course considered by means of a thorough analysis of the risks associated with localized corrosion (pitting, stress-corrosion, etc.) in order to choose, among the various types of steel, the type with the highest resistance to the most serious type of localized corrosion.

The bottles are essentially square-section boxes whose outer dimensions are 185 x 185 mm by a total 3,325 mm in height. They are built from stainless steel plates of AISI 316L, that are 4mm thick and folded with an internal curvature radius of the edges equal to 8mm. The plate is welded along the generant in the position of least stress, and submitted to heat treatment after welding. At the bottom the bottle is closed by a tight cover, which can be opened and closed under water and is clamped to a square flange welded to the shell. The lower plate is provided with a seat for the fuel assembly. The bolted cover of the bottle has a handle that can be folded back after handling.

The bottle interior is dry in order to prevent development of an "explosive gas mixture" due to water radiolysis; in addition a "feed and bleed" system for recurrent sampling and emptying of the bottles is available to check the internal atmosphere.

The bottles are also provided with an emptying system which consists of a sheathed stainless steel pipe located internally along the shell edge, with its upper end inserted in the flange. The pipe draws directly from the bottle bottom. The pumping process is carried out by sending compressed air through a fast coupling on the bolted cover and by discharging water through a remotely controlled thread plug on the flange connected with the above-mentioned pipe. In the lower part of the bolted cover a seat was provided for the ethylene-propylene O-ring seal. The bottles are designed to withstand a back-pressure up to 1.2 Kg/cm^2 at 150°C. CISE tests on a flange prototype proved that no release is recorded up to pressures far higher than the design pressure. The bottle is shown in Figure 2.

Licensing

The operating licence for the facility was obtained through a licensing process undertaken by FIAT. The following were the most important analyses performed:

Seismic and structural analysis: The following critical structures and equipment were analysed for seismic capability demonstration:

(a) storage racks;
(b) storage pool;
(c) storage building;
(d) cask-handling crane and related supporting structures.

All seismic analyses were based on the Newmark '73 response spectrum; for new structures and equipment, an asymptotic soil acceleration value of 0.19g was adopted.

Fig. 2. Bottle with irradiated fuel assembly.

(a) The racks are of the free-standing type, which means that they simply rest on the pool bottom and may therefore slide on it in case of an earthquake. The seismic input (motion of the pool bottom) was calculated with the "time history" technique. This input was used to check, not only the structural integrity of the racks, but also their stability against overturning and in addition to ensure that they will not knock against each other or collide with the adjacent pool walls if sliding occurs.

(b) The pool structure was analysed to check its capability of withstanding an earthquake without a sudden loss of water. The analysis was performed with the use of the three-dimensional finite element technique, arriving at stress determination by taking into account the soil-structure interaction.

(c) The storage building was analysed to check that, in case of an earthquake, it would not collapse on to the stored fuel. The analysis was performed with the same calculational technique used for the pool.

(d) The cask-handling crane was checked against overturning by means of a lumped-mass dynamic analysis, which gave the set of inertial forces generated by the earthquake.

<u>Nuclear analysis</u>. The array of the stored fuel assemblies was checked to ensure that the multiplication factor, in the most reactive condition, would never exceed 0.95. The analysis was performed with a two-dimensional diffusion theory computer code calculation, using very conservative assumptions, for example by considering the pool filled with unirradiated fuel at the highest enrichment, stored in an infinite array.

<u>Environmental impact analysis</u>. The radiological impact on the environment during normal operation of the facility and in case of malfunctioning or accident was analysed. During normal operation, the overall release of radioactivity to the environment was evaluated with conservative assumptions for all parameters involved and, in particular, for the extent of the faulty fuel rod fraction (1%). The results of the analysis showed that the radioactivity discharged by the storage facility to the environment amounts to 0.3% and 0.25% (for liquid and airborne substances, respectively) of the overall allowable limit for the Saluggia site. The accidents considered critical, although of a very low probability of occurrence, and whose radiological consequences were evaluated, are as follows:

(a) fuel assembly dropped into the pool;
(b) transport cask dropped during its handling inside the storage building.

The analysis results, which were based again on very conservative assumptions, show that the doses and contamination levels at the Saluggia site boundary are a few orders of magnitude lower than the corresponding emergency reference levels, so that no risk would result to the population following an accident in the storage facility. Also, some malfunctioning/outages of equipment and/or systems were considered from the standpoint of their possible radiological consequences.

The facility described in this paper was the first to operate in Europe for interim spent fuel storage independent of any reactor or reporcessing plant. It is likely to accomplish this task for another 10 to 20 years. Following the completion of the adaptation works described above, the storage facility was ready for operation in January 1984. Four months later ENEA-DISP released the operation licence.

During three years in which the assemblies have remained in the pool, the performance of all the facility auxiliary systems has been completely satisfactory and no undue release of radioactivity, either into the pool water or through the ventilation stack, was detected. In particular, the liquid waste discharged to the nearby river amounted to less than 1% of the admissible radioactivity discharge limit, and the gaseous effluents through the stack released an amount of radioactivity between 0.1 and 2% of the discharge limit, depending on the radionuclide species. Figure 3 shows racks installed in the pool with a fuel assembly being lowered down to its storage position.

The pool water activity level has always been well below the licensed limit of 10^{-3} Ci/m^3, with a maximum total beta-gamma activity of 3.6×10^{-5} Ci/m^3, (about 1/30 of the allowed limit). The water chemistry was also well within the design parameters.

A corrosion control experimental program is now being performed, with metal specimens immersed in the pool water. These specimens are periodically inspected for early detection of any possible corrosion problems that could arise in the different structural materials present in the pool, namely, the SS cladding and other structural parts of the Trino V assemblies, and the SS of the racks and of the bottles containing the Garigliano fuel.

Fig. 3. Installed racks in Avogadro pond with fuel assemblies.

THE TRANSPORT CASK

Description

Transport was carried out by a special cask called AGN1, which was designed and built for ENEL by AGIP Nucleare; this container can be used either dry or full of water.

The main characteristics of this cask, which is shown in Figure 4, are as follows:

The container consists of a hollow cylindrical carbon steel body, the outside and inside of which is lined with stainless steel. The lid and bottom consist of 330mm thick carbon steel forgings with a stainless steel lining. The lid is fixed to the container body by 32 DN 56mm studs. The container is sealed by two (elastomeric) snap rings between the head and the body. The space between these two rings allows the tightness tests to be carried out. Neutron shielding is provided by a stainless steel annular finned jacket, which is filled with a mixture of water and ethylene glycol. The outer maximum dimensions of the body are: diameter 1,880mm and length 4,120mm; the inside cavity dimensions are: diameter 840mm, length 3,380mm. Shielding is ensured by the 324mm carbon steel, the 11mm of stainless steel lining, plus the 95mm of water, if any. Two impact absorbers in the form of semi-elliptical carbon-steel shells are fixed at both ends of the container body by means of conic flanges and semibails.

The container is provided with a type of basket, where seven PWR or BWR fuel assemblies can be accommodated. This basket consists of 14 stainless-steel disks utilized as support and spacers for the seven longitudinal boxes of borated stainless steel sheet.

These boxes have a square internal section and can accommodate both non-bottled PWR fuel assemblies and bottled BWR fuel assemblies.

Fig. 4. AGN 1 cask.

During transport, the cask is laid down horizontally on a supporting structure which is attached to the semi-trailer platform and is protected by a slotted cover in stainless steel sheet.

The package was accepted by ENEA-DISP as of the B(M)F type, and, before shipment, the assemblies must have had a cooling period of at least 365 days. The total maximum authorized decay power is respectively equal to 10Kw for Garigliano assemblies (BWR) and 21Kw for Trino Vercellese assemblies (PWR).

<u>Utilization</u>

The package use is envisaged to be in the conditions:

- "dry" (1) for Trino assemblies with a total thermal output
             lower than 6Kw;
         (2) for Garigliano assemblies as dry capsules with an output
             of at least 10Kw;

- "dry" and "wet" for Trino assemblies with a total thermal output
             ranging between 6 and 10Kw;

- "wet" for Trino assemblies with a total thermal output
             ranging from 10 to 21Kw.

The maximum accepted activity is respectively equal to:

- 2.2×10^6Ci for the Garigliano assemblies and

- 5.6×10^6Ci for the Trino assemblies

The maximum burn-up of the assemblies that can be carried ranges between 25,500 and 35,000 MWD/t according to whether they are Garigliano or Trino uranium oxide or mixed oxide elements.

The container, which has an empty weight of 53,380 Kg, can weigh up to 57,890 Kg with all the assemblies bottled and the liquid of the neutron shield present. The package is anchored to the semi-trailer by means of a carbon-steel anchor structure. The anchor structure is clamped to the semi-trailer by a 48 bolt system and lock plates.

In the case of the design of the anchor system and its clamping to the semi-trailer platform, the provisions of 10 CFR-71 were applied; the minimum yield strain must not be exceeded at any point of the structure subjected to the following loads:

(a) a force in the direction of the longitudinal axis of the container equal to 10 times the container weight at the maximum load;
(b) a force in the vertical direction equal to 2 times the container weight at the maximum load;
(c) a force in the transverse direction (perpendicular to the above-mentioned) equal to 5 times the container weight at the maximum load.

Loads applied are considered both separately and in the combination (a) + (b) and (b) + (c).

The project and the manufacturing proposals obtained ENEA's conformity approval; the Provincial Traffic Control Authority recognized the fitness of the anchor structure.

The transports were organized in compliance with the non-stop principle to ensure the shortest possible time of transit of the convoy along the road, and with the immediate intervention principle to cope with any type of incident or accident (electrical, mechanical or nuclear).

The carrier was selected through a technical-economical tender to make sure that the vehicles used would be the most appropriate and that the personnel assigned to the task would be highly qualified for the transport of irradiated fissile material, handling of extraordinary loads, driving and repair of special vehicles, and for the application of measures required to protect the workers and the population against the hazard of radiation.

The selected carrier prepared the required safety report and QA procedures and worked out, with ENEA-DISP, the detailed program of actions to be taken in both normal and emergency conditions.

The convoy for the transport of the AGN-1 container, as shown in Figure 5, consisted of the tractor and the semi-trailer with an escort of (a) a standby tractor, and (b) a safety escort vehicle with two auxiliary escort vehicles, specially equipped to carry out "non-stop" transport and provided with the following equipment:

- Radiophone and VHF transmitter-receiver
- Two revolving searchlights set up on the tractor roof
- Bathrooms with a decontamination shower
- Air-conditioning and heating systems
- Six beds for personnel
- Sectionalized waterproof tank for foul water collection
- A special electric installation including an emergency
    3Kw-220V motor-driven generator and auxiliary batteries
- Back-pull hook
- Cooking range and refrigerator.

Fig. 5. Convoy for the transport of AGN 1 cask.

The safety escort vehicle carried a complete set of materials and equipment for the medical treatment of the personnel and any emergency situation.

Escort auxiliary vehicles carried fire-fighting equipment and the equipment to face any emergency concerning the means of transport (mechanical failures).

The tractors carried, in addition to halon and powder extinguishers, eight pneumatic sacks to remove the transport means in case of a skid or overturning, and the materials to delimit the area if emergency stops are required.

The convoy personnel included:
(a) two drivers (one driving) in the tractor of the semi-trailer;
(b) two drivers in the reserve tractor;
(c) two electricians with experience in repairing the vehicles in the convoy
(d) a highly skilled engineer with specific experience in spent fuel transportation and in health physics;
(e) up to five assistants for the safety and radio-protection of the convoy (one of whom was an expert in health physics);
(f) four radioprotection employees (one of whom was an engineer with experience and qualification in field health physics).

The personnel from (a) through to (e) were hired by the carrier; the persons under (f) were ENEL employees who worked at the station and travelled in a fully equipped motor-home. The personnel had their dosi eters and protective clothing if required in case of need. The convoys were escorted by the police which ensured both the use of the road and security.

The transport of the 322 assemblies was carried out in 46 successive trips with a frequency of a trip every two weeks; the sequence had to be interrupted owing to the unavailability of the road in the periods coinciding with public and summer holidays. The whole campaign took place in slightly more than two years, that is, from September 1985 to December 1987.

The average activity of the package was equal to about 180,000 Ci with a peak of about 200,000 Ci. The assemblies have a total residual activity of about 8.3×10^6 Ci. During these transports no incident nor accident occurred either to the means of transport, to the container, or to the auxiliary equipment. We believe that these results were achieved because of the scrupulous observance of the procedures and the QA program prepared for this campaign of transports and because of the professional qualification and training of all personnel involved.

ACKNOWLEDGEMENT

The authors wish to thank the personnel of the Garigliano Station for their assistance during the transport on the road and for the work during the preparation of the cask. The success of the whole program owes much to the collaboration and guidance of ENEA-DISP and of the police, without whose assistance the transports would not have been so smooth. FIAT, AGIP NUCLEARE and BORGHI provided information and photographic documentation for the preparation of this report.

TRANSPORTATION OF NUCLEAR MATERIALS
IN THE FEDERAL REPUBLIC OF GERMANY

G. Schwarz, F. Lange

Gesellschaft für Reaktorsicherheit (GRS) mbH
Schwertnergasse 1
D-500 Koln 1
Federal Republic of Germany

ABSTRACT

Currently a project is underway in GRS at the request of the Federal Government Ministry for the Environment, Nature Protection, and Nuclear Safety, with the objective of conducting a study to determine the magnitude and characteristics of shipments of all radioactive materials within the Federal Republic of Germany and of performing an assessment of the radiation exposure associated with the transportation of radioactive materials. For a specific subset of radioactive materials, i.e. fissile materials and non-fissile bulk quantity sources of radionuclides, the appropriate information has been compiled and is subject to further analysis. The conduct and results of the study for normal transport operations are described along with the major conclusions drawn.

INTRODUCTION

At the request of the Federal Government, Ministry for Environment, Nature Protection and Nuclear Safety, the "Gesellschaft für Reaktorsicherheit (GRS)", Cologne, is currently conducting a study related to the transportation of radioactive materials in the Federal Republic of Germany (FRG). The primary objective of the study is to provide an up-to-date overview of the magnitude and characteristics of shipments of all radioactive materials in the FRG, and an assessment of the radiological impact associated with these shipments. Assessment of the radiological impact refers basically to both aspects of concern with transportation of radioactive materials, i.e. the impact from normal (incident-free) transportation as well as the radiological impact from potential accidents.

The basic imformation required for studying this subject can be broadly divided into the following categories:

- quantity and characteristics of the radioactive materials shipped
- characteristics of the individual consignments
- mode, route, and number of packages and shipments of radioactive materials
- transport procedures and handling data.

Gathering this level of information has been and will be done by exploitation of various sources pertinent to the study area. The most relevant sources are questionnaire surveys, site visits, and examination of other appropriate data bases.

While the research project is still underway with results preliminary in various ways, for a specific subset of radioactive materials the relevant information and results of the study are fairly complete. This category of radioactive materials comprises shipments of fissile materials and bulk quantity sources of non-fissile radionuclides. Bulk quantity sources of non-fissile radionuclides are defined throughout this study as shipments of non-fissile radionuclides in quantitiies in excess of values given in the International Transport Regulation ADR, App. A, No. 2450 (5).

This paper is an account of work describing the research effort, data bases, and results of the study in terms of relevant transport statistics for shipments of fissile and bulk quantity sources of radioactive materials shipped in the FRG for the calendar year 1986 and the radiological consequences attributable to normal transport operations.

TRANSPORTATION DATA ACQUISITION AND DATA HANDLING

The Atomic Energy Act of the FRG and related regulations generally require approval by competent authorities of any transport of fissile materials and bulk quantity sources of non-fissile radioactive materials (FBRAM) on publicly accessible transportation routes. In addition, licensees, i.e. the consignor or carrier, are generally required to notify, with very few exceptions, the competent authority of any individual shipment of these materials. Thus, for this category of radioactive substances a fairly comprehensive and complete body of information exists in the FRG. The notification records presented to the competent authority were available for this study.

The transportation-related information reported to the competent authority generally includes the following items:

- address of the transportation licensee holder
- approval certificate number
- quantity and type of the radioactive material shipped
- type of package and package identification
- carrier of the shipment
- mode, date, and route of transportation and type of vehicle used.

Data specially related to describe the

- radiological package characteristics and
- transport and handling procedures

are less frequently reported on a routine basis. However, to the extent required and possible, such data were selectively requested from consignors, carriers etc. or derived by other appropriate means. Complementary data have been introduced into the data base where required on a judgemental basis. Examples are the presumed area of application of the nuclear material, the road travel distances and in rare cases the number of individual shipments of FBRAM's where these were not explicitly specified. In all cases where personal judgement has been exercised generally average transport conditions were assumed.

Not included in the database are shipments of minor quantities of fissile materials, i.e. miniscule amounts of activity per shipment or materials of low specific activity (< 74 Bq/g), which are generally exempted from approval and in even greater quantities exempted from the notification procedure. Transportation data of uranium ore shipments were also unavailable at this early stage of the study.

This body of information forms the principal basis for studying the transportation of FBRAM in the FRG and the radiological impact associated with these shipments. The information has been compiled and implemented in a computerized data base for further analysis. Examination of the data and conduct of the statistical analysis of the transportation data was facilitated by using a standard statistical program package.

RESULTS OF THE ANALYSIS OF TRANSPORTATION DATA

After appropriate editing and examination of the transportation data the computational analysis has been performed using a standard statistical software package (SAS, 1985). The results of this effort are summarized in a variety of tables presented subsequently. We note explicitly that the values presented should be considered as best estimates. The number of digits shown in the tables do not indicate the accuracy of the data acquisition procedure but are rather the result of rigid application of arithmetic. There is some uncertainty left in these figures due to shortcomings in the data acquisition process, e.g. incomplete data, and potential bias in the reporting procedure.

Table 1 presents the summary statistics of all shipments of fissile and non-fissile bulk quantity sources of radioactive materials (FBRAM) reported to the competent authority. The total number of shipments in calendar year 1986 is approximately 1676 comprising 8950 packages.

The estimated cumulative activity content shipped with these packages amounts to 23630 PBq (639 MCi). Most of this activity, however, has been carried by only a few packages. The average activity per shipment is approximately 14,1 PBq (0.38 MCi) with values ranging from very low values to a maximum of 670 PBq per shipment. The net weight of the nuclear material shipped via all transportation modes is about 6670 Mg.

The cumulative road travel distance of all shipments within Germany is about 260000 km. Travel distances for other transportation modes were not readily available for this study but will be included at a later stage.

The statistically relevant transportation quantities, i.e. the tonnage-road travel kilometer and the activity-road travel kilometer have been totalled to be $6,7.10^6$ Mg.road km and $1,9.10^6$ PBq.road km, respectively.

The average T.I. per package has been computed to be 0.6, the average number of driving and escort personnel on the road vehicle while carrying nuclear material was about 1.6 persons per shipment with individual values ranging from 1-4 persons per shipment. We note that both quoted values are based on limited data.

Further important characteristics of shipments of FBRAM are given in Table 2. Shipments, packages etc. are classified according to the transportation relationship, the transport mode, type of road transport vehicle, origin or end use of the radioactive material shipped, category of material, and the transport group.

Table 1. Summary Statistics of reported Shipments of Fissile and Bulk
         Quantity Sources of Radioactive Materials in the Federal
         Republic of Germany in 1986

Estimated number of shipments	1676
Estimated number of packages shipped	8950
Estimated cumulative activity shipped	23630 PBq[a]
Average activity per shipment	14.1 PBq
Estimated cumulative mass of fissile and radioactive materials shipped	6670 Mg
Average mass per shipment	4 Mg
Average T.I per package	0.6 [c]
Estimated cumulative road travel distance of all shipments	260000 km
Average road travel distance per shipment	155 km
Estimated Becquerel – Road Kilometers of all shipments	$1,23.10^6$ PBq.km
Estimated Tonnage – Road Kilometers of all shipments	$6,73.10^6$ Mg.km
Average number of driving and escort personnel on road vehicle per shipment	1,6 (1-4) [b] [c]

a) 1 PBq = 10^{15} Bq
b) In parenthesis: Range of values reported
c) Based on limited data

    <u>Transportation relationship</u>: The majority (63%) of shipments are domestic transports, i.e. shipments having their origin and destination within the FRG, while 37% of the shipments cross the national border. With respect to the net weight and amount of activity shipped, however, most transports of FBRAM are international.

    <u>Transportation mode</u>: Motor freight and rail as sole use account for approximately 83% of all shipments, 60 % of all packages, and almost 100% of the cumulative activity of nuclear materials shipped. The remainder of the shipments involve combinations of two or more transportation modes including road, rail, air and sea.

    <u>Road transportation vehicle</u>: The table indicates that light and heavy trucks are the preferred road transportation vehicle of nuclear material. 96% of the road travel distances are attributable to truck traffic.

    <u>Origin/End use</u>: Examination of the table reveals that the majority of nuclear shipments, of the number of packages, and of the activity shipped are attributable to the nuclear fuel cycle. Nuclear

Table. 2: Characteristics of Shipments of Fissile and Bulk Quantity Sources of Radioactive Materials in the Federal Republic of Germany in 1986

	Estimated Number of Shipments	Estimated Number of Packages	Cumulative Activity shipped PBq [a]	Cumulative Mass shipped (net) Mg	Road Travel Distance of Shipments km
Total	1676	8953	23630	6670	260000
Transportation relationship					
Domestic	1062	3499	1470	1358	120000
International	560	4471	22100	3360	127000
Transit	54	983	48	1947	12000
Transportation mode in the FRG					
Road	1300	5229	1250	2000	193600
Rail	65	143	16570	679	-
Rail/Road	21	64	5727	233	800
Air/Road	104	303	<<	13,5	8900
Sea	28	792	-	2218	-
Other combined modes	158	2422	84	1523	56400
Road transportation vehicle					
Truck	1499	7752	7060	3190	252100
Light vehicle	58	62	<<	0,12	6700
Unspecified/other than road transportation	119	1139	16567	3477	900
Origin/End use					
Fuel Cycle	1663	8940	23630	6670	257700
R & D	12	12	$2 \cdot 10^{-5}$	$6 \cdot 10^{-5}$	1800
Unspecified	1	1	$6 \cdot 10^{-4}$	$5 \cdot 10^{-6}$	200

a) $1 \text{ PBq} = 10^{15}$ Bq

Table. 2: Characteristics of Shipments of Fissile and Bulk Quantity Sources of Radioactive Materials in the Federal Republic of Germany in 1986 (continued)

	Estimated Number of Shipments	Estimated Number of Packages	Cumulative Activity shipped PBq	Cumulative Mass shipped (net) Mg	Road Travel Distance of Shipments km
Commodity					
• Fuel Cycle					
– Unspecified	141	424	1.6	0.4	19300
– Pre-fuel material, UO_2, UNO_3; etc.	218	3253	0.014	267	19500
– UF_6	231	1969	0.062	5074	58600
– Fuel elements, rods; nonirrad.	816	2127	154	1079	128300
– Fuel elements, rods; irrad.	103	107	23300	246	11700
– Pu and Pu-containing substances	55	261	214	0.40	8600
– Residues with further use	5	23	<<	0.002	100
– Wastes	84.	719	4.1	0.27	9900
– Emptied transport flasks	10	57	0.025	$4 \cdot 10^{-5}$	1650
• R&D	12	12	$2 \cdot 10^{-5}$	$6 \cdot 10^{-5}$	1800
• Unspecified	1	1	$6 \cdot 10^{-4}$	$5 \cdot 10^{-6}$	200
Transport Group					
5	2	5	<<	0.14	800
8/11	918	2378	<<	1159	120000
9/10/11	513	2701	22875	1394	100000
unspecified	243	3869	752	4115	39000

material shipments solely related to research and development occur less
frequently.

<u>Commodity type</u>: Nuclear material shipments are classified in
eleven commodity types. The commodity type of each shipment has been
determined on the predominant form of the radioactive freight. Therefore
small quantities of materials other than that specified may be included
in each commodity type category, notably small quantities of plutonium
and other actinides.

Commodity types accounting for the largest number of shipments are
fresh fuel element and rods, uranium hexafluoride (UF_6), pre-fuel
materials, and unspecified shipments with 49%, 14%, 13%, and 8%,
respectively. The commodity type accounting for the largest fraction of
the total activity shipped is, however, irradiated fuel elements and rods
with 98%.

Further analysis reveals that in various categories, notably in the
unspecified shipment category, for Pu and Pu-containing substances, and
for waste and emptied transport flasks, most of the individual shipments
carry only minor (less than 1 kg) quantities of nuclear material.
Examples are the numerous sample quantities of nuclear materials that are
shipped between nuclear installations and research and production
facilities.

RADIOLOGICAL IMPACT

The INTERTRAN code developed under the auspices of the International
Atomic Energy Agency (IAEA) has been used in assessing the radiological
impact of transporting nuclear materials in the Federal Republic of
Germany. The code calculates the radiological impact from both normal
(incident-free) transport operations and potential accidents during
transportation in terms of collective and individual doses to the
population at risk. Details of the computational models and the type and
format of the input information required for code applications are
described elsewhere (IAEA 1983).

Emphasis in this paper has been placed on the radiological impact
associated with normal transportation operations of FBRAM. The
incident-free dose model of the INTERTRAN code calculates external
radiation doses to various population subgroups resulting from the moving
and stationary phases from the transport operations under consideration.
The radiological impact is specified in terms of the annual expected
collective dose equivalent in man-Sv per year.

The radiological characteristics of shipments or packages were, to
the extent possible, derived from this study or have been adapted from
other sources (Lange, 1987). The input imformation relevant to the
transport and handling procedures reflect the experience gained from site
visits or are best estimates.

The radiologically most relevant population subgroups considered in
this study are:

- crew and escort personnel
- handlers
- population on and surrounding the transportation link.

The annual collective dose estimates resulting from <u>road</u> transport
operations of fissile materials and non-fissile bulk quantity sources of
radionuclides in the FRG in the calendar year 1986 are as follows:

	man-Sv/year
Crew and escort personnel on road vehicles	0,1
Handlers	0,17
Population on and surrounding the road transportation link	0,02
Total	0,29

CONCLUSIONS

The figures indicate that a major fraction of the total annual collective dose from transportation operations of nuclear materials can be attributed to the population groups actively involved in the transportation process. The exposure of members of the public is significantly lower.

Perspective about the calculated collective dose can be drawn from an examination of current levels of exposure of today's society. The collective effective dose in the FRG from ionizing radiation of natural origin and medical exposures is about 120 000 man-Sv/yr and 90 000 man-Sv/yr, respectively.

We note that at this stage of the study the figures presented above do not include the dose fraction associated with rail, air, and sea transportation. By the nature of the study the values should be qualified as rough estimates.

ACKNOWLEDGEMENT

The authors gratefully acknowledge the generous support and assistance of Prof.W.Collin and Mrs.R.Hoflich in providing the transportation data for the project. Discussions with Dr. U. Alter have been very valuable in completing this study. We also appreciate the effort of the numerous contributors in handling the data and the computational support provided by Mr.A.E.Meltzer.

REFERENCES

International Atomic Energy Agency, 1983, INTERTRAN: A system for assessing the impact from transporting radioactive material. IAEA-TECDOC-287, Vienna 1983.

Lange, F., 1987, Risk assessment with INTERTRAN for a subset of transports of radioactive materials in the Federal Republic of Germany (IAEA-SM-286/14) in Packaging and Transportation of radioactive Materials (PATRAM 86), IAEA. Vienna 1987.

SAS Institute Inc., 1985, SAS User's Guide: Basics, Version 5 Edit. Cary, NC: SAS Institute Inc., 1985

MR.M.S.T.PRICE, ATOMIC ENERGY ESTABLISHMENT, WINFRITH for Paper 4:1
Since the inception of Nirex, I have promoted the idea that Nirex should
tuck in alongside the CEGB so as to take advantage of their purchasing
leverage, because of the large amount of coal transport carried out by
British Rail for the CEGB. We heard yesterday of market forces possibly
affecting rail costs (ie. reducing). Is the costing information given in
the paper still relevant?

A further point is that we looked at sea transport of whole boilers
some six or seven years ago in connection with the initial studies on the
possible disposal to sea (which was an option at that time) by AEE
Winfrith working together with Burness, Corlett and Partners and other
AEA sites. On safety grounds we chose to rule out a roll-on/roll-off
(Ro-Ro) ship design. Is the Ro-Ro ship design what I might call a
'normal' Ro-Ro ship design or is it, what I might call, a safer Ro-Ro
ship such as the Swedes are using for shipments through the Baltic to
Cherbourg? How many watertight compartments could be flooded at the same
time?

MR.D.BENNETT, UK NIREX LIMITED:
I am not sure I can answer your first question on rail charges and
whether if we go along with the CEGB we might have more leverage. We
certainly didn't get anywhere near the stage of negotiating what we would
pay to British Rail for shipping low-level waste, perhaps they might have
more to say on that. I think that the charges for low-level waste
transport by rail could be perhaps revised downward, as I suspect Ian
Braybrook will talk about later, if a type of operation other than block
trains is used, that is if a Speedlink type system is used. So I suspect
that by using charges for dedicated trains we were at the top end of the
possible spread of charges for rail transport. The question arose of what
type of Ro-Ro ship could be used for transporting whole boilers, and four
options that were identified - three were variations on barges e.g. push
barge, towed barge, a barge carried on a barge carrier. We didn't look in
detail at the Ro-Ro ship option but it would probably be a Ro-Ro ship
carrying the heat exchanger mounted on the multi-wheeled land transport
vehicle, actually on deck. I am not sure of the Swedish design you
mention, but I would imagine that it is that kind of scheme. Is that what
actually happens in Sweden?

CHAIRMAN:
Could I just answer the question on the Ro-Ro ship, I think you are
talking of the Swedish ship which carries spent fuel packages and not
cars. I think when you question the integrity of any Ro-Ro ships then the
integrity of the package is the important factor not the requirement for
a special ship.

MR.J.W.STEPHEN, JAMES FISHER & SONS PLC:
I will be giving my paper in Session 4, which will answer Mr.Price's question. It will be on the movement of abnormal loads and it does address the problem of Ro-Ro conventional type vessel vs. a normal cargo where the cargo is rolled over the deck onto the weather deck.

MR.C.CHAPMAN, CEGB NPTC for Paper 4:1
(a) Did Nirex consider "floating" the "old" boilers from/to Bradwell - the method used in 1957 during construction?
(b) Was the pre-war railway route proposed to Bradwell (as a potential holiday resort) considered in the study for a rail link from Southminster?

MR.D.BENNETT:
In the first case, no. Though I wasn't born in 1957 I have seen a video of the shipment of boiler shells to the Bradwell site and so we were aware of that. The problem in trying to get them out by the same method is that I believe they were shipped as empty shells and were filled with boiler tubes on site. If you try and move the boiler without first stabilizing the boiler tubes in some way you create problems, as, when you start to tip a boiler, you get mass movement of tubes within and that can lead to handling problems. So the proposal is to fill the boiler with a light-weight grout before moving it, then it would sink if pulled out to sea.

In the case of the route you spoke of to Bradwell from Southminster, I confess I'm not aware of that. Because of recent industrial develop- ments around the rail head at Bradwell I know it wouldn't be possible to reinstate the same route, you would have to by-pass it to some degree. I don't know whether the identified route shares any of the original lengths and so I can't answer that question.

MR.I.K.BRAYBROOK, BRITISH RAILWAYS BOARD:
I would like to thank Mr.Bennett for his invitation to comment on the aspect of rail pricing. There have been allegations of subliminal and even direct advertising here at this conference, so I think Mr.Bennet has carried out a master stroke in actually bringing negotiations into this forum - assisted by Mr.Price! I would certainly not want people to leave with the idea that rail transport is necessarily dramatically more expensive than road and I think it is right to put the records straight and say that the costing information used for the original Nirex study is now out of date because it was recognized by all parties, once the announcement of the cancellation was made on the 1st of May 1987, that there was no point doing any further work. More recent discussions have identified many ways, some of which I will elaborate later, of probably reducing the overall price of rail. As far as the earlier rail link is concerned I am afraid I cannot shed any further light on that either.

MR.C.R.H.STRANG, STRACHAN & HENSHAW:
I think some of my queries have been answered but I was astonished by the figures given by Mr.Bennett concerning the cost of transport, rail and sea being quoted at £300 per cubic metre. In answering one of the earlier queries Mr.Bennett has now cast some doubt on the validity of his figures. Surely in this type of feasibility study the accuracy of the figures is essential?

MR.D.BENNETT:
I agree with that, accuracy of the figures is essential. I am not sure how I can validate those figures. At the time of the study we were happy with the basis on which those figures were derived. I would be interested to know why you are astonished by the figures.

MR.C.R.H.STRANG:
I am astonished because we have a rail network in position whereas
building special boats and special slipways at various positions around
the country would entail considerable extra expense to build up the sea
transport network and therefore I was surprised that the price quoted was
the same.

MR.D.BENNETT:
The reason for it is that the figures for sea transport of c. £300 per
cubic metre did not assume that various slipways were built around the
country, but that in fact existing ports are used for despatch ports. So
we are talking about the construction of one un-loading facility at the
repository site which would cost, we believe, of the order of about £5 -
6 M. The other thing to be borne in mind is that although the capital
cost of sea transport would be high, in that there would need to be a
harbour built at the outset and several ships purchased, the costs that
we derived were averaged over an operational period of about 40 years for
a repository. If you do the sums and assume paying British Rail charges
over 40 years rather than running your own vessels which are bought at
the outset, those are the figures that are arrived at.

MR.BANKS:
I would like to discuss some of the comments made by Mr.Chapman and
Mr.Bennett on Bradwell. I was a co-author of the paper written by Burniss
Corlett and also a member of the site construction staff at Bradwell some
30 years ago. It is true what Mr.Bennett says, the heat exchangers were
floated round with floatation collars. They were brought in at Bradwell
water-side, a trailer was taken out at low tide on to a slipway, frogmen
went down at high tide, secured the boilers to the trailer, then taken up
into the site. At that time they weighed 800 tons. The internal pipework
was then fitted and, due to thermal cycling, it had to be hung rather
than fixed. During the decommissioning period it was then proposed to
lift out the boilers in the vertical position, fill them with
light-weight aggregate concrete, put them on a tailing down device, on to
a trailer, take them out to a standard 300 x 90 North Sea oil barge.
Four would be put on each barge, which would then be towed out to sea.
The method of disposal at sea would be to tip them two at a time off the
side of the barge in order not to upset the stabilty of the barge at sea.
I understand, though, that this method is no longer acceptable and that
land disposal facilities are being looked at. I hope that ties it up with
Bradwell.

MR.F.JOHN L.BINDON:
I would like to just add a comment to what Mr.Banks has said because I
had the good fortune to be at Bradwell in 1959 when the station was first
being constructed and for the benefit of overseas delegates, I would like
to say that Bradwell has had a magnificent safety record ever since the
first reactor went critical in August 1961.

OPERATING EXPERIENCE IN NUCLEAR TRANSPORT
FOR THE FRONT END OF THE FUEL CYCLE

M.A.Simpson

British Nuclear Fuels plc
Springfields Works
Salwick
Preston
Lancs. PR4 0XJ

ABSTRACT

    The fuel and enrichment divisions within BNFL are involved in some
4,000 lorry journeys per year covering the transport of non-irradiated
fuel elements as well as the feed materials and intermediate products of
the front end of the nuclear fuel cycle. The range of materials being
carried means that a variety of methods of carrying and containment are
required.

    Further, the number of receipt and delivery points means that the
facilities need to be simple and the techniques employed chosen to give
maximum flexibility. There is thus much scope for improvement or
optimisation of routine transport.

    For the non-routine, simple statistics imply that traffic incidents,
involving front end nuclear materials, will occur during transport
operations. Because of the public perception of risks, it is necessary
for operators to improve upon the law of averages by careful procedures
and the training of operators, including sub-contractors. Contingency
plans have to be laid to cover the cases where even these procedures
prove inadequate. These plans require international collaboration since
the consequences of an incident will not be restricted to carrier.

    The record of transport for the front end of the nuclear fuel cycle
is a good one, but the scale of operations and the quality of the
achievement are not necessarily appreciated by the general public. There
needs to be a constant striving for excellence in procedures and a
willingness for international collaboration if the nuclear transport
industry is ₊to build upon its good record, and retain the public
confidence necessary for continued operations.

INTRODUCTION

    Transportation for the front end of the nuclear fuel cycle can often
seem to be the poor relation of irradiated fuel transport. Crashing a
train to demonstrate the safety of the movement of irradiated fuel is an

event which can catch the imagination of the media since the design and construction of the containers required sets challenges for engineers and the existence of fleets of ships designed specifically for their movement give it a touch of glamour. In contrast the movements of materials at the front end of the fuel cycle can seem dull and routine. The unhindered movement of uranium in all the forms necessary for the front end is however a vital part of the fuel cycle. Ensuring that it remains dull and routine is a demanding challenge which must be met by all the fuel manufacturers and enrichers throughout the world. This paper reflects upon how one operator, BNFL, has met those challenges.

FUEL CYCLE OPERATIONS

BNFL offers a complete fuel cycle service. Through its operating divisions it covers all aspects of the fuel cycle from the receipt of uranium ore concentrate (UOC) through to the production of metal or ceramic oxide fuel elements including the enrichment processes required for such elements. It also encompasses the reprocessing of elements after irradiation and the return of the products of reprocessing into the fuel cycle. Two of its divisions, the Fuel Division and the Enrichment Division cover the front end of the fuel cycle. The Fuel Division is centred at Springfields, near Preston, where UOC is converted to either the metal fuel elements required for the Magnox reactors or ceramic oxide elements for a variety of reactors. The site also exports natural uranium hexafluoride as an intermediate product for enrichment plants as well as enriched uranium dioxide powder and pellets as intermediate products in the enriched cycle. The Enrichment Division is centred on Capenhurst, near Chester. BNFL is a partner in the URENCO enrichment organisation and the site at Capenhurst houses one of that company's three centrifuge enrichment facilities.

The UK has no commercially mined uranium deposits and therefore BNFL relies entirely on imported uranium which it receives from sources worldwide. The majority of the uranium hexafluoiride (hex) conversion is for export and the material is transported to enrichment facilities in the United States, France and Russia as well as to the URENCO factories. URENCO itself is an international partnership with world wide contracts. The conversion of enriched hex to powder is achieved at Springfields through the integrated dry route (IDR) process. Half of the material from this facility, some 3,500 tes, has been exported to destinations throughout Europe for use in BWR and PWR facilities. BNFL have also supplied pellets as an intermediate product. In addition to the Magnox and AGR fuel elements distributed to reactors throughout the UK, BNFL have supplied magnox elements to Italy and Japan, fuel elements to the Fast Reactor Station at Dounreay in Scotland, PWR assemblies to the United States, BWR assemblies to Europe, as well as the SGHWR fuel elements to the reactor at Winfrith.

To cover these activities at the front end of the fuel cycle BNFL is a transporter in its own right with its own prime movers and a variety of standard and purpose designed containers. It also makes extensive use of transport sub-contractors. It delivers to the majority of European countries as well as to the United States, Russia and Japan. It makes use of road, air and sea transport, both load-on/load-off and roll-on/roll-off. In all, the front end transport movements involve some 4,000 lorry journeys per year, carrying total receipts and despatches from the Springfields and Capenhurst sites of over 15,000 tes of uranic products per annum.

DETAILS OF TRANSPORT MOVEMENTS

For the front end movements the Company makes use of three of the four categories of packages defined in the 1985 edition of the IAEA Regulations for the Safe Transport of Radioactive Material, namely, Excepted; Industrial; and Type A categories. Table 1 provides a breakdown of the total movements by package type. Before reflecting on operating experiences it may be useful to review the types of movement in more detail. The details of each type are given below. A summary of these details is given in Table 2.

Uranium Ore Concentrate (UOC)

Uranium ore concentrate, for customers contracting for conversion services, is received on the Springfields site in standard 20 or 40 ft ISO containers. The concentrate itself is contained in 45 gallon (210 litre) mild steel drums. The drums each contain approximately one third of a tonne of concentrate and are secured within the container by wooden retraints. The customers are widespread throughout Europe the United States and in the Far East. The ore is received either directly from mines throughout the world or from customers' intermediate stores. Receipts arrive in the UK either in freight containers, on ocean-going vessels, or on roll-on/roll-off ferries. Containers are collected from the port and generally transported by road to the Springfields site, although there are occasional deliveries by rail.

Natural Uranium Hexafluoride

Natural uranium hexafluoride is transported in the internationally recognised 48 inch UF6 cylinders (Figure 1). The cylinders are bolted to metal stillages which in turn are bolted to the framework of freight containers. For load-on/load-off ferries to the USA the freight containers employed are 40 ft ISO containers specially modified to accept the fixing arrangement of the metal stillages. For deliveries within Europe, BNFL employ in addition, its own specially modified 20 ft containers which are carried on roll-on/roll-off ferries. The Russian transport regulations require the used of a special metal stillage known as a Mafi flat. These are secured to flat bed trailers for their journey to the ocean-going vessel.

Enriched UF_6

Enriched UF_6 is contained in the internationally standard 30 inch cylinders. These in turn are enclosed in protective shipping packages (PSPs). The cylinders in their PSPs are exported from the Capenhurst site or received at Springfields, bolted to the base of 20 or 40 ft flats or ISO containers. Up to six cylinders can be carried on a single trailer. The cylinders can be despatched either on the flat on load-on/load-off ferries or on roll-on/roll-off ferries. Again receipts from the USSR make use of the Morflot Mafi flat with two PSPs per flat.

OU_2 Powder

The container used by BNFL for shipments of powder is a cylindrical metal drum with bolted lid which contains some 24 Kgs of uranium dioxide (the 1610 container). The metal drum is itself enclosed in a protective outer box which is of a solid wood construction, again with a bolted lid (the 1660) (Figure 2).

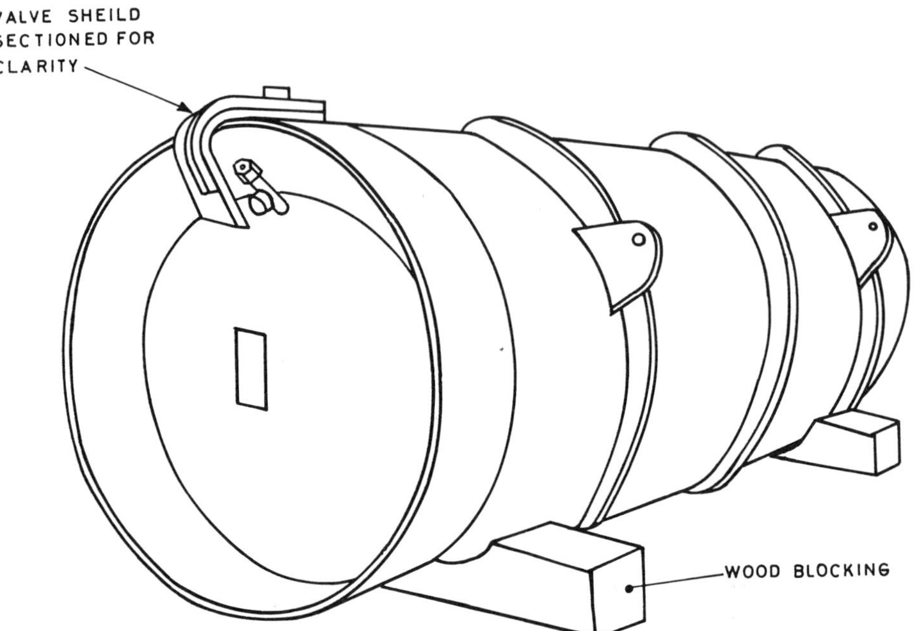

Fig. 1. UF$_6$ cylinder – 48Y model.

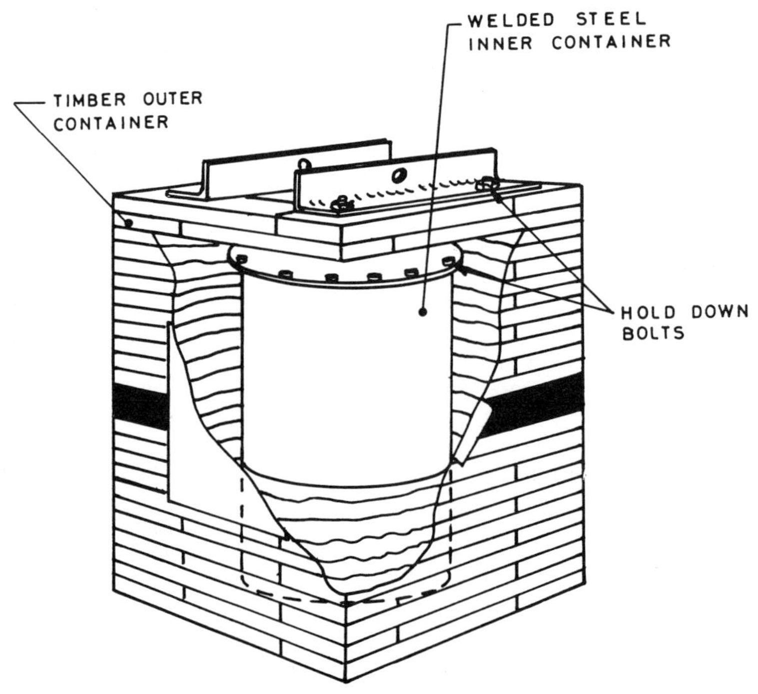

Fig. 2. Type 1660 container (outer), type 1610 (inner)
Uranic powder transport arrangement.

Over two hundred of these boxes are loaded into a 40 ft side loading
sliding door freight container in stillages. Alternatively some
customers have required that BNFL make use of containers which they have
supplied. These have included the US design of container, the BU5 or BU7
where an inner metal drum is contained within an outer metal package; and
more recently a CE250 container where a number of plastic bottles are
held within a double length metal drum placed in pairs on wooden
stillages. These again are transported in side loading freight
containers.

In all cases the loads are secured by strapping. Standard end
loading ISO containers can also be used for particular routes by
sub-contractors. The method employed on these occasions would be for
containers to be loaded from the end with the use of ramps. The
resultant load is secured by timbering.

OU_2 Pellets

Pellets are transported using a variation on the powder packaging
technique. For example the CE250 containers can be used for pellets. In
this case the plastic bottles containing powder are replaced by metal
suitcases holding the pellets. Pellets can also be wrapped in packaging
material and transported in the cylindrical drums associated with BNFL
containers or the BU5/7 container.

Magnox and AGR Fuel Elements

Magnox and AGR fuel elements are much shorter than the water cooled
reactor elements. This allows the elements to be packed easily into
metal boxes, either the small Magnox fuel boxes containing anything from
three to twenty Magnox fuel elements laid on their side (Figure 3), or
the larger AGR fuel element boxes containing eight AGR fuel elements with
a central absorber for criticality safety (Figure 4).

The boxes are secured to 40 ft trailers and taken by road to UK
stations. In addition, deliveries of Magnox fuel elements have been
made to Italy and Japan where standard ISO containers are used, the boxes
being secured within the containers by timbering.

Other Fuel Elements

The Springfields site produces SGHWR and PFR breeder fuel elements
as well as BWR and PWR fuel elements. Each of these elements have
purpose designed fuel element transport containers ranging from a single
tube type container for PFR breeder assemblies through boxes containing
one or two fuel elements to the perhaps more internationally familiar
cylindrical boxes containing pairs of PWR elements.

In addition to the requirements for the containers and radiation
safety, the elements would normally be packed along with instrumentation
such as accelerometers to gauge the stresses and strains that the
elements would be subjected to during transit to ensure that the elements
themselves have been unharmed by the movement inherent in transport.

In addition it is necessary to move numerous surface contaminated
items for waste storage, incineration or disposal. The quantities of
uranium involved can be minute. Nevertheless, since the load will carry
a radioactive symbol, for public relations reasons it is essential to be
seen to be handling such material as carefully as genuine uranic loads.
Enclosed containers or specially procured enclosed skips are needed for
this task.

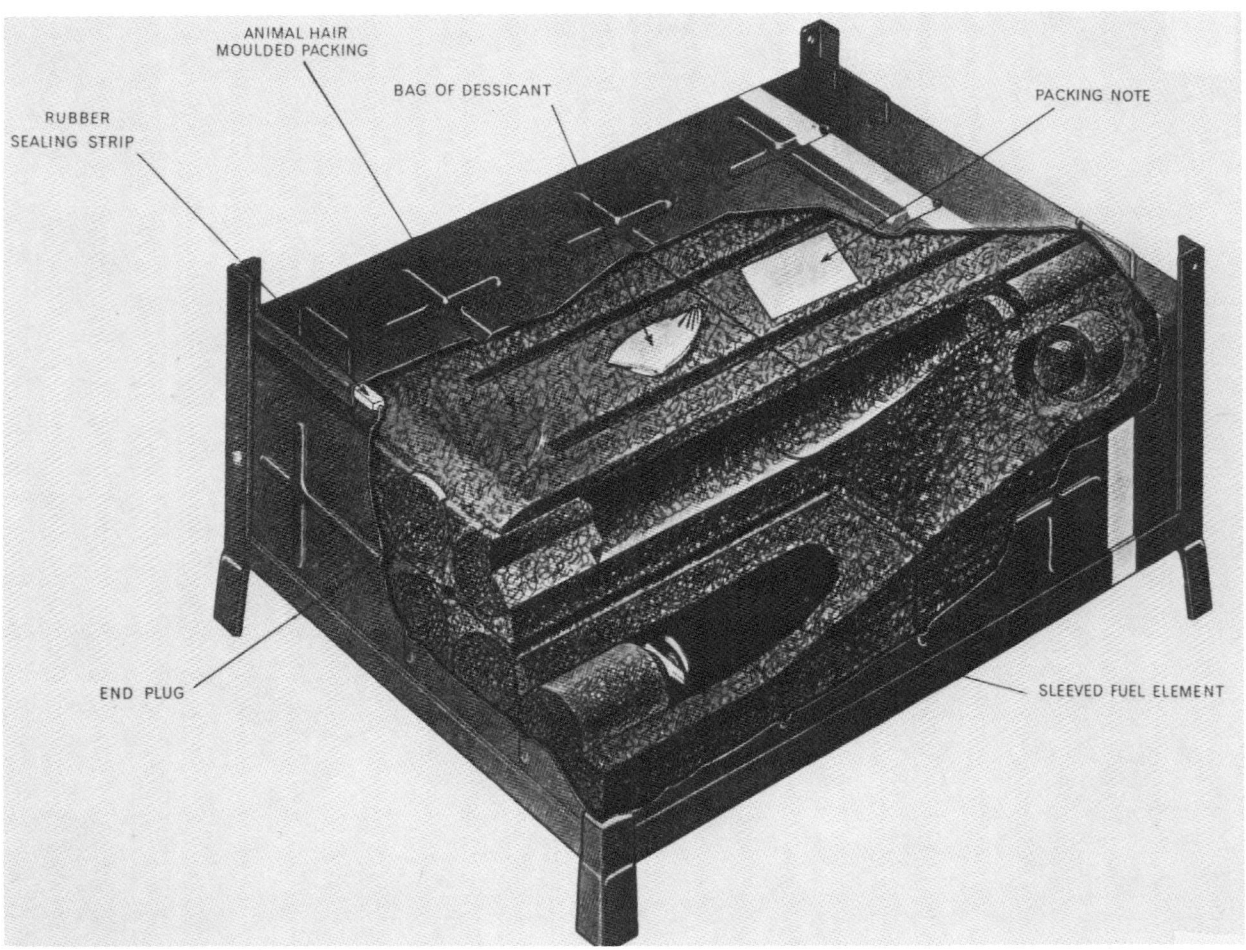

Fig. 3. Tokai Mura F.E. container exterior cutaway to show moulded packing.

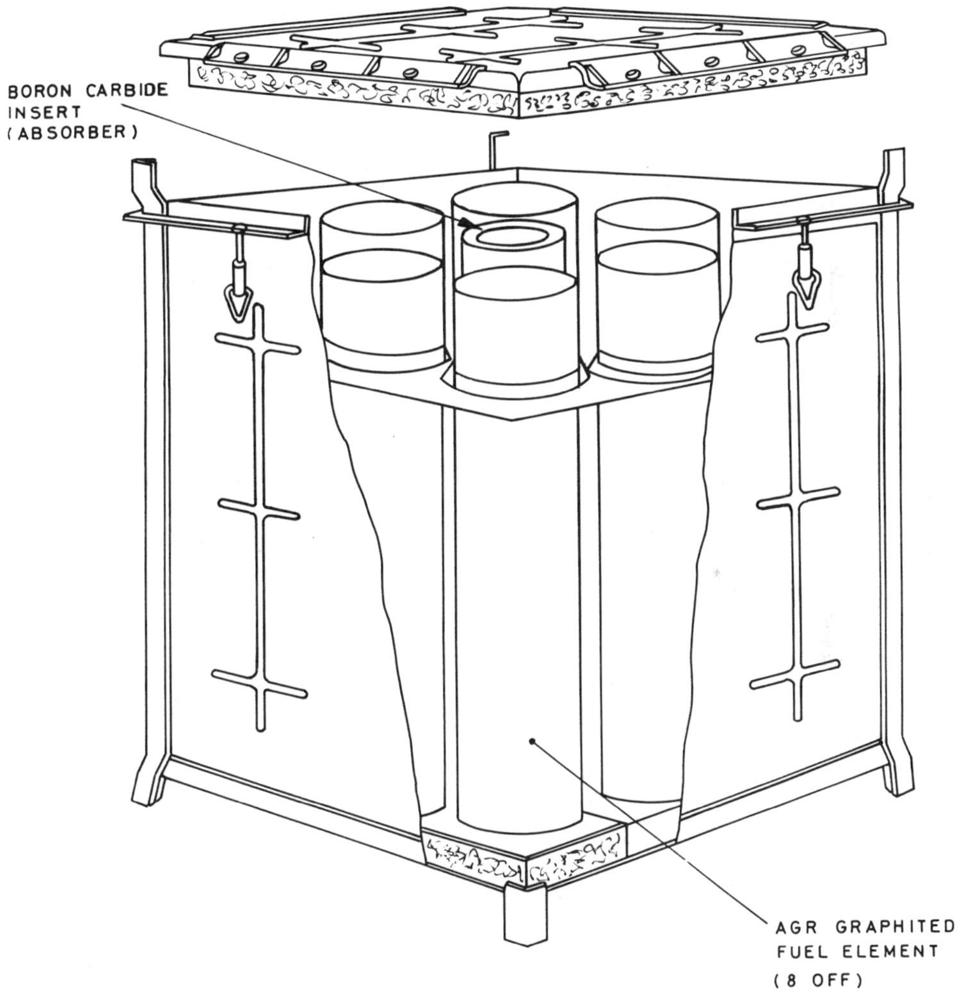

Fig. 4. AGR fuel element packing (type 1642).

DIFFICULTIES ENCOUNTERED IN FRONT END TRANSPORT OPERATIONS

The IAEA Regulations on the Transport of Radioactive Materials define logically the attitude and containment required for the safe transport of radioactive materials. The degree of containment is defined by rational assessment of the hazards presented by the radioactive material contained and the need to protect the general public against the risks from radiation both during normal operations and in accident conditions. The radiological impact of the materials involved in the front end of the nuclear fuel cycle has been assessed and can be shown rationally to be small compared with the possible impact of irradiated fuel. This is reflected in the requirements for packaging.

The public perception of the danger involved in anything labelled 'radioactive' can be very different from the scientific evaluation. further, the consequences of an incident involving material from the front end of the nuclear fuel cycle are more likely to cause political and public relations damage than physical harm. Thus adherence to the minimum standards called for by international regulations, while perfectly adequate for public safety, may provide inadequate protection for an industry so vulnerable to public opinion and political decisions.

During the course of its transport operations BNFL have been called upon to provide information and assurance in a number of situations. for example a shipment of food was held for a period by Saudi Arabian authorities when it had been discovered that it had travelled on the same ship as a consignment of natural uranium fuel elements destined for Japan. The dockers at ports and the Port Authorities themselves have required assurances about the safety of the materials that they were handling. Local Councils have been alerted by concerned inhabitants that lorries have been seen with radioactive symbols passing through their region and have demanded reassurances about the nature and potential impact of the loads. Organised demonstrators have taken action to protest about movements and have threatened interference with loads ranging from graffiti to physical damage. It is clear in all these cases that simple compliance with minimum legal requirements is only a starting point and that an assurance that operations were proceeding in line with regulations will not be adequate.

Given the number of movements simple statistical arguments will show that there will be traffic incidents involving vehicles carrying nuclear loads. There will not be the same feeling of assurance given by design considerations alone such as can be achieved by witnessing trains running into irradiated fuel flasks. Indeed it is paradoxical that while demonstrating the safety of irradiated fuel transport, it may have served to confirm the public impression that all nuclear movements are inherently very dangerous and require exceptional measures. For the front end of the fuel cycle it is essential that operating practices overcome or improve on the laws of averages and that adequate physical and public relations' responses are available to give assurances in advance and reassurances in the event of incidents.

RESPONSES TO THE PROBLEMS

From the experiences of BNFL there are a number of areas where operating practice either has been or needs to be improved in order to meet the demands of customers or the general public for safety and security in front end transport.

<u>Logistics and Design</u>

The requirements for testing type A packages are laid down in the
IAEA regulations and containers have been designed, tested and approved
for enriched fuel and powder movements. Hex, both natural and enriched,
is moved in containers fabricated to a standard international design with
a well established record of performance, including withstanding
immersion in the North Sea. For movements of low specific activity (LSA)
material, the requirement in the IAEA regulations is simply for
industrial containers. Uranium ore concentrate and other natural or
depleted powders are moved in the standard 45 gallon drums widely used
throughout industry. Specifications for the standard and quality of
drums are built into supply contracts and the system of double
containment with the drums in shipping containers has proved effective
for most circumstances over a period of many years.

The greatest area of variability rests in the movement of powder
where there has been less pressure for standardisation. BNFL cope with
three different types of containers, as described, which has made it
difficult to streamline or optimise the handling techniques. Further,
the practices at receiving plants can impose conditions which are more
stringent than the transport regulations. Powder containers are
necessarily filled in an area where uranium is exposed to the atmosphere
and where there is inevitably some degree of airborne contamination. The
containers are loaded in outer packages and transported to the receiving
plants where the bottles are off loaded-in many circumstances into what
is regarded as a 'clean' area. The level of contamination allowed in
these plants can be an order of magnitude less than the contamination
allowed on the outer surface of a package in the IAEA transport
regulations. For example the transport regulations call for the surfaces
of the outer container to be clean to 4 Bq/sq cm, equivalent to one
thousandth of a gramme of uranium. For full loads the outer container
can be interpreted to be the freight container, in which case there is
some relaxation on the packages themselves. BNFL have been asked by some
customers to achieve a cleanliness of 0.4 or even 0.04 Bq/sq cm on inner
containers within outer packages within freight containers.

This adds importance to the need for containers to be designed to
allow for ease of cleaning but also for plant designers to consider the
logistics of loading and off-loading material. Naturally, checks are
taken on containers to ensure compliance with contamination level before
they leave the site; though once again the method of obtaining results
can be a matter for discussion between parties. Obtaining representative
readings from, for example, unpainted wooden pallets and indeed cleaning
these, are problems which have to be overcome but could be avoided by
design.

Finally the security of any load in transit depends upon the
adequacy of the tie-down arrangements. In this regard, in the UK, BNFL
work to a Code of Practice agreed for the UK which sets standards beyond
the minimum legal requirements. (AECP 1006, 1979) This in turn has led
to sub-contractors when tendering for work, submitting designs of their
tie-down arrangements for consideration by BNFL in order to ensure
compliance with these standards.

<u>Handling Arrangements</u>

In moving over 15,000 tes of uranium in and out of its factories
BNFL have to handle packages ranging from gramme or kilogramme sample
quantities up to single cylinders containing 12 tes of uranium
hexafluoride.

A good number of the packages are in the IAEA category of Industrial Container. While these containers are robust enough for normal handling and minor incidents, their integrity could not be guaranteed in extreme accident conditions. As a general principle then, BNFL have adopted the policy of double containment wherever possible. For such movements it has procured 40 ft trailers with sliding side doors which give the advantage of a fixed containment while in transit but at the same time offer the flexibility of side loading which gives the maximum access for fork lift trucks. Where endloading is necessary, and in particular for UOC shipments, BNFL have recently built a dedicated off-loading facility with an unloading bay and docking ramps.

For hex movements the storage raft is served by Goliath. Goliath is a crane with a lift capacity of 20 tes running on rail tracks; the crane itself runs on a beam which has a span of 48 metres. This single crane is used to manoeuvre the stocks of natural and enriched cylinders on the 150 metre long storage raft.

With fuel elements the principle of double containment is already incorporated in the cladding of the element and the outer container of the box. The principle has been taken further by provision of an enclosed slide side container with specially designed tie-down points for the fuel boxes.

Operating Practices

Even with the improvements in handling, at the end of the day, the operation is vulnerable to operator error. Operator training, clear instructions on how things should be done and clear definition of responsibilities between the different sets of people involved in the process (packers, loaders, and drivers), all play a role in minimising risks of error. The system of training and instruction and the checking of loads within BNFL all form part of the Company Quality Assurance system with its audit procedures.

In any operation there will always be peak work loads, and because of this there is the necessity to employ sub-contractors. The control exercised here is less direct but procedures such as the use of approved contractors, the knowledge in advance of the drivers to be used and the necessary instruction for drivers who have not been regularly involved with nuclear loads, are all important features in the satisfactory movement of material. Modern training techniques, such as the use of video, moving in the future to interactive videos, are ideas which are being tried as a means of improving the training which can be given in short periods of contact.

A good proportion of movement however may be conducted by customers who can elect to deliver or collect their own materials. Any carriers employed by these people would necessarily need to comply with UK regulations. As already pointed out however, these regulations can sometimes be a bare minimum of what is needed and BNFL seek through contact with customers and by formal clauses in contracts to require extended co-operation between customer and BNFL over the transport arrangements to be employed in the UK.

Response to Incidents

BNFL at its Springfields and Capenhurst Works have always received excellent co-operation from the Police, Fire and other emergency services. By regular contact and their involvement in training

exercises, a good deal of knowledge and expertise has been built up. The
regional nature of these services however, makes it difficult to extend
the same degree of co-operation and training along the entire length of
all journeys.

In the UK the NAIR scheme exists to provide specialist advice to
emergency services on radiological incidents. For immediate responses
however, information available at the scene of the incident is important.
Apart from the training of the drivers, BNFL ensure that each load
carries with it information on the nature and potential hazards of the
material and the action to be taken in the event of incidents. For this
it has extended the use of the TREM card system required for hazardous
chemicals, to cover all its shipments. In addition each tractor unit
carries with it an emergency kit to cope with the immediate actions that
would be required for a breach of containment or other incident. More
extensive emergency equipment is held in readiness at the BNFL sites and
again each tractor unit carries with it a fireproof plate with a contact
number from which immediate advice or physical assistance can be sought.

The Company has demonstrated its readiness to respond to calls for
help moving quickly when necessary to the scene of the incident even
where the help has been simply to give reassurance. Recognising that
some locations may be remote, arrangements to allow for the use of
helicopters to provide more rapid transport have already been made.

<u>International Co-operation</u>

International co-operation is important because of the wide
variation of transport regulations between nations and a difference in
the expected actions to be taken in the event of incidents. National
knowledge and advice is essential in coping with these situations. A
delivery to a customer in one country can involve the movement of the
material through one or two neighbouring countries in order to achieve
the delivery.

The requirements for advance notification, escorting and
communications between the carrier and the point of delivery, can be
different in each of the countries through which the vehicle passes
whilst the planning and preparation for such movements is something which
requires experience and a good list of contacts.

The scale of operations means that it is equally likely that
incidents could happen abroad, as well as in the UK. Experience in the
UK has demonstrated that an incident involving a foreign carrier will be
treated in the minds of the general public as if it were the
responsibility of the domestic nuclear industry even where the legal and
practical responsibilities are clearly with the carrier. Because of the
importance of maintaining public support for nuclear transport BNFL has
shown it willingness to extend its response capability to cover movements
to and from its factories even when those movements are the
responsibility of another operator.

Discussions with other European operators, partners, customers, and
suppliers, have shown that the public response will be similar in other
countries. Through its contacts BNFL have detected a very welcome
willingness on the part of the international nuclear community to supply
assistance and support where necessary: a willingness stemming from a
common appreciation of the nonpartisan impact of bad publicity. For
example different operators in Germany are recognising the need for them
to have some common response capability to incidents. In contacts with
customers and overseas transport contractors, BNFL has sought to

establish, for all its overseas trips, a national contact to provide on
the spot assistance. This assistance can in the first instance be simply
providing information and advice, particularly where the language barrier
presents a problem; coping with the local media; and moving on from
there to practical assistance such as providing replacement vehicles or
assistane in recovering loads. To date by good fortune and good practice
there has been no requirement to test the system in the event of a
genuine nuclear incident. However, work continues to strengthen and
formalise the contacts.

Areas of improvement can be an agreement between operators
comprising knowledge of the emergency equipment being carried by vehicles
and that which could be provided from responding sites, and establishing
firm points of communication to ensure that speedy contact can be made.

Contact with the Media

The experience of the Mont Louis incident demonstrated the need for
a co-ordinated and unified response to the media. This requires
information being prepared in advance to guide responses and agreed in

Table 1. Number of Package Shipments occurring in a 12 month period

(a)

	Type of movement		
	UK internal	Export	Import
Exempt	54	153	11
Industrial	15,500	2,600	16,000
Type A (including Enr Hex Cylinders)	1,360	13,290	420
All Types	16,914	16,043	16,431

(b)

	Type of movement			
	Road	Rail	Air	Sea
Exempt	218	–	143	21
Industrial	34,100	600	1	18,600
Type A (including Enr Hex Cyclinders)	15,070	17	–	13,710
All Types	49,388	617	144	32,331

Table 2. Summary of Movement Details

Type of Load	Method of Transport	Package Category	Load Category	Container	Method of Carrying
UOC	Road/Sea	Industrial	LSA	Mild Steel Drum	ISO container
Natural/Depleted Uranium Hexa-fluoride	Road/Sea	Industrial	LSA	48 F & Y Hex cylinder	Modified ISO container
Enriched Uranium Hexafluoride	Road/Sea	Type A*	Fissile III	30 B Hex Cylinder in PSP	Flat bed trailer or modified open top ISO container
UO_2 Powder	Road/Sea	Type A	Fissile II	1610 metal drums in 1660 outer container 3 or 5 kg pails in BU5/BU7 30 lt plastic bottles in CE250 outer	Enclosed slide side freight container or ISO container
Magnox fuel elements	Road or Road/Sea	Industrial	LSA	Metal boxes BNFL type 0012 or 0013	) Flat bed or) enclosed) slide side) freight
AGR fuel elements	Road	Type A	Fissile II	Metal boxes BNFL type 1642	) container)
Other fuel elements	Road Road/Sea Road/Air	Type A+	Fissile II	Purpose designed boxes as appropriate	Flat bed or open top ISO container

*currently carried by special arrangement + PFR breeder elements require only Industrial containers.

advance to ensure that the information being passed to the media is consistent. It was some time after the Mont Louis incident before a clear appreciation was arrived at on the effect of water on hex in a cylinder and, again, some time before that appreciation could be translated into a presentation of the style which could be appreciated by the media.

If the media response to incidents is to be minimised or turned to advantage then quick accurate responses and responses in a style which can be easily publicised are required. It is interesting that in the incident of the food containers in Saudi Arabia, referred to above, references to radiation level, IAEA regulations, and the like left both the shipping company and the Saudi authorities unmoved and uninformed. It was pointed out that had the food been air-freighted out then it would have been exposed to slightly enhanced levels of radiation. Even standing the food next to the fuel containers would only have resulted in irradiation equivalent to the air-freight. It was only when the issue was put into these terms that there was a hint of movement in the situation.

CONCLUSIONS

The record of transport for the front end of the nuclear cycle is a good one. The scale of operations involved and the quality of the records of achievement are points which are not fully appreciated by the general public. Acceptance of these movements is something which has to be worked at by slow and persistent education wherever the opportunity presents itself.

There needs to be constant striving for excellence in procedures and operating practices in order to minimise the risk of publicity which could be used to jeopardise public confidence. Given the scale of movements, however, some incidents will be inevitable. In order to mitigate the effect of such incidents it is vital that industry is prepared in advance to respond swiftly both physically and with information.

Further, the response needs to be geared to satisfying the public perception of the risks rather than the straight scientific evaluation of the potential harm. In view of the interdependence of the nuclear industry internationally it is also essential that international contacts are maintained and extended to ensure that the response is not only immediate but co-ordinated and unified. By such measures it should be possible for the nuclear transport industry to build on its good record and maintain the public confidence necessary for continued operations.

REFERENCES

AECP 1006 (1979), Securing Radioactive Material Packages to Conveyances: AECP 1006 -June 1979.
UKAEA, SRD (1987), The Consequences of the Release of Radioactive Materials from Typical Cargoes carried in the English Channel and the North Sea: SG2(87 10 SRD Culcheth - May 1987.

SEA TRANSPORTATION OF HEAVY PLANT FOR THE POWER GENERATION INDUSTRY

J.W.Bristow, T.W.Stephen

James Fisher & Sons plc
P O Box 4
Fisher House
Barrow-in-Furness
Cumbria LA14 1HR

INTRODUCTION

The regular shipment of heavy and voluminous electric power generation plant for power stations, around the coast of the United Kingdom by sea and river, has been effectively and efficiently carried out over the last two decades.

The location of power station sites alongside, or close to, rivers, estuaries and the coast, makes sea transport a convenient, efficient, and environmentally attractive, operation. The sea shipment mode of transport minimises land transport operations thereby avoiding many problems associated with road transport such as traffic congestion, strengthening of roads, bridges and so forth. Units which cannot be moved by road because of their weight or size, can conveniently be moved by barge or ship provided suitable loading and off-loading facilities are available or can be erected. The plant or equipment is shipped as a complete unit which enables manufacturers to apply quality assurance and testing procedures to the products prior to leaving their works thus avoiding problems which could arise with "on-site" assembly.

This paper addresses, in broad terms, some of the operational aspects of heavy load vessels as well as design considerations pertinent to the sea transport of heavy and sensitive items.

REQUIREMENTS OF THE NUCLEAR INDUSTRY

It has been apparent since the advent of nuclear power station construction that, within the construction period, there will be a requirement for large heavy indivisible loads to be transported from their point of manufacture to the construction site. In order to achieve the necessary quality assurance, design criteria and finished tolerance together with factory testing of certain equipment, each plant item is preferably completely constructed and fully tested at the factory/ construction yard and fully tested, avoiding the costly prefabrication and assembly on site constraints. It is in considering the above points that transport modes need to be addressed.

By their nature nuclear power stations require cooling water and

The map of the United Kingdom (Map No.1) showing the extensive waterway network illustrates how coastal and inland waterway transport can be used to advantage in this country.

The designers of large items of plant need to take into account at an early stage the method to be adopted for transport. The design should allow for such items to travel the short distance from manufacture to the nearest appropriate port and the on-carriage by sea, bearing in mind that sea transport will require the particular items to be safely secured on board the ship. In some instances, ships will require to fix additional points to secure particular loads, depending not only on their weight but also their dimensions. The general term "sea fastenings" can be defined as points on the structure from which adequate wire/chain or purpose designed equipment can be attached without unduly stressing the item of plant to a point on the ship's structure. It is therefore important that designers and manufacturers get together early to plan the overall transport operation with the shipowners.

The availability of suitable ships is a further consideration at the planning stage. They must be able to serve both the manufacturing outlet port and the port closest to the construction site. To overcome the latter it is prudent to consider a purpose-built jetty at the site for this purpose. In nearly all cases the transport of equipment in the U.K. from manufacturing point to ports is by heavy load transporter vehicles. It is therefore both cost effective and efficient, particularly with the absence of fixed heavy capacity cranes at ports, to utilise a heavy load roll-on roll-off (Ro-Ro) vessel. This concept allows the vehicle, together with its load, to be driven on board the ship, remain with the ship to its destination, and then be rolled off direct to the nearest port to the nuclear power station or to a purpose-built jetty at the site.

HEAVY LOAD ROLL-ON, ROLL-OFF SHIPS

Ro-Ro ships built for the specific purpose of transporting heavy and voluminous loads, on a commercial basis, came into operation in Europe about the mid-sixties. America also had in service heavy load carrying Ro-Ro ships, some of which were converted army landing craft. These ships were used for transporting heavy and bulky refinery units from manufacturers' works to operational sites.

Since the mid-sixties the number and type of heavy load ships has multiplied considerably. Today, vessel versatility is the keynote, not only for accommodating various sizes and masses of units, but also for flexibility of loading and unloading. Systems of loading include float-on, float-off (Flo-Flo), roll-on, roll-off (Ro-Ro) and lift-on, lift-off (Lo-Lo). Flo-Flo vessels can lift and transport single loads of over 8000 tonnes, and if this figure is compared with the 100 ton heavy derricks installed in heavy lift ships about three decades ago, it will be readily realised that great progress has been made in the shipment of very heavy and bulky cargoes.

HEAVY LOAD VESSELS: KINGSNORTH FISHER AND ABERTHAW FISHER

The two heavy vessels, "Kingsnorth Fisher" and "Aberthaw Fisher" (Photo.1) were built to the requirements of the CEGB for transporting electrical plant from manufacturers' works to power station sites in the U.K. Both vessels came into service in 1966 and are owned and managed by

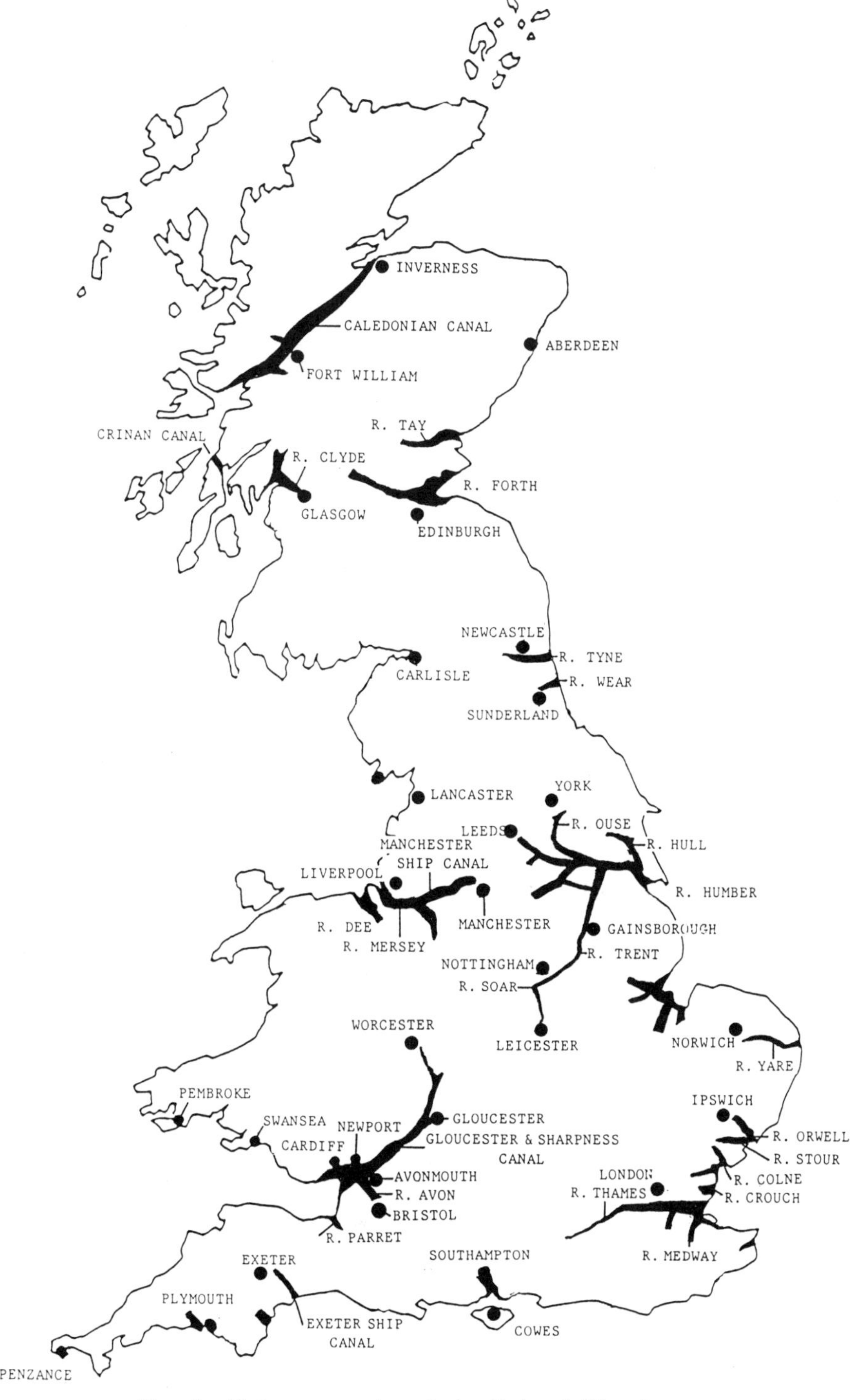

Map 1. Waterway network in United Kingdom.

Photo. 1. Aerial view of Kingsnorth Fisher.

James Fisher and Sons at Barrow-in-Furness who operate the vessels on charter to the CEGB. They were named after coal-fired stations in South Wales and on the River Medway. Map No. 2 shows locations of suitable berths for heavy load Ro-Ro ships.

These vessels were the first of their type, and the method of loading and unloading units, using a hydraulically operated roadway and an elevator for taking the units into the hold, is unique.

Figures 1a, b, c and d show the sequence of loading.

Firstly, the roadway is raised to the level of the quay and the load is hauled aboard using the vessel's 50 ton winch as illustrated in Figure 1a.

In the second and third operations the roadway is lowered to the level of the lift platform (Figure 1b) and the transporter is hauled forward until the load is central with the lift platform (Figure 1c).

The fourth operation involves the disconnection of the front and rear bogies of the transporter from the girders, once the load has been jacked down on to the timber bearers, and the load is then lowered into the hold where it can remain on the hatch for the sea voyage.

These ships have been designed to carry loads on deck, provided the plant can be transported in exposed conditions. They were originally designed to carry single pieces of electrical plant, on girder frame trailers, having a net weight of up to 300 tons. Since the vessels came into service, however, the growth in weight of electrical plant has necessitated a greater lifting capacity for single loads, and loads of up to 470 tons can now be shipped (subject to certain unit loading conditions being met).

Map 2. Locations of suitable berths for heavy load Ro-Ro ships.

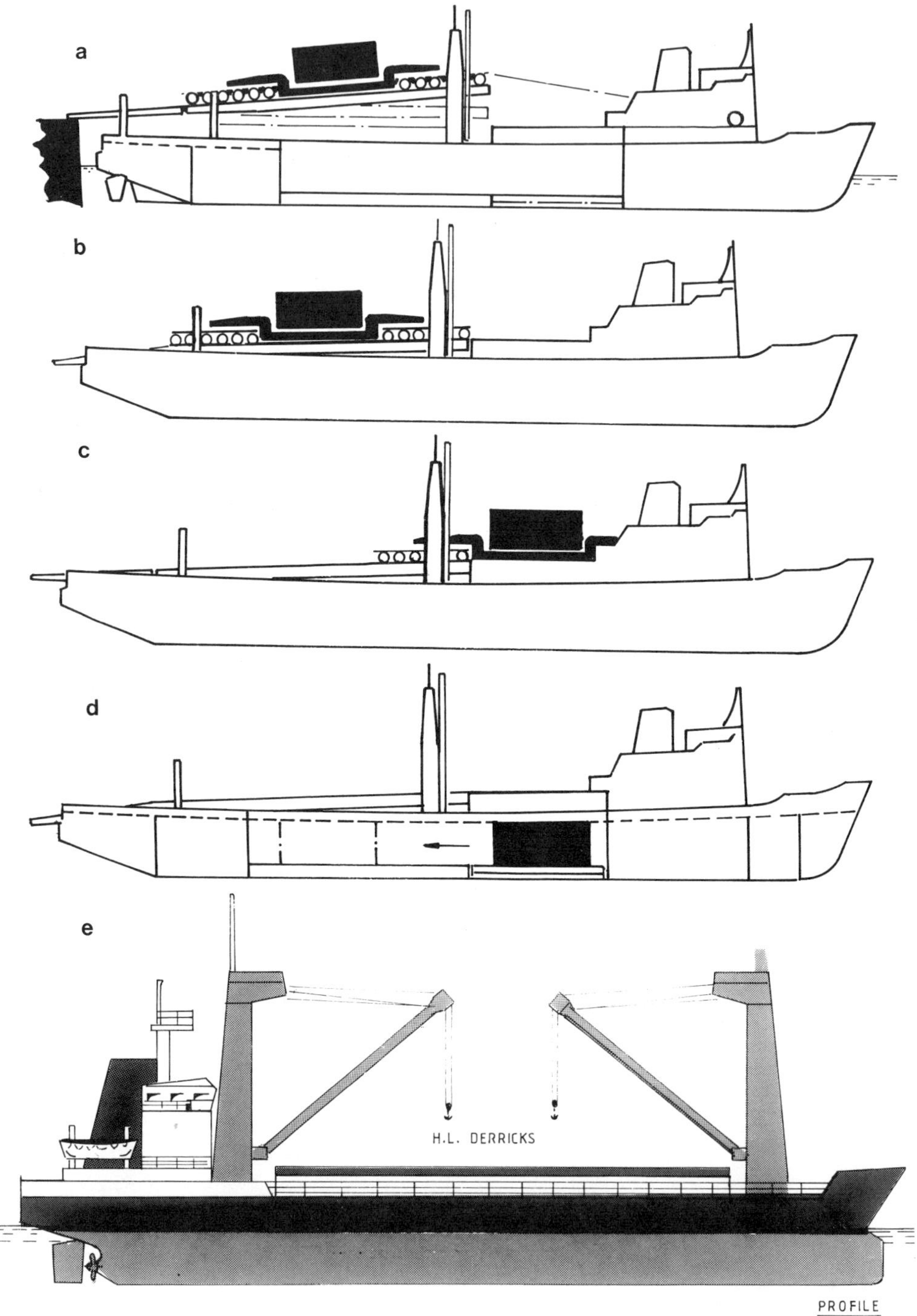

Fig. 1a, b, c and d. Method of loading a transporter onto HLV's
MV Kingsnorth/Aberthaw Fisher.

Both ships are under the British flag, manned by experienced British crews, fully versed in the safe and efficient carriage of large heavy indivisible loads.

Their shallow draught and engine/bow thrust configuration make them ideally suitable for the varied coastal regime of the United Kingdom. Vessels have, over the years, been deeply involved with the new generation of power stations, delivering plant items through small fishing harbours, for example Port Penrhyn in the Menai Straits, North Wales, in connection with the pumped storage power station at Dinorwic. In this operation large tunnel lining sections, penstocks, turbines and generating equipment were successfully delivered on time through this small harbour, which dries out at low water. Other examples are: Combwich Quay, River Parret, for Hinkley Point Power Station; a purpose-built breakwater and berth constructed on the shore line for Torness Nuclear Power Station; Heysham Harbour for Heysham Nuclear Power Station; and Folkestone Harbour for the cross-channel link transformers at Sellindge. These are but a few, all of which were jointly considered at an early stage with design teams, hauliers and shippers, taking into consideration that weather and tidal factors, good communications and co-operation from the various harbour authorities are essential.

The ships have the flexibility to cater for tidal range and shallow waters. On average, from docking, a heavy load of 470 tons can be safely delivered ashore in 45 minutes. Photographs 2 and 3 show an advanced gas cooled reactor boiler unit being off-loaded at Heysham.

ENVIRONMENT ISSUES

Nowadays, environmental issues are another significant factor at the early planning stage and have a bearing on the location and design of off-loading facilities. Sea-borne transport certainly assists in minimising the disruption to the general public, reducing delays in traffic flow, particularly at holiday periods, or eliminating the need for major road improvements which may be required in respect of heavy loads.

DEVELOPMENT AND DESIGN OF THE HEAVY LOAD VESSEL

In this paper it is fitting to discuss the origin of the heavy load vessel (HLV) and its subsequent development up to the present day, identifying at the same time the pros and cons and characteristics of the various types.

The geographical locations of the plant manufacturing sites and nuclear and fossil fuel power stations impose, in many instances, constraints on purpose-designed heavy-load-carrying vessels in regard to their size.

At the outset it is important to identify all restricting factors which are likely to affect, or influence, design criteria. In this respect a detailed study of sea, river and canal routes from loading points to discharge sites should be undertaken to ensure the movements of the proposed vessel will not be inhibited or prevented by water depth, river or canal width, bends, clearances below bridges and lock lengths and widths.

Because they have to operate in narrow, shallow and tortuous waterways, as well as in coastal waters, heavy load vessels rarely exceed

Photo. 2. Kingsnorth approaching berth at Heysham.

Photo. 3. Kingsnorth off-loading advanced gas cooled reactor
component at Heysham

Table 1. Typical Plant items which are shipped on heavy load vessels

Plant Item (Tonnes)	Net plant weight and dimensions			
	Weight	Length (m)	Width (m)	Height (m)
Heavy Electrical Plant				
Grid Transformer	141	8.3	4.5	4.9
Generator Transformers (3 x 1 single phase units)	188	6.3	4.5	4.9
Generator Stator Inner Cores	272	10.9	3.9	3.8
Generator Outer Casings	127	11.0	5.4	4.9
Generator Rotors	76	*	*	*
Turbine Plant				
HP Turbine Modules	179	6.7	4.0	4.9
LP Turbine Modules	167	8.0	6.9	5.6
Moisture Separator Reheaters	240	22.0	5.2	5.6
Condenser Nest Modules	235	17.3	6.6	6.0
Condenser Neck Modules	107	20.0	6.6	6.3
HP Heater Modules	175	17.2	6.0	6.5
LP Heater Modules	130	15.5	6.0	5.0
De-aerators	105	35.0	5.7	6.7
Turbine House Crane Beams	106	4.3	4.2	2.0
Turbine House Crane Trollies	65	9.5	6.6	3.6
Nuclear Plant				
Reactor Pressure Vessel	358	12.0	6.8	6.3
Reactor Lower Internals	127	14.0	4.6	5.0
Reactor Upper Internals	69	6.0	4.5	4.3
Pressurizer	100	14.5	2.5	3.0
Steam Generators	372	21.3	5.8	5.8
Polar Crane Beams	195	44.0	4.0	4.5
Polar Crane Trolley	120	16.0	7.0	3.0

3,000 tonnes deadweight. It should be noted that the deadweight is the total cargo weight of the vessel including fuel, fresh water and stores. It does not necessarily signify that the vessel can carry a single piece of plant of that weight.

Table 1 is a list of typical plant items which are shipped on heavy load vessels.

Heavy load carrying ships can be divided into categories as follows:-

(i)   Lo-Lo type (Figure 1e)

(ii)   Ro-Ro types (Figure 2a & b)

(iii)   Ro-Ro, Flo-Flo, Lo-Lo (Figure 4)

The general characteristics of the above types of vessels, and basic differences, are broadly discussed as follows:-

<u>LIFT-ON, LIFT-OFF TYPE</u> (Figure No. 1e)

Before Ro-Ro H.L.V.'s came into operation commercially, this was the sole means of handling heavy unit loads. Such loads were shipped on general cargo vessels which were equipped with a heavy lift derrick. Stability of these vessels was not a problem when loading or unloading heavy units, because the deadweight was in the order of 10,000 tons and the mass of the unit (up to 100 tons) was small by present-day standards. These deep draughted vessels were only suitable for plying between major ports which had deep water access.

For vessels up to 3000 tonnes deadweight any overside heavy lifting operations, using either the vessel's own derricks or cranes, would require ballasting to maintain a safe stability margin and a satisfactory angle of heel. The amount and location of water ballast depend on the particulars of the vessel's hull geometry, its displacement and centre of gravity (vertical), the mass of the units being lifted and the height and outreach of the derrick or crane.

Some advantages of using derricks or cranes for handling heavy items are:-

- Quay threshold height and ship's freeboard are not critical as in the case of Ro-Ro vessels.
- Items can be loaded relatively quickly to their final stowed positions in the hold.
- They are independent of port cranage and Ro-Ro facilities and therefore only require a suitable berth or apron to load from.
- High concentrated loadings on the quay apron are avoided.
- Derricks can load items direct on to pontoons or barges for inland waterway shipment.

Some disadvantages of derrick ships are:-

- Weight of derricks and stiffening increases the vessel's lightship - and therefore reduces its deadweight.
- Deck and hold space is lost by way of derricks.
- Height of derricks could prevent vessel from operating in rivers and canals with overhead obstructions.
- Stability has to be carefully monitored when lifting items.
- Capital costs for installing derricks are high. Maintenance costs are also high.
- Statutory requirements call for testing at regular intervals.

ROLL-ON, ROLL-OFF (Gearless) VESSELS - (Figure 2b)

The origin of the Ro-Ro vessel (general type) was probably the LST's which were used for military operations.

The arrival of the Ro-Ro ship concept in Northern Europe in the early 1960's engendered great interest among the shipping fraternity, who were quick to recognise distinct advantages over conventional vessels - mainly in the areas of rapid loading and unloading and flexibility.

In tidal ports, link spans were constructed to allow these vessels to load and unload their wheeled cargo at any state of the tide; thereby keeping down-time in port to a minimum.

In impounded ports, the vessel's own ramp accommodated height differences between quay and deck to facilitate loading.

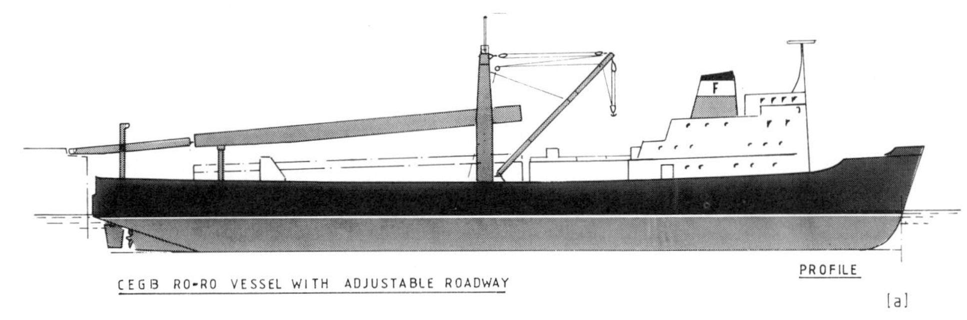

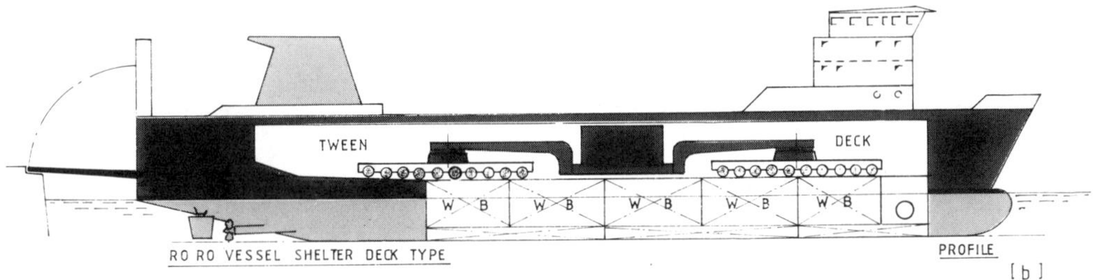

Fig. 2. (a) Profile of HLV's Aberthaw/Kingsnorth Fisher
(b) Profile of typical Ro-Ro Vessel.

In addition to the transportation of commercial freight on roll-on, roll-off ships, it was recognised that this method of loading, using wheeled vehicles, could be applied to the shipment of heavy and large indivisible units. Over the last twenty years or so, purpose designed Ro-Ro heavy load ships have been operating with great success.

Advantages of the Ro-Ro heavy load ship can be summarised as follows:-

- Both transporter and load can be shipped as a combined unit.
- The combined unit can be speedily loaded and unloaded.
- During loading/unloading operations, the vessel's stability loss is small compared to lift-on, lift-off operations.
- Provided the beam of the vessel is not designed to accommodate a particular width of unit, a Ro-Ro ship's beam may be less - for stability reasons - than the beam of a derrick ship.
- Air draught is generally less than a derrick ship's air draught.
- Vessel's strength can be up-graded to carry heavier units more easily - and at less cost - than a derrick ship.

Considerations affecting the use of the Ro-Ro vessel are:-

- Vessel's Ro-Ro deck level and quay level have to be compatible.
- For very heavy loads going over the stern or bow, high ballasting rates are necessary to maintain a satisfactory trim of the vessel.
- Quay strengthening may be required to accommodate high concentrated link span loading on apron.
- To encompass a wide range of quay heights and/or tide levels, long link spans and ramps and provision of water ballast are necessary.
- Unless an elevator is provided (an expensive item) in the vessel, the "hold" space will be in the form of a tween deck and this would result in a vessel with a high freeboard to accommodate the maximum height of cargo.

FLOAT-ON, FLOAT-OFF HEAVY LOAD VESSELS - (Not illustrated)

This type of vessel is effectively a self-propelled floating dock which came into operation about the mid 1970's. Heavy lift operators saw a need for such vessels and had several of this type of vessel built. Their principal role is in the movement of very heavy floating plant such as barges, dredgers, jack-up rigs, offshore modules, ships etc.

Cargoes are floated into the vessel's hold or on to the submerged pontoon's deck - usually in an estuary or a sheltered location with sufficient depth of water - following which the vessel is de-ballasted to lift the cargo out of the water. A stern door renders the hold watertight and enables the vessel to transport the cargo under cover.

ADVANTAGES OF FLOAD-ON, FLOAT-OFF CONCEPT

-       Vessel can carry ultra-large floatable unit of cargo that cannot be lifted by derricks or sheerlegs.
-       No special loading facilities are required for loading or unloading, just sufficient depth of water.
-       Can be designed as a "mother" ship from which a shallow water type self-propelled pontoon would operate.

Very few obvious disadvantages exist with this type of vessel because it can be adapted to combine the benefits of float-on, float-off with roll-on, roll-off and lift-on, lift-off.

COMBINATION OF FLOAT-ON WITH ROLL-ON AND LIFT-ON - (Figure 3)

All these operational functions can be carried out satisfactorily with this type of vessel. The lift-on facility, either in the form of derricks, cranes or gantry, can be an expensive addition and a thorough analysis of this mode of cargo handling must be made to ensure that the facility is necessary and that the cost is justified.

Likewise, the float-on concept of loading could be a very expensive facility if used for a small part of its operating life.

The previous brief comments on heavy load vessels are broad generalisations which are intended to highlight fundamental design differences - and consequential merits and drawbacks - of each type.

To design an efficient and cost effective vessel for a particular purpose, it is necessary to identify a range of factors which would be influential in formulating the correct concept.

Inputs and information required:-

-       Is a "float-on" capability required?
-       Will the vessel be expected to load or discharge its cargo from a beach?
-       Is it probable that the trailer and load may have to be trans-shipped on to a barge for inland waterway operation?
-       Port particulars including:-
        - Ports vessel will operate from;
        - Quay heights above water level;
        - Whether ports are tidal or impounded - or both;
        - Facilities for Ro-Ro vessels;
        - Maximum load capacity of quay apron;
        - Water depths.

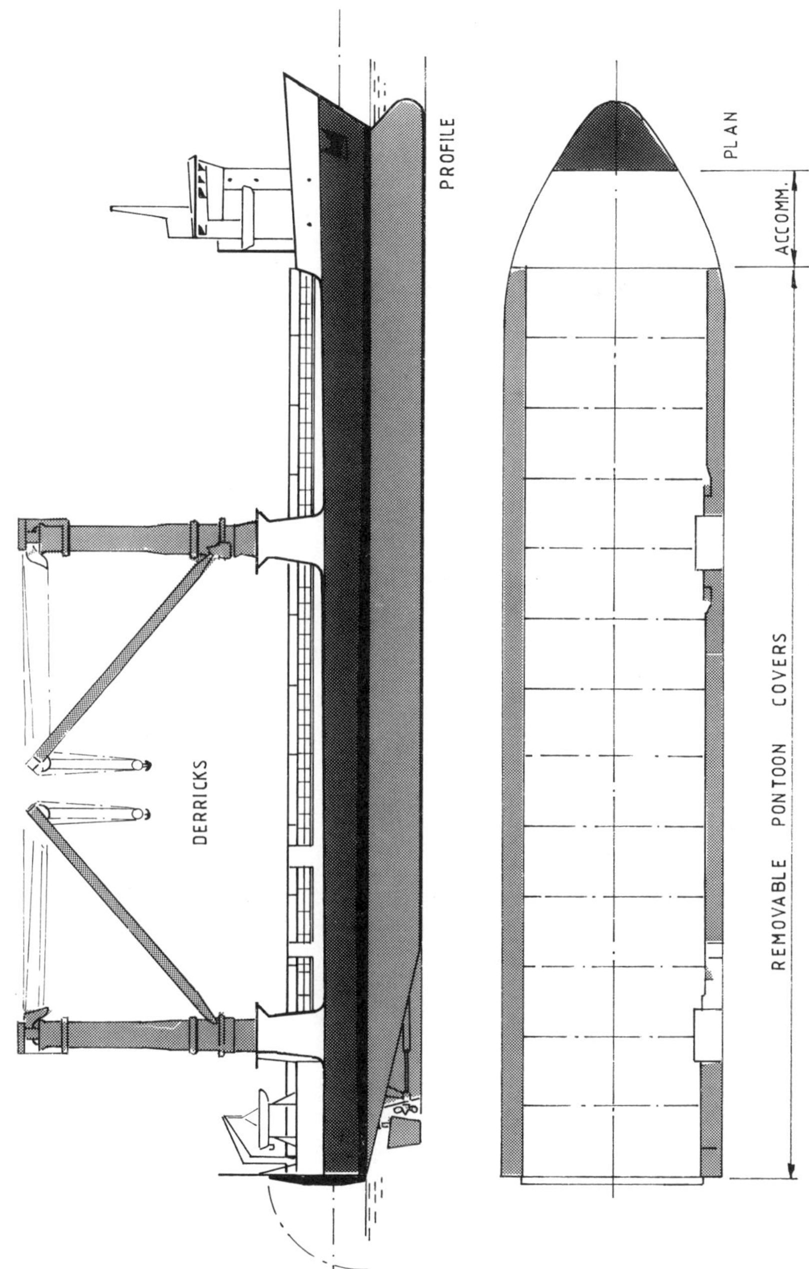

Fig. 3. Profile of Ro-Ro, Flo-Flo type vessel.

        - Access clearance for trailers behind Ro-Ro berth.
        - Areas where vessel is likely to be beached or to load aground.

-     Particulars of Transporters:-
      (a) Maximum length and width of Transporter.
      (b) Maximum laden mass of Transporter.
      (c) Bogie lengths.
      (d) Maximum wheel loads.
      (e) Maximum ramp angles and girder clearance.
-     Indication of vessel's operating life.
-     Indication of maximum 'g' forces plant can withstand.
-     Any constraints, in addition to existing constraints, on the vessel's principal parameters, e.g. draught, beam or air draught, would have to be determined.
-     Information on ports of shipment from which the vessel would operate, and additional operating requirements (e.g. beach loading/unloading capability), would be required.

VESSEL CONCEPT

The correct concept of vessel, and size, at the outset is of paramount importance to ensure that it will operate effectively and will be cost effective throughout its predicted operating life.

To achieve this end, it is necessary to obtain, from the client, relevant information or guidelines on the specific requirements for the vessel in regard to load particulars (sizes, masses and numbers), method of transporting the load, frequency of movements, berth particulars and loading/unloading facilities, if existing.

The various types of H.L.V. already described, apart from the ABERTHAW/KINGSNORTH FISHER vessels, have been designed to provide a world-wide service for a wider and more general heavy load sector of business covering miscellaneous items such as process plant equipment, barges, pontoons, dredgers, electrical plant etc.

As well as dealing with a wide variation in heavy loads, the vessels have to have a capability for loading and unloading these units from quays of widely varying threshold heights. These vessels do not usually ply between defined ports - where the port's loading facilities are known during the design stage.

DESIGN CONSIDERATIONS

It is important to distinguish between two fundamentally different types of Ro-Ro ships.

Figure 2a an elevation of the Kingsnorth Fisher/ Aberthaw Fisher type where the piece of equipment is rolled over the stern and on to the roadway situated on the weather deck. A lift is provided, to move units that require protection from the elements, from the weather deck into the hold where they are stowed.

Apart from the loading system and hold transfer system, this vessel is similar to any other single deck cargo vessel in regard to hull layout.

Figure 2b illustrates an elevation of a vessel with a tween deck. The equipment is loaded through stern doors, or bow doors if fitted,

374

directly into the hold or between deck space. With this type of vessel it is vital to preclude water from the tween deck otherwise the vessel will be rendered unstable. At present this type of vessel has no compartments in the tween deck space to reduce free surface effects of any water on the deck, however caused, which adversely affects stability.

Transverse bulkheads which subdivide the tween deck into compartments are not practicable for long trailers.

A major problem with the shelter deck concept is that it has a much reduced loading/unloading range by virtue of the location of the deck in relation to the waterline. Its capacity is restricted to the length of the tween deck and as the transporter would take up most of the length, especially the very long transporters, very little space would result for carrying multiple under deck cargoes.

The height of the tween deck would have to accommodate the highest units which require to be stowed under cover, and if this is added to the freeboard deck height the depth of the vessel would be such as to confine it to loading only from ports with the highest quays. The loading arrangement and deck level must be such as to permit loads to be rolled on board or discharged within the specified range of quay heights above waterline.

A vessel of the type described above would have a high windage area and, as a result, would require powerful and effective steering gear and thrusters to maintain its course in confined and tortuous channels and rivers under storm conditions. The part of the vessel below the freeboard deck would comprise the machinery space and ballast tanks.

The single deck type of vessel with a hold (Figure 2a) has the advantages of being able to load over a wider range of quay heights and also to maximise on hold and deck capacity. The "Kingsnorth Fisher" and "Aberthaw Fisher" have been designed on this principle, except that the variation in quay heights is allowed for by adjusting the height of the roadway which accommodates the trailer and load.

HULL FORM

The hull form design is critical for the satisfactory performance of a heavy load vessel. A shallow draft design requires a hull having a high block co-efficient and low length to beam ratio whilst low resistance properties call for a low prismatic co-efficient. Also, to maintain a satisfactory stability standard with cargo on deck, it is necessary to have a vessel with a low length to beam ratio (Figure 4).

The locations of the longitudinal centres of buoyancy are critical for establishing a satisfactory service trim for good sea-keeping properties and good hydrodynamic qualities, and for loading and unloading trim.

Excessive metacentric height results in high rolling accelerations which will induce high 'g' forces in the cargo, especially when the cargo is situated far from the centre of lateral resistance of the vessel.

In this respect, the geometry of the hull form, as well as the distribution of mass, determines the motion characteristics of the vessel and it would be prudent, in fact essential, to address this particular aspect of hull design in order to ensure that, after taking all other hull design factors into consideration, the effects of vessel motions on

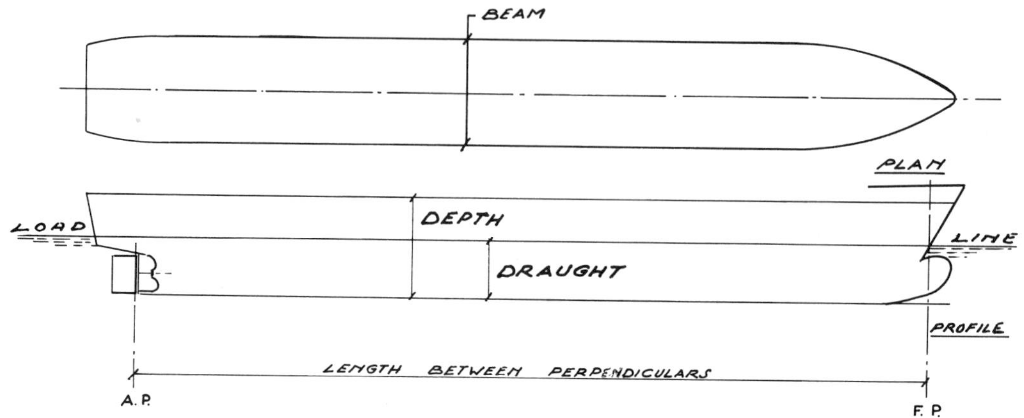

Fig. 4. Definition of ships' principal particulars

the structure and cargo will be mitigated when the vessel is operating in rough seas.

This aspect of hull design involves a study of wave histograms and probability distribution for the vessel's operating zones (U.K. coastal waters) from which "response amplitude operators" (RAO's) are computed for the vessel. This information produces vessel excursions and accelerations - at any point in the vessel - in the six degrees of freedom (roll, pitch, heave, yaw, surge, sway) (Figure 5).

A high block co-efficient and low length to beam ratio conflict with the characteristic features of a good hydrodynamic form but are necessary to meet a shallow draft criterion and minimum stability standard. The end result has to be a compromise but most of the above factors can, with careful form design, be integrated to give an optimum hull form that would meet the necessary and desirable standards of stability, shallow draft and hydrodynamic performance.

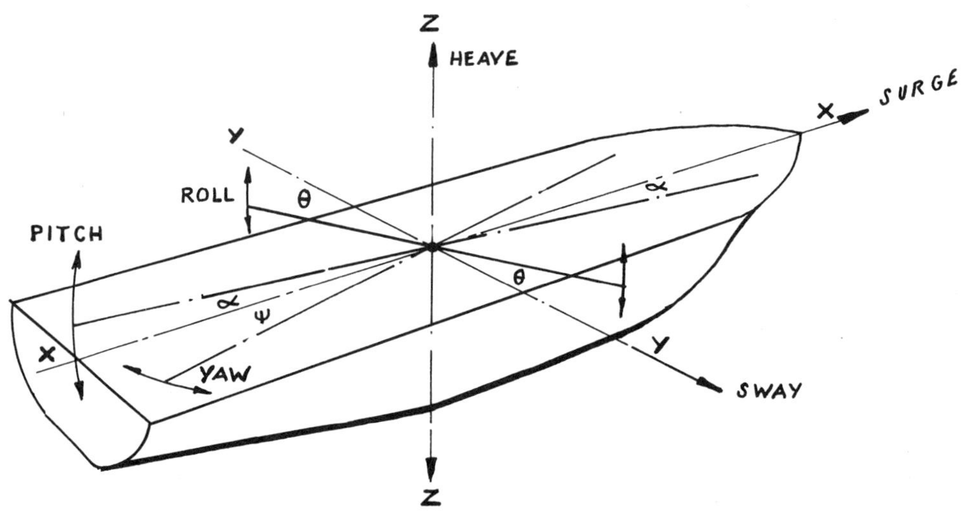

Fig. 5. Definitions of six degrees of freedom.

STRENGTH

Vessel's strength can be divided into longitudinal, transverse and local strength. The longitudinal strength is maintained by the hull girder and must withstand the worst loading conditions envisaged. The high concentrated loads to which this type of vessel will be subjected must be resisted by the hull. In this respect, it is anticipated that the bending and shearing stresses would be low due to the high modulus of the hull, especially if longitudinal bulkheads forming the ballast tanks are fitted. Local strength applies to structures such as the lift platform, hatch covers, link spans, winch and hydraulic ram reactions etc., in the case of the Kingsnorth and Aberthaw Fisher.

Most strength calculations (hull in particular) would be carried out using the finite element method, which is used by Lloyd's and other classification societies.

STABILITY

All sea-going vessels must conform with minimum legislative stability standards. In the United Kingdom these standards are contained in The Merchant Shipping Load Line Rules 1968 - Schedule 4, Part 1, Paragraph 2.

The minimum stability criterion has been formulated to ensure that a vessel, when heeled from its vertical position due to wind and wave action, will not capsize. The vessel must always eventually return to its original upright attitude. Stability standards and good seamanship must be combined to avoid a capsize situation occurring.

Stability Conditions are prepared in the Stability Booklet to cover various loading conditions of the vessel. The curves of righting lever against angle of heel are drawn for each condition and the areas under the curves are calculated up to various angles. These areas must not be less than the areas given in the regulations for the respective angles.

These curves are known as statical stability curves because they do not take account of dynamic conditions.

SHIP MOTIONS

In the design and operation of heavy load vessels, it is of paramount importance to address the problems associated with ship motions to ensure that heavy units of plant being carried will not break their lashings and also to ensure that sensitive plant, such as transformer cores and windings, are not damaged. It is equally important for manufacturers of heavy plant to be aware of the nature and order of magnitude of forces attributed to ship motions in order that they can be accommodated in the design of the structure.

Ship motions include what are known as the "six degrees of freedom" and they are defined as follows (Figure 5):-

"Roll" occurs about the vessel's longitudinal axis.
"Pitch" occurs about the vessels transverse axis.
"Heave" is the vertical movement of the vessel
"Yaw" is the rotation about the vessel's vertical axis.
"Surge" is the forward movement of the vessel.
"Sway" is the sideways movement of the vessel.

Generally, only roll, pitch and heave need be considered for determining ship accelerations.

When the ship is rolling the maximum deceleration and acceleration forces on a piece of equipment occur at the end and beginning of a roll, respectively. The magnitude of the acceleration forces is related also to the location of the piece of equipment in relation to the centre of oscillation of the vessel. The farther away the equipment is, the greater the 'g' forces. This means that units carried on deck will be subjected to higher 'g' forces than units in the hold.

Compounded to the roll 'g' forces are the static force components at the angle of heel being considered and also the probability of increased 'g' forces due to sway.

Pitching motions are not so severe as rolling motions but the combined effect of pitch and surge could be significant.

Synchronous rolling motion with a train of waves would lead to excessive excursions and high static component forces due to the mass of the unit.

Forces due to heaving motions can also be significant and must be considered along with the other motions.

From the ship design aspect, the naval architect has to examine the stability of the vessel with the maximum number of heaviest units on deck to determine the beam of the vessel. In most cases, heavy load vessels have a small length to beam ratio for this reason. The wide beam reduces the draught, but it also results in reducing the rolling period when the heavy loads are in the hold. Such a situation leads to higher 'g' forces on the units under certain sea conditions. Very often topside ballast tanks are fitted to increase the rolling period and reduce the 'g' forces. Both Aberthaw and Kingsnorth Fisher are fitted with topside tanks for this reason.

DAMAGE SURVIVABILITY OF HEAVY LOAD SHIPS

At present there are no mandatory regulations for the subdivision of heavy load ships or, for that matter, general dry cargo ships. Chemical carriers and oil tankers are required to meet stringent damage survival standards in accordance with MARPOL requirements. These requirements are intended to prevent, or mitigate the effects of, pollution at sea.

Dry cargo vessels carrying sensitive cargoes may be required to conform with survival standards of foreign countries if the vessels operate in their waters.

The International Maritime Organization (IMO) are currently drafting proposals for the damage survival of dry cargo ships over 100 metres in length, and it is anticipated that these proposals, when finally agreed, will become mandatory for new ships in about 5 years' time. At present it is proposed to exempt ships of below 100 metres subdivision length. The subdivision length can be defined as the greatest moulded length of that part of the ship at or below the deck or decks limiting the vertical extent of flooding.

These draft proposals for ships over 100 metres in length will involve extensive subdivision of the vessel's hull and will introduce problems in regard to cargo disposition in the ship's hold.

To summarize, any degree of subdivision for a one or two compartment standard would be at the request of the owner, as no mandatory requirements exist at present.

CONCLUSION

It is considered that water-borne transportation of heavy, voluminous and sensitive cargoes will continue to be the most effective and efficient means of moving such plant, with the added advantage of avoiding populated areas. Nuclear plant can in many cases be loaded straight from the manufacturers' works, or close to them, and shipped directly to a purpose-built berth, adjacent to the station, where the plant can be rolled straight into the power station.

Co-operation by all interested parties - manufacturers, designers and transport operators - is essential to ensure, inter alia, that the plant is safely delivered to the station and unaffected by the voyage.

RAIL TRANSPORT OF VERY LARGE NUCLEAR COMPONENTS

M.W.Snow, D.J.Bargh

Combustion Engineering Inc.
100 Prospect Hill Road
Windsor
CT 06095
U.S.A.

ABSTRACT

Overdimensional or overweight nuclear products of Commbustion Engineering (C-E) have been shipped by rail, over the water and on road conveyances. Constraints imposed by the structural and dimensional limits of a particular route have strongly biased the choice of method and in one case, even justified the design and purchase of a unique railroad transport.

C-E acquired the 36 axle Schnabel (US Patent Nos. 4,041 879; 4,080,905; 4,083,311) rail car in order to move items weighing more than 800 short tons over long distances by railroad. Doing so takes advantage of existing rights-of-way and more robust bridges and route structures than exist for highways. Railroad dimensional clearances are generally greater than those for highways, and the schnabel car was designed to utilize that situation. It can raise, lower, and horizontally move the load (carried in between two similar car halves) for avoidance of obstructions which encroach upon the transport route. The schnabel car's integral power systems make it possible to reach directly and pick up a load from, say, the deck of a barge. The car's structural flexibility is such that it can alternately roll onto and off a floating barge or other seagoing vessel to load or offload cargo.

The 36 schnabel car is the world's largest rail vehicle and has thus far operated successfully over eight of North America's Class 1 railroads. Because it is so large, it includes several extraordinary onboard systems, such as an alarm system to warn operators if one side of the car becomes more heavily loaded than the other, and a "bootstrap" rerailing system which can, if ever needed, reach down to derailed wheelsets and lift them back onto the track.

With fewer nuclear components to be delivered, C-E's emphasis is now shifting to use of the 36 schnabel car for the shipment of spent fuel and dismantled nuclear plant components. In decommissioning activities, the use of this equipment permits minimal civil construction and environmental impact, and minimal disruption to other on site facilities. This is so because most nuclear plant sites already include rail installations on which the loaded car can depart with virtually no additional construction. In the case of spent fuel shipments, the schnabel car's high carrying capacity makes fewer individual shipments possible for a given quantity of fuel and provides a more massive, protected and controlled shipping method than do most alternatives.

BACKGROUND SITUATION

     Combustion Engineering's (C-E) large plant in Chattanooga, Tennessee is ideally located for water shipment of large fabricated components like nuclear pressure vessels. A gantry crane with capacity of 800 short tons is situated such that barges of up to 55ft width can be directly loaded from the shop door. Early nuclear plants had been conveniently located on navigable waters, and C-E's delivery obligations for large vessels were met with safe arrival of barges at customer-provided berths. Customers were then responsible for barge unloading and transfer of components to final positions within their plants.

     As the number of nuclear plants constructed in the United States grew in the 1970's, however, more and more of them were sited remote (hundreds of miles) from navigable waters. With the trend toward remote siting of plants came increased risks associated with long overland delivery routes and customers' bid specifications began to place those risks with suppliers. In C-E's case, not only was overland delivery distance of great concern, but also, C-E's design employed steam generators which were very large at about 70 feet in length , 21 feet in diameter (Figure 1) and 800 short tons in weight. Reactor vessel size was dimensionally similar to steam generators, but the reactor's weight was considerably less, at about 500 short tons.

     The problem of overland delivery of mammoth vessels was greater for C-E than for any of its competitors, because competitors' nuclear plant components were smaller than C-E's due to process and design philosophy differences.

     C-E utilized contacts in the transportation industry to estimate overland transportation costs for nuclear components at the proposal stage. However, the vessel sizes and weights were so much greater than transportation companies had previously carried over long distances that much uncertainty was associated with those estimates. Doubt existed also as to whether delivery of the large vessels to some proposed plant sites was even physically possible. The delivery problems to these distant points not only stretched the state-of-the-art for transportation equipment, but also required rapid definition and resolution, since shipping schedules were tight.

     The magnitude of this concern led to the establishment of a task force to investigate the following options for solution of the overland delivery problem:

(a) modified plant design involving smaller steam generators,
(b) field fabrication of large vessels from smaller parts, and
(c) development of optimized water and highway, railroad or airborne
     methods to deliver the major vessels to plant sites.

     The task force concluded that engineering and licensing commitments were too far along to accommodate a plant redesign (option a) utilizing smaller steam generators and that field fabrication (option b) would double the cost of the large components.

     Option (c) was thus chosen for further development, and in order to minimize the overland transportation problem to the extent possible, it

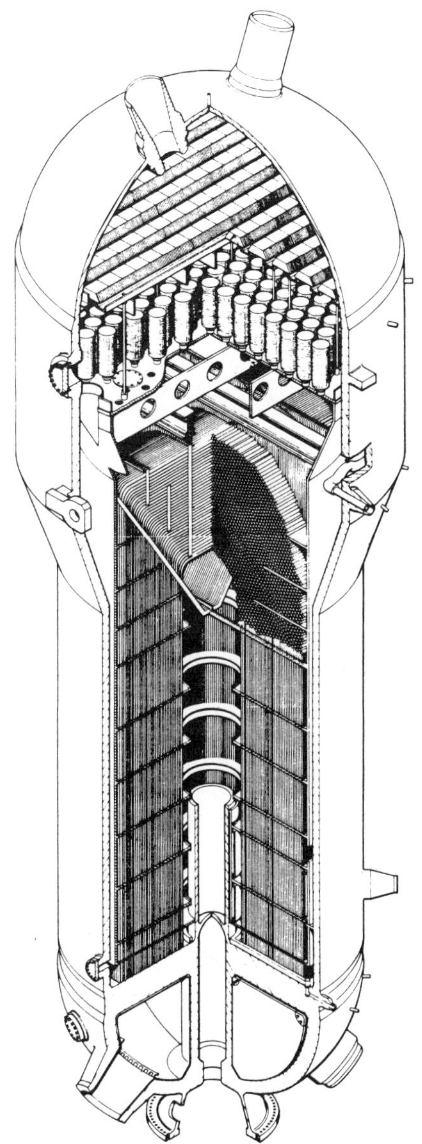

Fig. 1. SYSTEM 80 Nuclear steam generator.

was decided that all shipments of components would be made from
Chattanooga by water to the most practical point of transload to highway
or railroad. It was also concluded that over-the-road and railroad
routes and equipment should be studied further to determine applicability
to the problem at hand.

TRANSPORTATION OPTIONS

Results of studies indicated that delivery of very large loads to
some plant sites would be more economically accomplished by higway routes
while the same deliveries to other plant sites would be less costly by
railroad. The routine loads carried by most railroads travel across

structures with load bearing capabilities which were found to be generally adequate to support C-E's large load configurations.

This is not the case with highway construction. Highway culverts and bridges must thus usually be bypassed or reinforced at shipper's expense. Highways also have many low overhead clearances and many states will not grant permits for heavy and/or overdimensional loads via highway if another transport method is available. Where civil construction outside of existing highway right of way is required, environmental impact must be considered, and land must be leased or purchased from private owners. Traffic control must be acquired and coordinated with local law enforcement authorities. Once a road route is chosen and developed for delivery of oversized loads, an economic and physical commitment exists which, in the case of multiple product deliveries, biases subsequent operations toward road as well, even though all products are not necessarily best transported that way. Security is not passively inherent in over-the-road operations as it is in the case of other transport modes. Both the highway route and the transported cargo are more visble and accessible to saboteurs, and security must be addressed specifically. Multi-wheeled over the road transporters are hydraulically suspended and steered, requiring frequent routine maintenance service with a high spare part usage rate. Although the wide footprint of such a transporter provides a great degree of lateral stability, the road width is sometimes exceeded and road shoulders must be stabilized and strengthened to support the loading.

The relatively narrow gauge of U.S. rail systems combined with high centers of gravity for oversized loads makes it necessary to plan and program every increment of the route and mandates slow, controlled progress which is imcompatible with the railroads' preferred operational style of rapidly moving large blocks of freight. Some railroads are thus reluctant to consider non-routine operations on their systems. Others will permit such traffic only with special handling and assessment of train delay and/or special train charges. The general condition of United States' railroad plant was known to lacking at the time of this study due to years of "deferred maintenance". Also, a limited number of routes was becoming ever fewer, due to abandonments of unprofitable lines or those requiring large remedial maintenance programs by the owning railroads. Each rail route considered for overweight or over densional operations had to be thoroughly researched not only from the standpoint of clearance obstruction but also to confirm load bearing competence of rail installations and substructures and to establish those actions necessary to prepare and protect that route for the extraordinary deliveries.

Because of the limited demand for specialized rolling equipment, railroads were unwilling to invest in it themselves, and any capital costs would thus be borne by the shipper. In studying available equipment, C-E found that Westinghouse Electric Co. owned a railroad schnabel car (so-called because it used two beak-like "schnabels" to grasp the load and carry it between two car halves) which was used for delivering 500 short ton reactor vessels. Modification of that car for 800 short ton steam generator loads was impractical ,however, since with only 22 axles to spread the load over the railroad, weight concentrations would have been excessive. Looking to Europe, C-E found that West German and Swedish railroads operated 32 axle schnabel cars, also limited to 500 short ton loads.

Working with railroad equipment vendors and the Association of American Railroads however, C-E became confident in the prospects for successful application of a schnabel design to the rail transportation of

800 - 900 short ton steam generators, and the decision was made to
proceed further toward acquisition of such a transporter.

SPECIFICATIONS FOR ALL TRANSPORT

     Because a schnabel car carries its load suspended between two car
halves, the maximum cross sectional area which the carrier and its load
presents during rail movement is a function of the dimensions of the load
itself. A schnabel car is capable of laterallly and vertically shifting
the load in order to avoid some of the obstacles which encroach upon the
track. Schnabel equipment had been widely used in Europe for electrical
transformer and generator component deliveries in the 500 short ton
range, at speeds of 60 KM/hr. Such operations had even included
intermodal exchanges of schnabels and loads (Figure 2) from rail to
over-the-road conveyances. Looking at the specific problem at hand, it
was determined that in order to deliver 800 short ton components over
American railroads at low enough axle loads to be compatible with existing track and structures, a 36 axle configuration would be necessary.
This choice was based also on dynamic studies whic indicated that at 10 -
20 miles/hr, staggered jointed rails could induce natural "rock and roll"
frequencies in a loaded 32 axle car which would be dangerous.

     Maximum operating flexibility and stability were provided for,
despite the car's great length, by choosing 2-axle trucks, 38-inch
diameter wheels and 6-foot axle spacing. On other cars, 3-axle trucks
had been seen to be susceptible to derailments on tight curves and to
concentrate wheel loads in a shorter track length.

     As was previously mentioned, inherent in the schnabel car's design
is the ability to raise, lower and laterally re-positon its load.
Previous cars ad accomplished this to a limited degree, but C-E wanted
the absolute maximum practical, in order to lift and side-shift loads
around a great variety of obstacles.

     In order to conform as closely as possible to routine operating
parameters of the North American railroads, C-E adopted the car design
requirements of the Association of American Railroads (AAR) for this
project.

Empty Schnabel Car

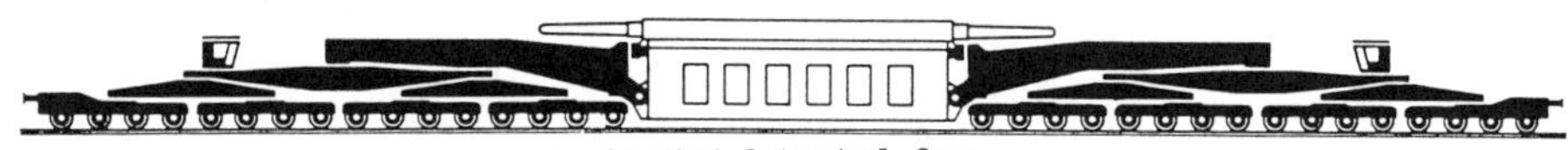
Loaded Schnabel Car

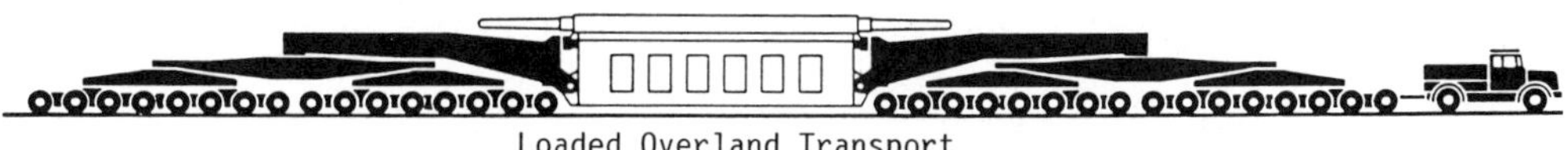
Loaded Overland Transport

Fig. 2. Rail to road Transload of large load.

A major impact of these criteria was to set a maximum gross weight on rails of 2,520,000 lbs (70,000lb/axle) and to limit permissible schnabel car weight to 742,000lbs.

Working with individual railroads to determine the most restrictive operating environment which the car would encounter, C-E was able to identify worst case horizontal and vertical curves and track superelevations. But railroads only catalogue horizontal and vertical clearance dimensions to a limited distance from track centerline and top of rail, because of the fixed dimensions of most rolling stock. Recognizing the need to quickly and accurately assess the physical clearance situation confronting any potential schnabel-carried load over many possible specific rail routes, C-E developed a clearance template for use in field surveying. This device, mounted on a railroad inspection vehicle, can simulate any load configuration and is easily positioned where the real load would be as it travelled a route carried by schnabel car to indicate interference with route-adjacent obstructions.

Meetings with the AAR's car construction committee identified railroads' concerns over such a large car's high center of gravity, its stability in travelling through reversing superelevated curves and its potential for disruption of traffic in the event of derailment under load. In response to these concerns, as it developed specifications for the car, C-E included design requirements for audible and visible alarms which would alert car operators in the event of unequal side-to-side loadings of the car and for an ability to re-rail individual trucks and axles using systems incorporated in the car itself, if necessary. Static and dynamic test requirements were also specified in order to prove functionality and safe operation of the completed car. Instead of using outside contractors and expensive additional lifting and jacking equipment to off-load the large components from barges at water destinations, an objective was set to develop a capability to do this off-loading with the schnabel car itself. Figure 3 shows the concept which was developed to meet this objective: schnabel car halves are positioned on rails on either side of the barge slip, and the schnabel arms are swung over the barge and attached to the large components which are located at an angle to the barge centerline. Schnabel car hydraulics are then actuated to raise the loads, and the schnabel car is pulled forward on the rails to move the loaded component over land. An "HO" scale model was constructed and used to evaluate the ability of the car to function in this way, and computer modelling was developed in order to verify dynamic stability.

Because of the uncertainty over future markets for a schnabel car, another requirement added to specifications was to design for eventual reconfiguration of the equipment for other uses. All specification design requirements are summarized in Table 1.

Following a solicitation from a select list of qualified manufacturers, an order for the 36-axle schnabel car was placed with Krupp Industrie and Stahlbau of Essen West Germany. The car was fabricated and tested in West Germany, disassembled and delivered to the U.S. in less than 3 years from order date. Preparations for its first use began immediately

SCHNABEL CAR OPERATIONS

When an item is considered for transport by schnabel car, a number of tasks are undertaken to determine project feasibility. A clearance drawing is prepared which integrates the load with the schnabel car and

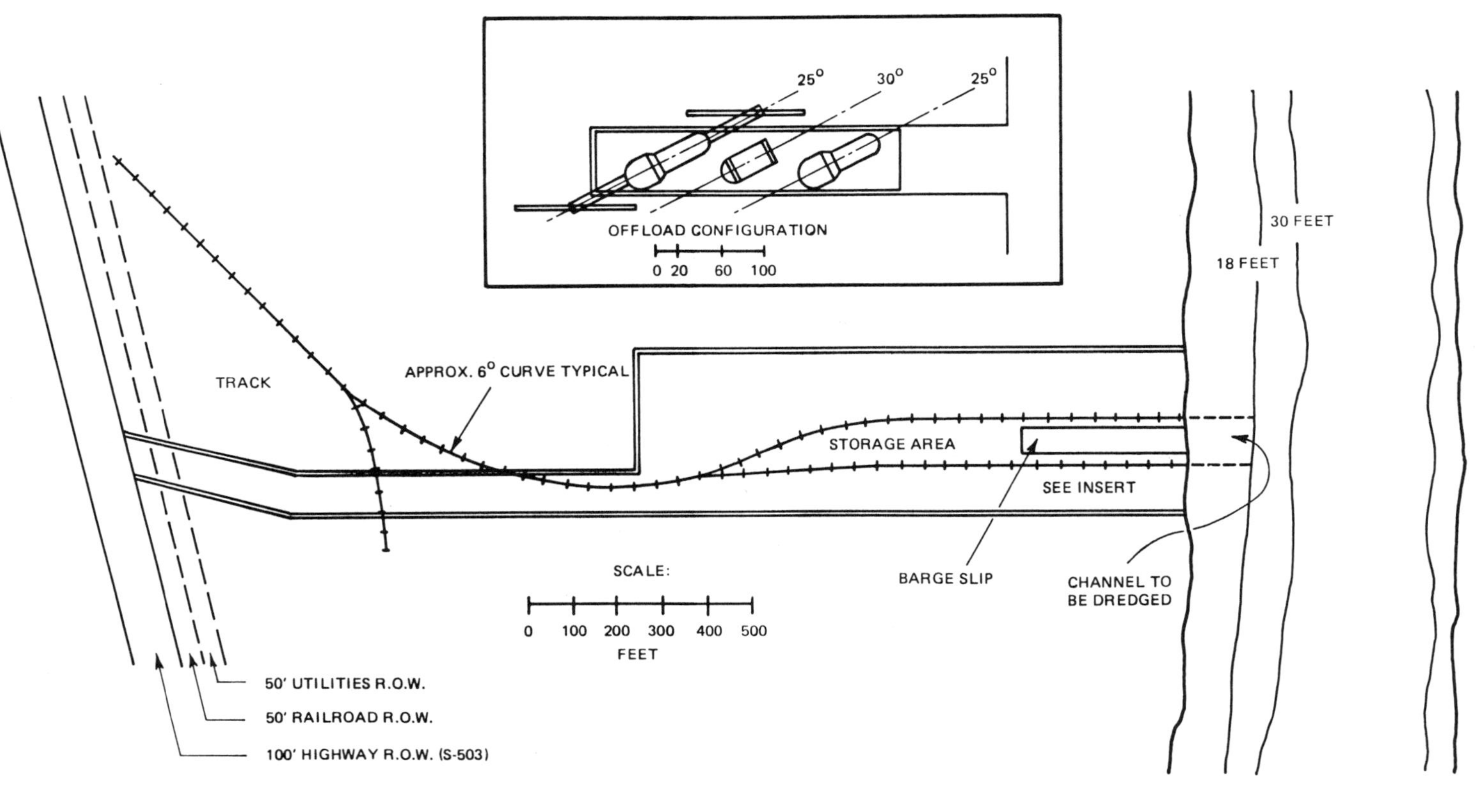

Fig. 3. Water-to-rail Transload facility.

Table 1. Specification Design Requirements for 36-Axle Rail Car

____

Railroad operating compability features
o   38 inch diameter wheels
o   7" x 12" AAR-G wheel bearings
o   6 ft axle spacing
o   unloaded dimensions 10'-8"W x 18'-0"H
o   191 ft radius (30°) horizontal curve navigability
o   2000 ft radius vertical curve navigability
o   self-contained re-railing devices
o   65,750 lb. maximum load per axle @ 812.5 short tons live load
o   standard railroad braking systems
o   wheel load monitoring system
o   all axles equally loaded
o   standard railroad couplers and cushioning devices
o   AAR approval of design

Loading Handling Features
o   self-powered ability to raise/lower load from top of rail to 44
     inches above
o   self-powered ability to shift load laterally to 20 inches either side
     of car centerline
o   ability to rotate schnabel arms 45° from car centerline while
     loaded
o   operability at -40°F
o   back up emergency power systems
o   lighting systems
o   ability to reconfiegure car for empty shipment or carriage of other
     loads

____

provides tabulated dimensional data for preliminary clearance evaluations
by railroads. This enables the railroads to identify the most major
clearance obstacles and to select a reference route. In order to further
advance the feasibility of a given route, C-E's clearance template is
usually taken into the field. A joint railroad/C-E party actually
travels the route with the template set up to simulate and capture
load/route clearance compatibility data. The resultant data are
converted back at the office into estimates of cost and schedule impact.
Some of the identified obstructions will be avoided by movement of load
position with the schnabel car's systems. Other will require removal or
reconstruction.

Once a route is chosen, a schedule is developed for the
implementation of permanent and temporary modifications to physical
interferences. A detailed procedure is developed which will be closely
followed by schnabel car operators and railroad operating personnel as
the route is travelled. In response to the railroad's concern over the
schnabel car's unfavorable impact on routine traffic on high volume
routes, the commitment has been made to provide accompanying operators
for the car whenever it moves.

Maximum loaded speed for the schnabel car is 15 miles/hr. The
loaded car travels in special train service with its own engine and
auxiliary cars (necessary to afford protection and minimize dynamic load
transfer to the cargo during transit). Spare parts and special tools for
the schnabel car and shelter for the operators are also provided on the
auxiliary cars.

Loaded operations have been conducted successfully over eight North American railroads. Some of the design features of the actual car which made for its flawless operation are the following:

Hydraulic Side Bearings

On standard rail cars, each two-axle truck is fitted with a flat center plate and a side bearing with gap on either side. When the car body is induced to lean by centrifugal forces or superelevated track, it closes the gap at the side bearing on one side which provides stability in the lateral plane. Since the 36-axle car must have several bodies (bolsters) piled on top of each other in order to equally divide the load between all axles, this side bearing gap is accumulated from bolster to bolster causing instability. Also, long bolsters require larger gaps due to track changes. The hydraulic side bearings (Figure 4) eliminate this problem.

Each bolster has a spherical center plate at each end allowing the load to be transferred without eccentricity to the bolster below. On either side of each center plate (except at trucks where standard side bearings are provided), a hydraulic side bearing is located with no gap. The side bearings are connected in parallel, i.e., side bearings on one side of each bolster are interconnected. This allows flexibility without loss of control. When the car is tilted to one side, pressure in the side bearings on that side increases, but no fluid is transferred to the opposite side, thereby preventing body roll which would move the center of gravity to a point of instability. When the car starts to negotiate a supperelevation, and only one end of a bolster is subjected to roll, fluid is transferred from one end of the car to the other preventing twisting in the structure or load, without adding instability in the lateral direction.

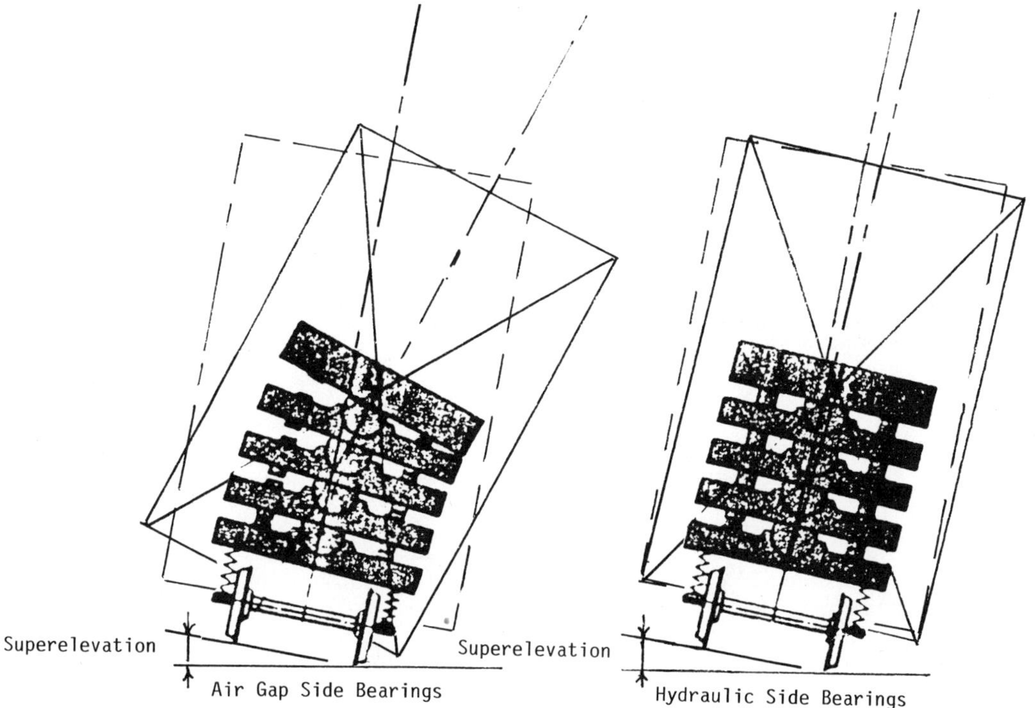

Fig. 4. Hydraulic side bearings stabilize leaning.

Initial pressure for each side of each bolster is provided by integral hand pumps and gages set according to the load to be carried.

Loading/Unloading Flexibility

Schnabel cars are "pinned" to their loads with large pins which form a tension force transmitting hinge between the two elements. Because of the magnitude of the loads anticipated, each pin in this case weighs 600 lbs and pin handling insertion and withdrawal was seen as a potential problem. Large distances on the load and schnabel arms, combined with the close tolerance required for the pin holes, made alignment a serious design consideration.

A schnabel arm manipulating system was incorporated into the car whereby two hydraulic-lift cylinders were installed on either side of each schnabel arm and used to manoeuver the arm (and thus the pin holes) very close to the required position in the vertical directions. Rollers installed under the cylinders on lower bolsters can move the arms in a lateral direction.

With these motions in combination and the main lift-shift elements on the other end of the schnabel arm, any motion can be achieved. A hand-operated hydraulic cylinder is provided for pin removal. A small jib hoist located above the pin supports the pin load. This combination of systems allows pins to be inserted or removed quickly and safely.

From an early observation of European schnabel car operations, a unique manoeuvering capability was discovered. When encountering a side clearance problem in double-tracked areas, operators would switch half the car to one track and half to the other thereby decreasing the load dimensions with relation to the outer sides of the two tracks. Considering this, a method of loading and unloading was devised using this unique capability and eliminating the need for heavy lift equipment. Also, where multiple lifts were required, it was seen that this would save on transloading costs. A track is laid on either side of the transloading area (barge slip, laydown area, or other mode of transportation) and the car halves are positioned independently adjacent to the load. The inherent ability of the schnabel car's lifting arms to swing away from its longitudinal axis was extended in this case so that the 36-axle schnabel car could lift or lower loads with its self-contained systems, thus avoiding the need for heavy rigging for barge offload.

On one project, a barge facility was designed and constructed to facilitate the schnabel car's use in barge offloading. Upon barge arrival, the two 800 ton steam generators and one 500 ton reactor were completely offloaded and barge released in five days.

On another project, it was decided that a deep-harbor floating barge offload of heavy cargo utilizing the lifting systems of the schnabel car would be developed. The schnabel car was shipped on the barge with the cargo, and upon reaching the end of the water voyage a comprehensive barge ballasting program was executed to maintain barge trim against rolling, lifting, tides, harbour swell and other dynamic forces. A pier was modified such that a structural rail bridge could support the schnabel car in various configurations from barge to pier. (Figure 5)

On a third project, five heavy vessels were each moved over an 865 mile route in the dead of winter, so that railroad substructure would be frozen and sufficiently strong. Operations were conducted at temperatures near $-30^{\circ}F$ and schnabel car cabins were extended and insulated to facilitate uninterrupted operation. (Figure 6)

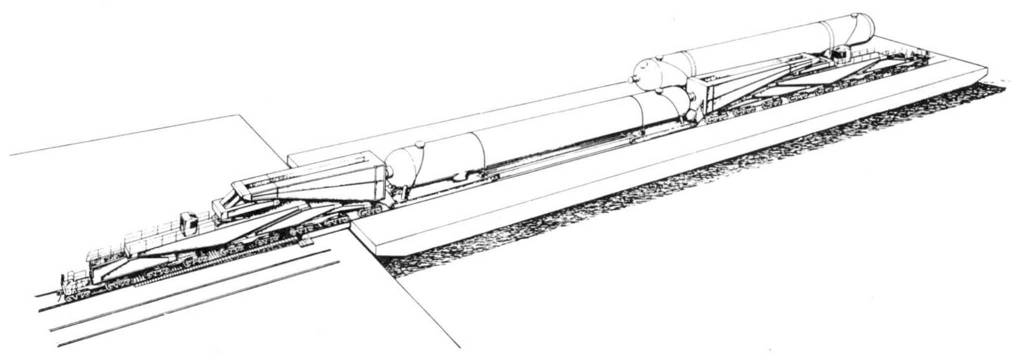

Fig. 5. Floating offload using schnabel car.

All the schnabel projects so far have involved much manipulation of the load by the car in obstruction avoidance. Avoidance of obstructions equals, of course, avoidance also of the cost of modifying, moving or reconstructing them. Because the car's systems are capable of moving what it is carrying · through such large spatial volumes, some very dramatic gains have been realized by manipulation of loads over, under, around and through obstructions.

FUTURE OPERATIONS

In addition to large non-nuclear vessel deliveries for petrochemical and energy industries, there are two potential nuclear areas where large rail schnabel cars may be used in the future. The conservative nature of railroad design results in a low incidence of problems, particularly where (as in the case of large loads carried by schnabel cars) progress is so carefully planned and controlled. Virtually all nuclear plant installations include railroad sidings for various in and out shipments. Those existing sidings should be used, if possible, for outbound shipments of spent nuclear fuel and of disassembled plant components to repositories rather than construct new and short-lived facilities to support very limited operations.

Decommissioning/Dismantling of Nuclear Power Stations

C-E's experience in overdimensional shipments has been that water transportation generally includes the most open clearance dimensions. It is therefore usually best to maximize the water-borne portion of any such transportation operation, if it involves considerable distance. A schnabel car could be used to haul a removed nuclear reactor, for instance, from the power plant's existing rail head to a point where, with construction of a simple bulkhead, it could be rolled on rails onto a barge or self-propelled ship. At a point closer to ultimate destination, a similar facility can again provide for roll-off and overland travel to placement for final burial. A continuous flow of activity results, with the removed component always mechanically supported by the schnabel car's lifting and transporting fixtures. Other advantages of this approach are its ability to be executed with minimal civil construction for minimal disruption of access to other facilities or work locations and its minimal impact on the environment.

Technology transference to othe decommissioning/dismantling projects is facilitated because existing infrastructure of a type common to all plant sites is used to the greatest extent possible.

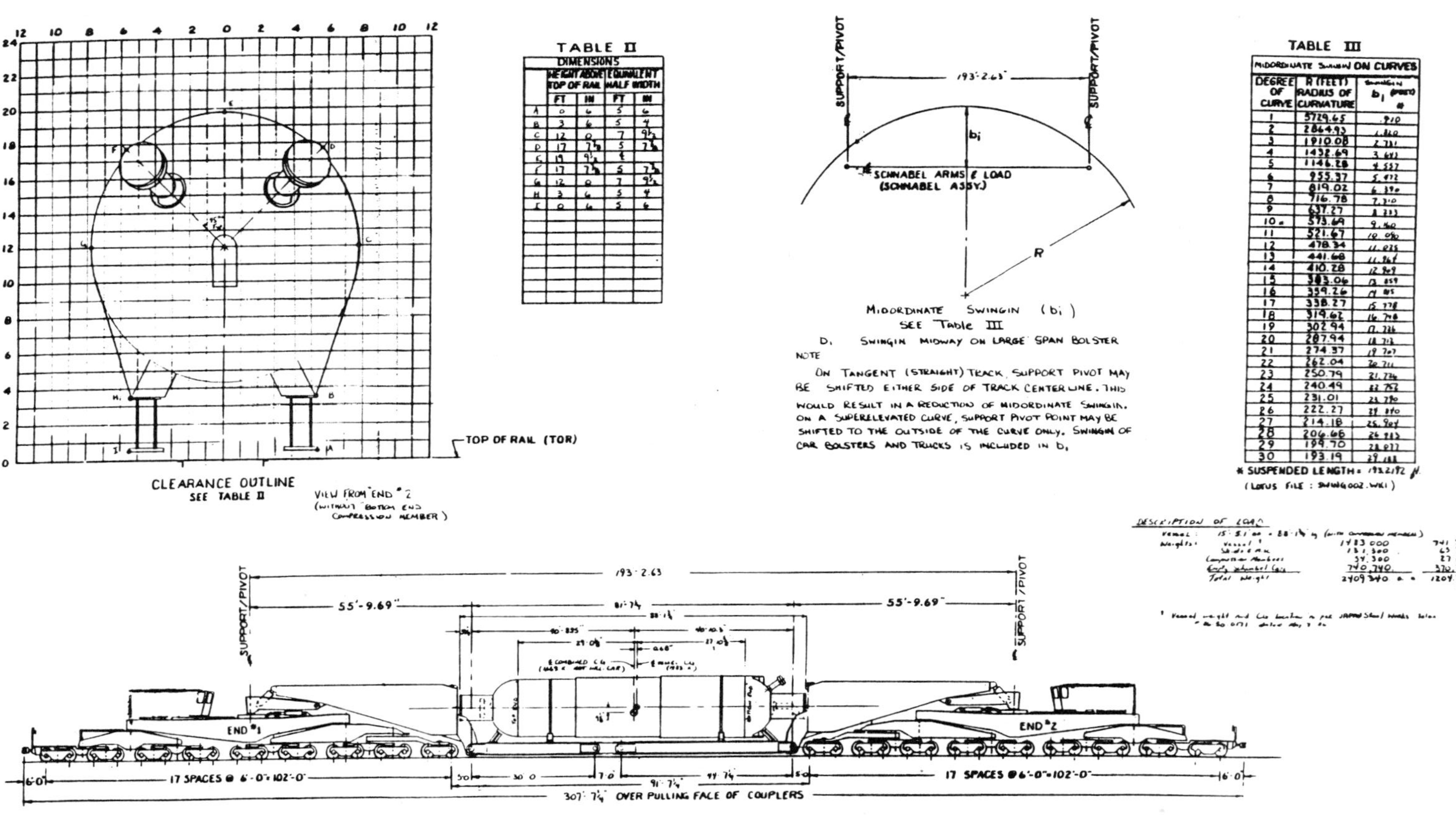

TABLE II

	HEIGHT ABOVE TOP OF RAIL		EQUIVALENT HALF WIDTH	
	FT	IN	FT	IN
A	0	6	5	6
B	3	6	5	4
C	12	0	7	9½
D	17	7½	5	7½
E	19	9½	[illegible]	
F	17	7½	5	7½
G	12	0	7	9½
H	3	6	5	4
I	0	6	5	6

TABLE III

MIDORDINATE SWINGIN ON CURVES

DEGREE OF CURVE	R (FEET) RADIUS OF CURVATURE	SWINGIN b_i (FEET)
1	5729.65	.910
2	2864.93	1.820
3	1910.08	2.711
4	1432.69	3.641
5	1146.28	4.557
6	955.37	5.472
7	819.02	6.390
8	716.78	7.310
9	637.27	8.233
10	573.69	9.160
11	521.67	10.090
12	478.34	11.025
13	441.68	11.964
14	410.28	12.909
15	383.06	13.859
16	359.26	14.815
17	338.27	15.778
18	319.62	16.748
19	302.94	17.726
20	287.94	18.713
21	274.37	19.707
22	262.04	20.711
23	250.79	21.724
24	240.49	22.757
25	231.01	23.780
26	222.27	24.810
27	214.18	25.904
28	206.66	26.912
29	199.70	28.022
30	193.19	29.111

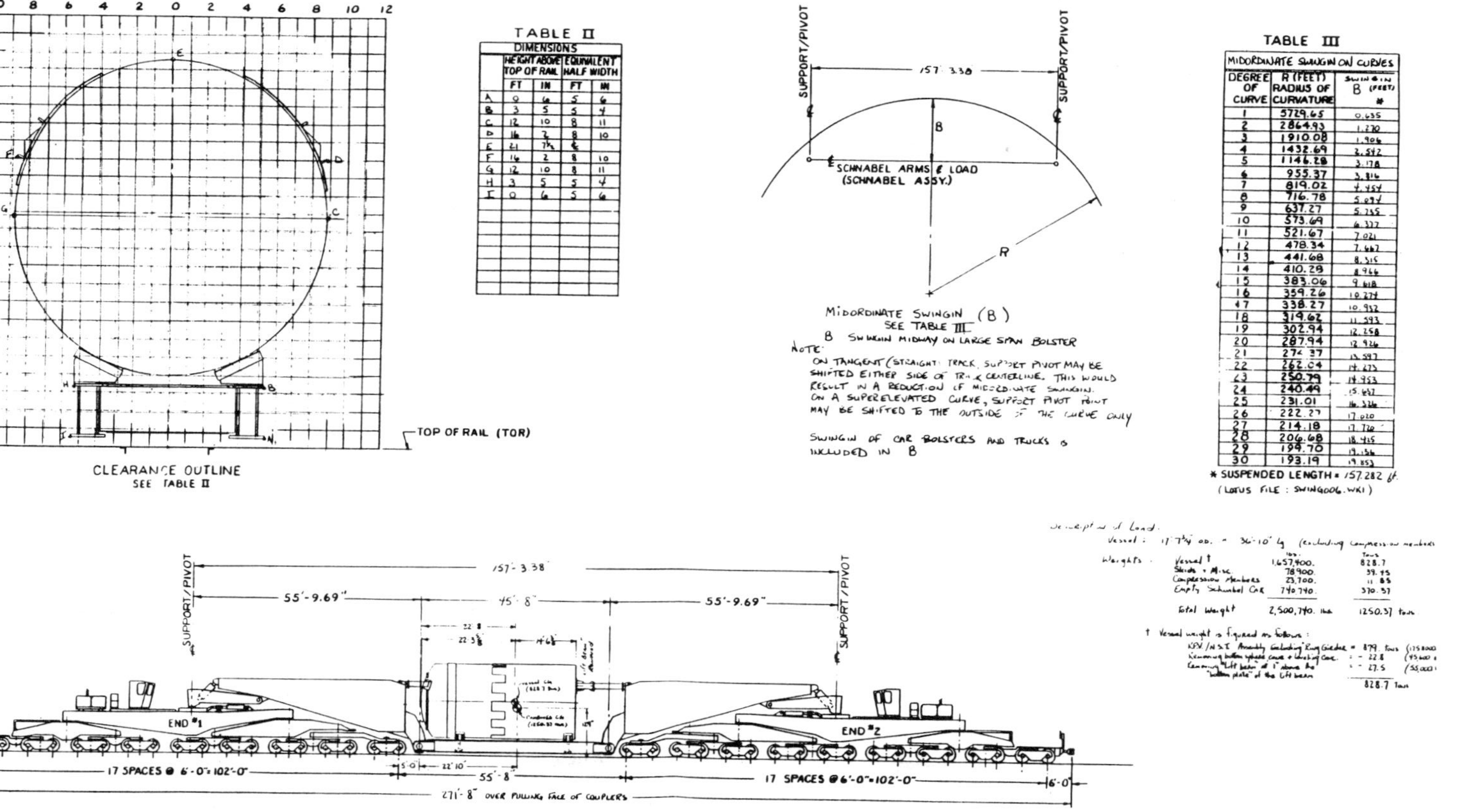

TABLE II

DIMENSIONS				
	HEIGHT ABOVE TOP OF RAIL		EQUIVALENT HALF WIDTH	
	FT	IN	FT	IN
A	0	6	5	6
B	3	5	5	4
C	12	10	8	11
D	16	2	8	10
E	21	7½	6	
F	16	2	8	10
G	12	10	8	11
H	3	5	5	4
I	0	6	5	6

TABLE III

MIDORDINATE SWINGIN ON CURVES

DEGREE OF CURVE	R (FEET) RADIUS OF CURVATURE	SWINGIN B (FEET) *
1	5729.65	0.635
2	2864.93	1.270
3	1910.08	1.906
4	1432.69	2.542
5	1146.28	3.178
6	955.37	3.814
7	819.02	4.454
8	716.78	5.094
9	637.27	5.735
10	573.69	6.377
11	521.67	7.021
12	478.34	7.667
13	441.68	8.315
14	410.29	8.966
15	383.06	9.618
16	359.26	10.274
17	338.27	10.932
18	319.62	11.593
19	302.94	12.258
20	287.94	12.926
21	274.37	13.597
22	262.04	14.273
23	250.79	14.953
24	240.49	15.637
25	231.01	16.326
26	222.27	17.020
27	214.18	17.720
28	206.68	18.425
29	199.70	19.136
30	193.19	19.853

* SUSPENDED LENGTH = 157.282 ft.

(LOTUS FILE : SWING006.WK1)

Fig. 7. Configuration clearance drawing – decommissioning operations.

Spent Fuel Movements

Design approaches for spent fuel transport have so far yielded rail casks which weighed no more than 125 short tons when loaded with 32 PWR fuel assemblies. This equates to less than 25 short tons of spent fuel per cask. With over 14,000 short tons of spent fuel already in temporary storage at the commercial nuclear plants of the United States and over 3000 short tons year expected eventually to be moved to a repository, a cask system with more capacity would seem to be attractive. Using the state-of-the-art which the 36-axle schnabel car represents, at least 6 times and probably 10 times as much spent fuel could be transported in a single shipment. With the primary objective of minimizing radiological exposure to personnel to as low as reasonably achievable and an absolute need to make such shipments in a climate of mounting public concern, the incentive to minimize the number of shipments is obvious.

A REVIEW OF RAIL TRANSPORT FOR THE NUCLEAR INDUSTRY

I.K.Braybrook[1], P.J.Abbott[2]

[1]Railfreight
Room 318
Euston House
24 Eversholt Street
London NW1 7DZ

[2]Director of Operations
British Rail
Paddington Station
London W2 1FT

ABSTRACT

The paper sets out to provide a comprehensive guide to the British Rail freight network insofar as the transport of larger items of radioactive material is concerned (freight traffic rather than parcels traffic). The content includes an explanation of the planning process required to gain access to the rail network, the options available for the provision of terminals, handling equipment and rolling stock and a basic introduction to the range of services available. A certain amount of technical detail is given for guidance purposes on the design of new rail links and of some items of equipment, in an attempt to assist transport planners and their consultants with their initial evaluation of competing transport methods and of the siting of new facilities.

A basic guide is also given to the regulations that govern rail transport and to the essential ingredients that enable British Rail to carry radioactive substances safely.

The paper will be of most value to agencies concerned with the following types of transport problem:-

- inland and international transportation of high level waste
- the decommissioning of nuclear installations
- the movement of low level waste to repositories
- intermodal developments for waste materials, such as containerisation.

INTRODUCTION

This paper sets out to be unashamedly educational in content. British Rail (B.R.) is no longer the system of memory or folklore: in recent years, there have been many changes, in outlook, in performance and to working practices. As a result few outsiders, even those with relatively recent 'inside' experience, have a clear idea of the potential

of the rail network and of the means of gaining access to it. This paper
is a guide to that network, written with transport for the nuclear
industry in mind, an industry for which the railways already play a
significant role and with which even closer links are likely to be
forged,as environmental issues assume greater significance. The paper
concerns itself with transport items large enough to be considered
'freight', rather than large packages or parcels, the minimum consignment
probably being the equivalent of a 20 tonne load by road.

In 1983 B.R. underwent an internal transformation, aimed at
improving efficiency and customer orientation and at ensuring that it
held its place as the least subsidised national rail network in the
world. The basis of this change was the creation of five separate
businesses:

- InterCity
- Network SouthEast
- Provincial
- Freight
- Parcels

each with its own clearly defined responsibilities, financial targets and
accounts. It bestowed upon railway managers 'bottom line' responsibility: Freight, Parcels and InterCity each have a government target to
trade profitably and achieve a given return on assets by 1989/90.

Railfreight is the trading name of the freight business and
similarly has been divided up into 5 customer orientated segments:

- Coal
- Metals
- Construction
- Distribution
- Petroleum

Because of the continued predominance of coal in the UK energy
market, Coal is the title chosen to describe the Railfreight department
that specialises in all aspects of transport relate to U.K. energy
production, including gas and nuclear, but excluding Petroleum.

The unique nature of the nuclear industry is recognised and a small
team has been established within Railfreight Coal to develop and maintain
a special expertise. In addition, key production and support functions
within B.R. have developed specialised cores of expertise to deal with
all aspects of nuclear traffic:

- engineering   : wagons and terminal equipment
- operations   : safety and transport requirements
- research   : radiological advice
- personnel   : industrial relations questions
- public relations : handling public concern

B.R. can truly claim to have grown up with the nuclear industry and
to have developed its methods and techniques to match the progression
from the earliest magnox stations and the first nuclear services by rail
in 1962, through AGR's to the next generation of PWR's. In addition to
the CEGB and SSEB, we also work, amongst others, for BNFL, NTL, PNTL, and
UKAEA. Flasks are carried to Sellafield from power stations and from
selected ports. Low level waste is carried to Drigg and special
movements of drummed waste for disposal are arranged when required.

B.R. is now preparing itself to serve the nuclear industry in the next decade and indeed into the next century; this includes meeting the challenges of power station decommissioning, serving new waste repositories and undertaking the movement of containerised waste. The purpose of this paper is to explain the process by which access to the rail network is obtained, the range of services on offer, and the management of transport safety.

ACCESS TO THE NETWORK

The Rail System

The potential network comprises the totality of operational rail lines in Great Britain. In theory, nuclear traffic could be handled over any stretch of line between Penzance and Wick. However, it is necessary to make certain distinctions between the various categories of radioactive materials, because practical and economic considerations do give rise to certain limitations.

For example, irradiated nuclear fuel is normally moved on dedicated services separate from other traffic, whilst material of low level radioactivity, including waste, can be attached to trains conveying 'general merchandise'. In both cases however, services may prove to be expensive to provide in parts of the country where there is little or no existing freight movement, because the traffic in question would have to bear the full cost of resource provision (locomotives and men) as well as some of the costs of local infrastructure (track, signalling, management control). The basic freight network is shown as Figure 1. Projected movements which involve transits long distances away from this 'core' network are feasible, but should be discussed at an early stage with Railfreight to ensure that the best deal can be put together, which usually means finding the best balance of volume and frequency. Services are also likely to be more costly to provide away from the core network.

Rail Terminals

The assumption is made that all major receiving points for nuclear traffic are already, or will be, connected to the rail network. Figure 2 shows all existing rail sites that handle nuclear traffic. These basic options are available for other locations:

-   direct rail connection at the loading point.
-   a dedicated road/rail transfer terminal as close as possible to the loading point.
-   a shared road/rail facility within reasonable distance of the loading point. (Subject to the nature of the traffic involved).

Which option is chosen will depend upon local geography as well as considerations involving volumes of traffic, public sensitivity, cost, existing road network and duration of contract.

Direct Rail Connection A dedicated rail loading point inside the plant or facility generating or receiving the traffic obviously represents the best option as far as rail logistics are concerned. However, other than in exceptional circumstances, such a facility is unlikely to be justified for volumes of less than one trainload approximately every week. Railfreight is able to offer specialist advice on the planning, construction and maintenance of such a facility, including track layouts, staffing levels and cranage facilities.

In certain cases, it may be necessary to consider the construction of a new rail line to link the plant with the rail network. Costs for constructing simple freight 'spur lines' are affected by a number of factors, including ground conditions, extent of earthworks, the number of structures required en route, level crossings and the cost of land acquisition. These costs are broadly in the order of £400,000 – £700,000 per single track kilometre at 1987 prices. Initial studies to assess the feasibility of such projects are relatively simple and usually involve site visits together with straightforward analysis of Ordnance Survey maps and data in order to establish the optimum line of route, taking into account costs, operational flexibility and environmental impact.

British Rail is also able to advise upon the appropriate statutory procedures for constructing a new link line. B.R. itself promotes a bill through Parliament each November for the authorisation of new or significantly altered rail lines. This process from submission to authorisation normally takes 12-15 months and confers on the British Railways Board all the rights required to run a statutory railway, including:

-    immunity from prosecution
-    the power to alter public rights of way
-    the right to alter the works of other statutory undertakings (gas, water electricity etc.)
-    compulsory purchase powers for land acquisition

<u>Dedicated Road/Rail Transfer Terminal</u>   If distance, cost or inhospitable terrain preclude the construction of a direct rail link, a purpose built road-rail transfer terminal connecting on to an existing line can be considered. The rail layout would be designed to cope with the volume of traffic and size of train predicted. British Rail would not own or operate the site, but would provide assistance in ascertaining the suitability of alternatives. B.R. itself has many parcels of land available that would be suitable for such facilities, although such land is now rapidly being disposed of by the British Rail Property Board, and it would be appropriate for any parties to register interest now, even if a facility may not be required for five or more years, in order that the decision may be taken whether or not to put a 'reserve' on any land. Because of the emotive nature of the traffic, good road links with any proposed transfer terminal are desirable and should take account of the size of road vehicle proposed.

<u>Shared Road/Rail Transfer Terminal</u> British Rail's existing freight network already makes use of several hundred privately owned terminals. Over one hundred of these are operated by companies which specialise in the transhipment from road to rail of all forms of general merchandise. Co-operation with such companies should not be ruled out, although will obviously depend upon the exact nature of the nuclear material being handled and its compatability with other traffics.

Low level waste is the most likely candidate for this type of solution, although even with this type of traffic, it would be prudent to plan to utilise terminals specialising in the handling of steel, bricks, paper etc. rather than foodstuffs. Many such terminals are already equipped with all the necessary handling equipment, ranging from overhead cranes to simple forklift trucks.

<u>Handling Equipment</u> The exact nature of equipment required will obviously depend upon the type and volume of 'containers' to be handled. A 'freightlifter' type vehicle (basically a large forklift truck) will

prove adequate for modest volumes of smaller ISO containers. A conventional mobile crane would probably cope with specialised flasks and containers, and drummed waste could be either craned or loaded by forklift. However, the preferred option for higher weights and greater volumes would be a fixed gantry crane capable of spanning at least one rail track and one roadway (8 metre span). Lifting frames and telescopic spreaders would also be required if different types of flask and container are to be handled with the same equipment.

British Rail is able to offer initial technical advice on the specification of equipment required and approximate costs involved. For example, a 15 metre span (2 rail tracks and 2 roadways) 120 tonne capacity gantry crane would cost approximately £300,000 at 1987 prices.

Wagon Standards

All wagons for use on the British Rail network must be designed, constructed and maintained to standards set and monitored by B.R.'s Director of Mechanical and Electrical Engineering. Full details of these criteria are known by all prospective U.K. based wagon builders. Advice should be sought from B.R. before tenders are discussed with overseas suppliers.

The most important criteria concern the weight, overall size, running speed and braking capacity of each wagon type.

Weight is important with regard to the forces transmitted through to the track when the wagon is in its most heavily loaded condition. This is measured in terms of axle weight, calculated by dividing the gross laden weight of a vehicle by the number of axles.

Whilst the maximum axle loading permitted is 25.5 tonnes (giving a gross laden weight potential of 51 tonnes for 2 axle and 102 tonnes for 4 axle vehicles), 22.5 tonnes is still recommended as a more flexible design standard, as there are many secondary routes which are unlikely to be upgraded in the foreseeable future. Because of the numerous localised lower limits that apply, special attention should be paid if the vehicles are intended for use in areas such as Scotland, Wales and indeed parts of Cumbria.

Specialised vehicles can be designed within these parameters to achieve greater overall weights as described in the section on specialised wagons, but these require a correspondingly greater number of axles.

The external dimensions of each vehicle and its load must conform to the B.R. Standard Freight Loading Gauge. The basic outline is shown in Figure 1 for guidance purposes, but once again contact with Railfreight is advisable at the earliest possible stage to ensure that the requirement can be complied with. Full details of permitted widths below 1065 mm above rail level (A.R.L). and for vehicles of great length are omitted from Figure 2 for simplicity. A minor relaxation to this gauge allowing an additional 75 mm width above 1065 mm A.R.L. is expected early in 1988. It should be noted that this gauge is more restrictive than that allowed on the mainland of Europe.

All vehicles should be designed to run at 60 m.p.h. in order to provide rapid transits and to fit in with the needs of a progressively higher speed passenger railway. However, lower speeds may e necessary for very heavy multiple axle vehicles. All vehicles are now built with air brakes as standard equipment.

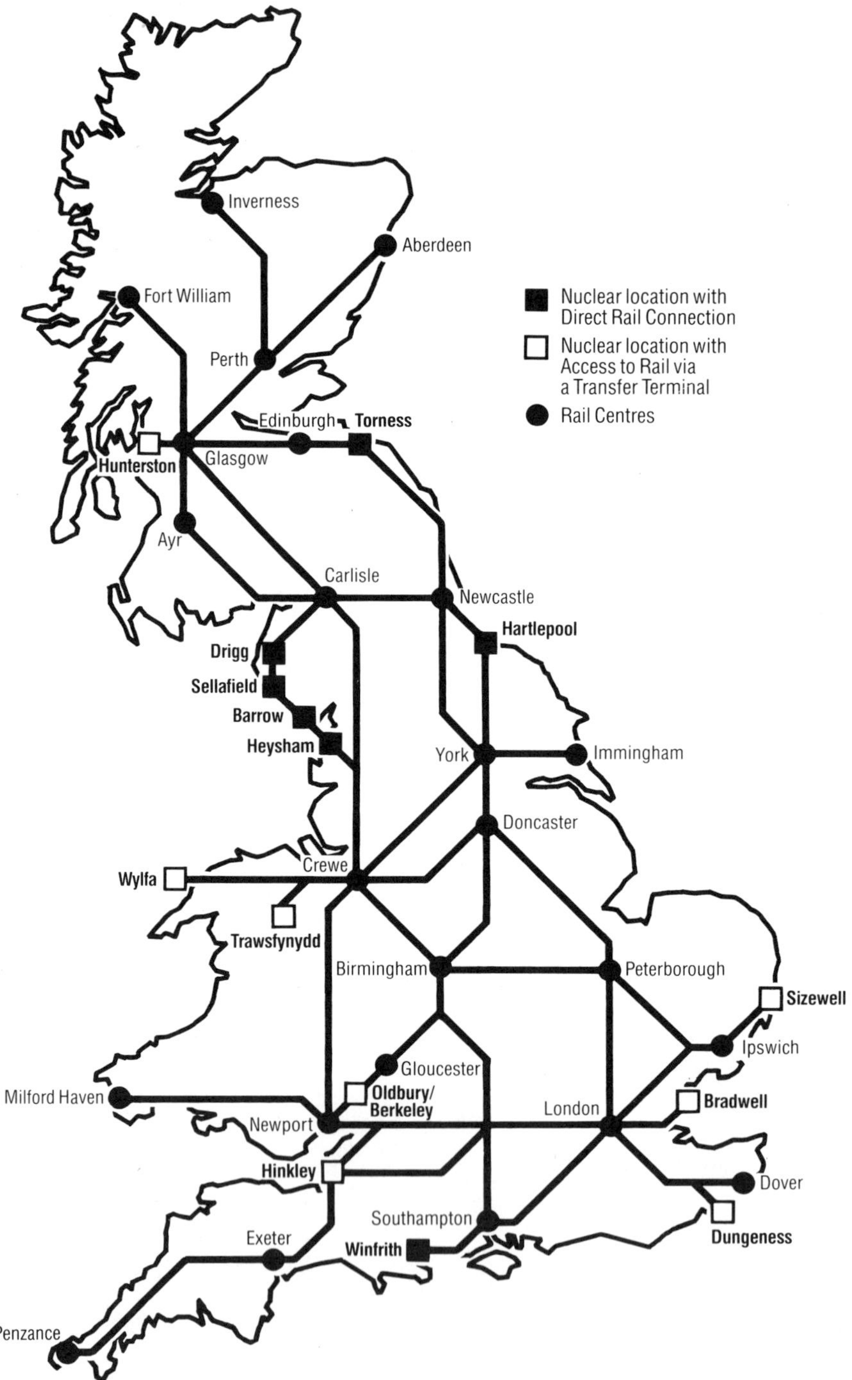

Fig. 1. Railfreight Core Network

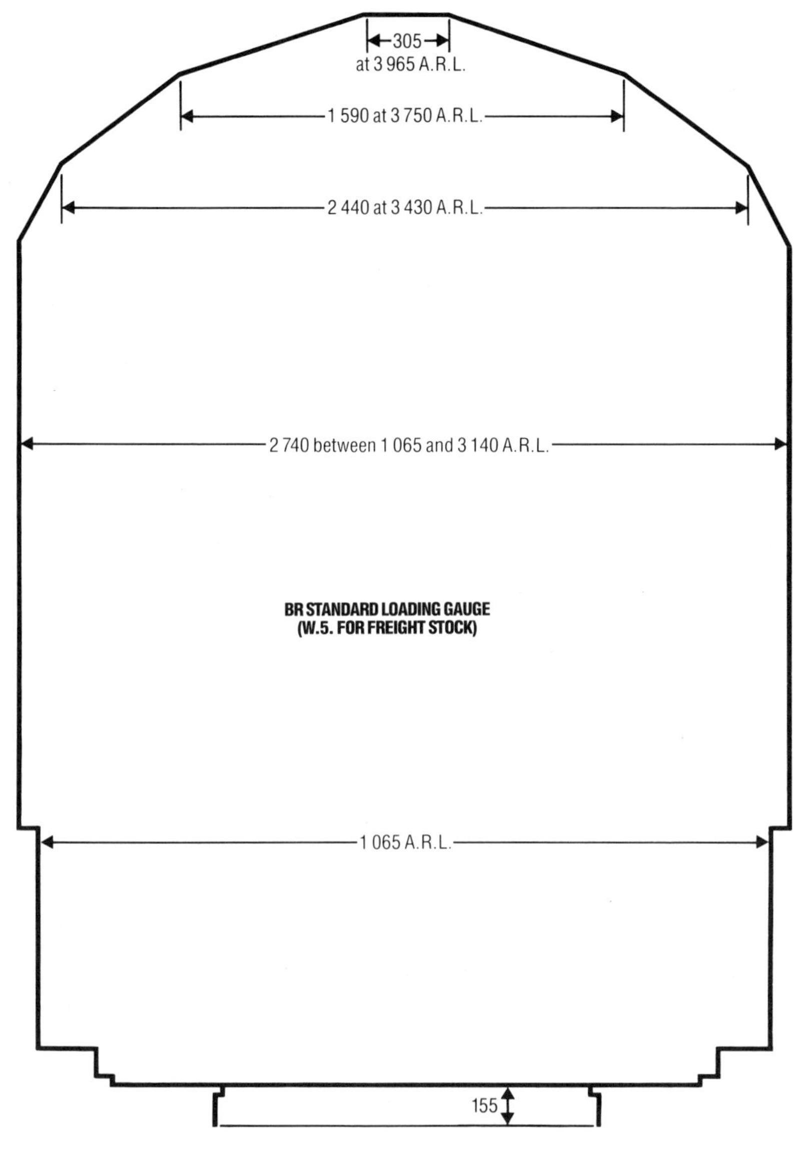

Fig. 2

Wagons

Specialised Wagons Specialised Flask Carriers, as operated for example for the CEGB or BNFL can be designed and built to suit specific requirements. Whilst standard wagon types are usually designed to weigh a maximum of either 90 or 102 tonnes (including approximately 60-70 tonnes of payload), total weight can be increased by multiplying the number of axles in order to restrict the total axle load to 25.5 tonnes per axle or other agreed maximum.

However, the heavier the vehicle, the more sophisticated it will be and inevitably more costly. A vehicle with many axles will also probably be of great length and so will be of restricted width (in order to allow for 'overthrow' when rounding curves) and could be subject to speed and route restrictions as well as requiring specialised trackwork within terminals.

BNFL/NTL already operate a small fleet of 8 axle (double bogie) well wagons for transporting specialised flasks. These are more than 25 m long overall, weigh 58 tonnes empty and can carry loads of up to 102 tonnes which can so add up to a total weight of 160 tonnes (equating to a 20 tonne axle load). Such vehicles as these require 'runners' or 'spacers' between each loaded wagon, in order to reduce the weight imposed on track structures.

Conventional Wagons Railfreight, along with many private owners and operators, run fleets of conventional wagons capable of carrying general cargoes (loose or palletised), bulk materials, liquids, powders etc. Within the standard loading gauge, covered and open wagons exist up to 20m in length and capable of carrying up to 56 tonnes payload. Many of these are also suitable for running directly to and from the Continent via the Dover-Dunkirk train ferry. (It should also be remembered when considering European transactions that many more plants, installations and factories are directly connected to the rail network on the continent than in the U.K.) Modern rail vehicles reflect the latest road based technology and are available with sliding roofs and doors, or curtain sides. Such vehicles are available either for long term contract hire or for 'spot lot' movements.

Where vehicles are considered for the 'ad hoc' transportation of, for example, drummed waste, account will have to be taken of the likelihood of contamination. Wherever possible, dedicated wagons are preferable for such traffic and there is perhaps scope for a central nuclear wagon provider, to whom any company could turn when wagons are required. Such a 'pool' would be operated in the same way as all existing fleets for non-nuclear traffics, and would combine economic provision with security of supply.

Container Carriers Until recently, containers to be transported by rail were the domain of Freightliners Ltd, a subsidiary of British Rail created by the 1968 Transport Act to cater for containerised transport. Freightliners operate from a countrywide network of strategically located terminals offering a total collection and delivery service between any two points.

In the last five years, the continued spread in the use of containers for inland transport has led to complementary rail orientated developments beyond the Freightliner network. Individual wagons are now available to carry 20 ft., 30 ft. and 60 ft. ISO containers between private terminals on general merchandise trains, as well as on dedicated 'block' service. Because of the restricted nature of British Rail's loading gauge (see Figure 1), specialised flatbed wagons are required to transport containers. These container carriers have a platform height of approximately 1 m, which enables 8 ft. high boxes to be carried on most lines. 8 ft. 6 in. containers are subject to further route restrictions but most major axes are now cleared and engineering work is providing an increasing route availability. 9 ft. containers are prohibited on this type of wagon.

If 8 ft. 6 in. containers are intended for general use and a very wide range of routes is required, then three options are available:

(i) A relatively simple wagon has been developed for the coal trade with a platform height of only 838 mm. The basic principle is that

the wagon and container design allow the wheel rims to protrude
through the wagon chassis and into the underside of the container.
Standard ISO containers require relatively little modifiction.

(ii)   Designs have been produced for a 'well' wagon, which would allow a
standard 8 ft. 6 in. container to sit in the well between the
axles. The main disadvantage is that not all the length of each
wagon can be used for loading, and so transport costs may be
slightly higher, particularly where overall train length is a
restricting fact as outlined in the section on train sizes.

(iii)   Trials are currently taking place with a 'small wheeled' bogie,
which combines the advantages of both (i) and (ii), in that
standard containers would be suitable and the whole wagon length
would be available for loading. Further progress with this design
is expected during 1988.

Intermodal wagon designs are also coming on to the market, giving
further potential for improved flexibility for road-rail combined
operations. Two basic applications already exist, one involving
specialised trailers that can run either on road or rail, and the other a
development of the system known elsewhere in the world as 'piggyback',
whereby road heavy goods vehicles can be carried on special rail wagons.

Wagon Provision

Wagon building is now a very competitive industry within the U.K.
and under constant pressure from Europe, particularly France and Finland.

Wagons may be owned by British Rail, or by the intended operator or
by a third party, who will be prepared to hire or lease wagons as
required.

Most wagon builders are prepared to hire and lease, as well as sell,
and specialist 'wagon providing' companies have also been established
over the last ten years, who are prepared, in addition, to offer wagon
management and inclusive maintenance contracts. The exact nature of
wagon provision will depend upon the circumstances of each individual
case, particularly volume and frequency of traffic, duration of contracts
and the need for specialised equipment (or possibly dedicated equipment
because of cross-contamination problems.)

Train Sizes

Traditionally, trains conveying nuclear traffic have not needed to
be of great weight or length, and consequently the type of locomotive
usually now employed would result in a limit of about 1,000 tonnes total
trailing weight, dependent, of course, upon the nature of the route
followed and the gradients involved.

At the other end of the scale, Railfreight now employs locomotives
able to haul up to 4,500 tonnes trailing weight, and indeed most new
types ordered over the next 5/10 years will be of the larger variety.

Similarly, train lengths do not constitute a restriction for
existing nuclear traffic and 300 to 350 m can safely be considered as a
working guide for most parts of the country. (This will obviously also
have an effect upon, and be affected by, terminal design). Where bulk
materials are concerned, trains are becoming longer, as well as heavier,
and 650 m is now being achieved on certain routes.

At the other end of the scale from full train loads, reference has already been made to 'general merchandise' services, which are run under the brand name of 'Speedlink' and comprise a large network of timetabled trains which run (usually overnight) between the major centres shown on the core network (Figure 2). Portions, which may range from one wagon of a particular customer's commodity to many hundred tonnes of varied goods are attached and detached en route. For example, a train running overnight from Dover to Glasgow would call at London, Birmingham and Warrington, where connections would be made to other parts of the country. Space can be booked in advance on these services in the same way as seats on a passeger train. They are already used for some of the more spasmodic flask movements.

SAFETY IN RAIL TRANSPORT

Introduction

The essential elements enabling the safe carriage of any dangerous goods involve knowledge of the properties of the product to be carried and of the hazard presented, so that the correct choice of packaging for transport conditions and the correct labelling of packages can be determined. Operating procedures must also be correctly applied to reduce the chance of an incident or minimise its effects.

Conditions and practices are subject to constant review to ensure safe movement within the total railway operation. As with any transport operator, the pressures for cost containment and reduction have to be considered. At the same time, the awareness and concern of the general public of the potential hazard presented by dangerous goods increases.

Acceptance

Dangerous goods are accepted for conveyance on B.R. under conditions outlined in the List of Dangerous Goods and Conditions of Acceptance (LDG) B.R.22426.

Customers forwarding dangerous goods enter into a contract with B.R. incorporating our General Conditions of Carriage and the Special Condit- ions relating to dangerous goods conveyance. Non compliance with these conditions may make the customer liable to B.R. for breach of contract.

For any dangerous goods, including radioactive materials, not previously carried on rail, the precise specification of the material is given specialist consideration to determine what conditions have to be applied for acceptance and carriage.

Classification

Radioactive Materials are defined as a material with any specific activity greater than 0.002 microcuries per gramme. B.R. bases its acceptance requirements on those agreed internationally for all modes of transport under the International Atomic Energy Agency (IAEA).

Safeguards

The achievement of safe working is aided by the application of a number of safeguards, as detailed below:

Packaging B.R. specifies acceptable forms of packaging in the LDG for the carriage of radioactive material. There are basically five standards of packaging which may be used, depending on the quantity and activity of the material to be conveyed. These are:-

a) Packaging used for 'exempt' quantities of radioactive material, such as luminous watches, only required to ensure that in normal transport there should be no leakage of radioactive material.

b) Industrial packaging, such as metal drums, used for the transport of uranium ores, also required to meet normal transport conditions only.

c) Strong industrial packaging used, for example, for the transport of certain radioactive wastes and which is required to meet tests comparable with those recommended by the United Nations.

d) Type A packaging used, for example, for the transport of radioactive isotopes, of a limited, but greater, activity than that allowed in the lower standard of packaging. These packages must meet certain performance test standards to take account of minor transport accidents.

e) Type B packaging, used for the transport of radioactive materials of high activity such as irradiated nuclear fuel from power stations. These packages are designed to withstand very severe accident conditions and must be capable of passing stringent tests. Type B packaging must be approved by the Competent Authority, the Transport Radiological Adviser at the Department of Transport. Traffic is not accepted until the appropriate Certificate of Approval has been issued.

<u>Special Wagons</u> Specialised or specially approved wagons may be required to complement packaging adequacy. The requirement and processes involved in wagon provison have already been mentioned.

<u>Labelling and Marking</u> Small packages must carry B.R. hazard diamond labels to designs based on U.N. recommendations.

Wagons must have B.R. dangerous goods labels indicating the principal hazard and the emergency code. This information is input to B.R. Total Operations Processing System (TOPSaW The emergency code contains the commodity U.N. number followed by two letters (Alpha Code) identifying the source of Specialist Advice, with full details maintained on a data base.

International Traffic

This traffic is subject to the following requirements:

- Convention concerning International Carriage by Rail (COTIF) including CIM (Goods)
- Annex 1 to CIM. (Regulations concerning dangerous goods, RID)
- Special Conditions relating to UK legislation.
- International Union of Railways (UIC) recommendations and requirements.
- Special requirements occasioned by differences between B.R., RID and Maritime (SOLAS) requirements.

The Channel Tunnel operators ar currently considering which dangerous goods will be allowed through the tunnel. The future of the existing rail ferry link is also under review.

Working Manual for Rail Staff (B.R. 30054) Integrity of packaging containment is the key to safety in the transport of radioactive materials. In addition, B.R. has operating procedures specifically applied to the transport of dangerous goods. These are contained in our Working Manual for Rail Staff, Part 3, known as the 'Pink Pages'.

This manual is for the use of all operating staff, as well as specific customers and the emergency services. It covers all aspects of identification, loading, marshalling and movement of dangerous goods together with emergency procedures.

Marshalling and Movement Requirements for adequate segregation or prohibition of incompatible goods on the same train are based upon considerations as to whether the interaction of one commodity with another could present a hazard.

Loaded flasks are, for example, prohibited from being carried on the same train as explosives, flammable gases, highly flammable and flammable liquids, spontaneously combustible materials and organic peroxide.

Segregation and marshalling requirements are verified through TOPS to ensure compliance. This check also ensures that wagons formed in trains are fit for the purpose in terms of, for example, braking capacity, weight and speed restrictions.

Although 'Driver Only Operation' is now the norm for freight train operation, guards are being retained on trains conveying flasks to assist in emergencies. The movement of flasks is also monitored to ensure adherence to service plan.

Emergency Procedures In addition to B.R.'s standard instructions for action to be taken in the event of an incident, special procedures also apply where radioactive materials are concerned. The National Arrangements for Incidents Involving Radioactivity (NAIR) scheme managed by the National Radiological Protection Board at Harwell, covers all forms of transport, except flasks.

For flasks, B.R.'s procedures are agreed with, and form part of the CEGB/SSEB Irradiated Fuel Transport Flask Emergency Plans. The procedures seek to differentiate between incidents and emergencies to ensure that the correct level of response is initiated in each case.

Incident Control Precise methods of incident control vary according to the problem. The Railway Manager and Emergency Services must quickly establish one control point through which all requests for action and recovery can be promptly handled with the knowledge of all participants. The Senior Railwayman must be identified as the Mishap Controller.

This provides a secure framework within which specialists can operate effectively, saving life, containing fire and hazard problems, and dealing with any load transfer requirements.

Liaison and Influences There is regular liaison, formal and informal, with those having an interest in the transport of dangerous goods.

Safety in the transport of radioactive materials, including routeing of flask movements, is an aspect of concern to many in contact with us.

It is vitally important that our attitude towards these concerns is both accurate and positive. The use of direct routes, with modern track and signalling, together with the other procedures described, have established an excellent safety record. We continually seek in our procedures to ensure that the risk to the general population is minimised, and in turn, our own staff must not be forgotten.

<u>Monitoring</u> For the movement of flasks, for example, a certificate is required from the consignor to state that the wagon and flask have been monitored and that external radiation doses do not exceed the specified levels. Our Radiological Protection Advisor and other qualified B.R. personnel carry out random checks of radiation levels on flasks, wagons and staff involved.

<u>Safety Performance and Review</u> B.R. constantly reviews all aspects of the movement of dangerous goods. This involves consideration of actual performance and of the potential impact on B.R.'s operating procedures caused by changes to external regulations.

We strive to ensure that the highest operating standards are maintained and at the same time keep the public, decision makers and our own staff informed of our performance.

CONCLUSIONS

Most of this paper has concerned itself with the mechanics of modern freight railway operations and with the basic planning issues that have to be considered by any party involved with the use of rail for transporting radioactive materials. However, it would not be right to close without going beyond the mere features of railway operation and raising some of the benefits offered.

At the simplest level, rail is obviously a transport method which must compete with road, sea and air. However, whatever traffic one is considering, rail scores highly because of its safety record, its internal security (ranging from its own police force to the fact that every consignment is monitored by computer for the whole of its transit) and the public acceptability of its environmental impact. Whether one is discussing 2000 megawatt coal fired power stations, the disposition of the country's steelmaking facilities or roadstone depots in Greater London, the railways are not simply a method of transport but the key to the existence of such facilities in their current form. Many such developments have only been able to take place because the railways provide safe and acceptable transport for the traffic involved. The same is true of radioactive materials, as even most anti-nuclear pressure groups will agree that if such substances do have to be moved, then rail is the least unacceptable means of so doing.

ACKNOWLEDGEMENTS

The Director of Freight and the Director of Operations are thanked for their permission to publish this paper.

CHAIRMAN:
I would like to thank all the presenters for the very interesting papers they have put forward, and in particular the reminder of the early pioneering days of transport and communications.

MR.S.F.JOHNSON, ABB ATOM, SWEDEN for Paper 3:6 and 4:5
ABB Atom means in detail the Swedish Fuel Fabrication Plant. I would like to add to the comments made on impediments mentioned by the authors of these two papers. Because of the trouble with Transnuklear in Germany at the end of 1987, the ferries from and to Sweden refused to transport radioactive material and Sweden was practically cut off for several months. We have started to transport by chartered ships and by air. I can give you an example - there were 21 drums, each containing 62 kg, U2 by a four engine DC8, and we are chartering cargo ships now which are going (instead of the direct shortest way round Denmark to a harbour in Germany which will accept this cargo). We have been talking a lot about the safe transport and the improvements in safe transport and so on, but this method that we have been forced to use, and still are forced to use, does not make the transport safer. We have to use what we can get, we have time schedules, our customers are waiting for their fuel, and we use the same way back, for those delivering fuel to Swedish reactors - they couldn't get in to Sweden and we couldn't get out. If I take the question and say "Why did the Germans over-react like that?" no one could answer it here, and if I ask when will the situation be normalized, I think nobody can answer that one either. It creates much trouble, and it is not good for the nuclear industry.

CHAIRMAN:
Thank you. I am certain that we are very much aware of the implications for transport, whether it be in Sweden or the USA, wherever we are all striving to make it a safe transport against quite a lot of environmental pressure. Everyone is trying to go forward to continue to have a very safe business.

MR.F.JOHN L.BINDON for Paper 4:8
I hope the delegates will forgive me for being a bit parochial here, if I the two authors to say what the differences are between transporting nuclear waste on electrified tracks as against un-electrified tracks? The reason I ask is because in the area of Great Britain where I live - North Wales - we have at present two Magnox nuclear power stations and we are hoping before the end of this century to have a PWR there as well. We would like our track from Crewe to Holyhead electrified but there are capital costs involved and we are also told it is not just a question of erecting an electrified over-head line from Crewe to Holyhead, but a question of the track as well. Does that mean that the standard of track on which we are transporting radioactive waste from Wylfa and Trawsfynydd

to Sellafield is of somewhat less good quality than that going from S.E.
England up to Sellafield?

MR.I.K.BRAYBROOK:
There is only one answer to that - it is safe.

MR.F.J.L.BINDON:
Well, could I then ask British Rail to make sure that they meet the
people who complain about the non-electrified tracks, and assure them
that the only reason is the capital cost of the overhead lines and
nothing to do with the track.

CHAIRMAN:
Closing remarks.

DELEGATES

PROFESSOR J.R.A. LAKEY, PRESIDENT, THE INSTITUTION OF NUCLEAR ENGINEERS

SESSION 1 - Chairman: Mr.P.T.McINERNEY, UK Nirex Ltd.
 Technical Secretary: Mr.R.W.T.SIEVWRIGHT, UK Nirex Limited

SESSION 2 - Chairman: Mr.M.S.T.PRICE, United Kingdom Atomic Energy
                  Authority, Winfrith Atomic Energy Establishment
 Technical Secretary: Mr.R.COLYER, Department of Transport

SESSION 3 - Chairman: J.G.TYROR, United Kingdom Atomic Energy Authority,
                  Safety and Reliability Directorate
 Technical Secretary: Mr.K.L.LEIGH, United Kingdom Atomic Energy
                  Authority, Safety & Reliability Directorate

SESSION 4 - Chairman: Mr.D.BIBBY, Nuclear Transport Limited
 Technical Secretary: Mr.D.BOYER, British Nuclear Fuels plc

GUEST SPEAKER: Dr. John GITTUS, Director of Communication & Information,
              United Kingdom Atomic Energy Authority

ABBOTT Mr.P., British Rail
AKIYAMA Mr. H., Mitsui Engineering & Shipbuilding Company
ANDREWS Mrs. C., Euratom Safeguards (CEC)
APPLETON Mr.P.R., UKAEA Safety and Reliability Directorate
ARITOMI Dr.Masanori, Tokyo Institute of Technology
ARMITAGE, Mr.J., Rolls-Royce & Associates Limited
ASANO Mr.Norio, Sumitomo Heavy Industries Limited
AVERY Mr.A., UKAEA Winfrith Atomic Energy Establishment
BAILEY Mr.J.R., John Brown Engineering & Construction Limited
BAKER, Mr.L., Project Management Services Limited
BENNETT Mr.D., UK Nirex Limited
BERNARD, M H., COGEMA
BEYEN, Mr.H., Reederei und Spedition "Braunkohle" GmbH
BIBBY Mr.D., Nuclear Transport Limited
BILLOWES Mr.A., Strachan & Henshaw
BINDON, Mr.F.J.L., The Institution of Nuclear Engineers
BIRKETT, Mr.W.C., Central Electricity Generating Board PITB
BLACKBURN, Mr.S., Nuclear Transport Limited
BLACKBURN, Mrs.S.M., The Secretary, The Institution of Nuclear Engineers
BLACKMAN, Mr.D., Department of Transport
BOND, Mr.M.,
BOSTEN Dipl.Ing. F., TüV Baden EV
BOURRELIER Mr.P., COGEMA
BOYD, Mr.R.A., Ministry of Defence (PE)
BOYER, Mr.D., British Nuclear Fuels plc
BRACEY Mr.W., Transnuclear Inc.
BRAGARD, Dr. J-C., London Weekend Television, "The London Programme"
BRISTOW, Cpt.J., James Fisher & Sons plc
BRAYBROOK, Mr.I.K., Railfreight
BURGESS Mr.M.H., UKAEA, Winfrith Atomic Energy Establishment
BURGOYNE Mr.W.M., South of Scotland Electricity Board
CAINE Mrs.M., Costain Engineering Limited

CARLTON, Mr.J.F., British Steel Corporation (BSC) Cumbria Engineering
CECCOLINI Mr.A., ENEL DAA
CHAPALAIN Mr.J-J., Mory Technologis
CHAPMAN Mr.C., CEGB Nuclear Power Training Centre
CHESHIRE, Mr.R.D., British Nuclear Fuels plc
CHETWYND Mr.M., Sigma Power Controls
CLARK, Mr.R.G., EPS Logistic Technology Limited
CODEE Dr.H., Covra N.V.
COLYER Mr.R., Department of Transport
COOPER, Dr.C.A., UKAEA, Winfrith Atomic Energy Establishment
CORLESS, Mr.J., Ministry of Defence (Navy)
COWNLEY, Mr. J., Hovair
CRUICKSHANK, Dr.A.D., 'Nuclear Engineering International'
CUNNINGHAM, Mr.B.F., Costain Engineering Limited
DAY, Mr. G., UKAEA HQ
DALY, Mr.G., British Nuclear Fuels plc
DEAN Dr.R.B., W.S.Atkins Engineering Sciences
DESNOYERS Mr.B., COGEMA
DIERCKX Mr. L., Transnubel SA
DONELAN, P., Ove Arup & Partners
DOUGALL, Mr.I., CEGB, Health & Safety Department
DRAULANS, Mr.J., Transnubel SA
DUVAUCHELLE, Mr.H., Electricite de France
DYBECK Mr.P., SKB
EAVES, Mr.I., B&R Taylor Limited
EMMISON Mr.J., British Nuclear Fuels plc
EVANS Mr.J.H., Ministry of Defence (PE)
FIELDS, Mr.J., Impell Corporation
FIELDS, Mr.J., British Nuclear Fuels plc
FINDLAY, Mr.J.R., UKAEA Harwell Laboratory
FISCHER Mr.L., Lawrence Livermore National Laboratory
FITZPATRICK, Dr.J., UK Nirex Limited
FLETCHER, Mr.J., BSC Stainless
FORGAN, Mr.R., Gravatom Projects Limited
FRADIN Mr.J., International Nuclear Transport Network
FRANCIS Mr.R.J., UKAEA Harwell Laboratory
GARDELLINI Mr.A., Nucleco s.p.a.
GARTHWAITE, Mr.M., James Howden & Company Limited
GEISER Mr.H., GNS
GELDER Mr.R., National Radiological Protection Board
GITTUS, Dr.J., UKAEA HQ
GOWING, Mr.R., British Nuclear Fuels plc
GRIMM Mr.P.D., United States of America Department of Energy
GUIDOTTI Mr.M., ENEA
GUSTAFSSON, Mr.B., SKB
HANSON Mr.A., Transnuclear Inc.
HARTLEY, Mr. G., South of Scotland Electricity Board, Torness PS
HARRISON, Mr. A., Forgemasters Engineering Limited
HAYLOR Mr. P., Technician - Audiovisual
HERON, Mrs.L., British Steel Corporation (BSC) Cumbria Engineering
HIGSON, Mr. J., United Kingdom Atomic Energy Authority
HIKOSAKA Mr.Junichi, Marubeni Corporation
HINTON, Mr.B., Impell Corporation
HOGSTON Mr.J.R., Rutherford Appleton Laboratory
HOLDEN Mr.G.V., Gravatom Projects Ltd.,
HOLLAND, Mr.A.W.J., Rolls-Royce & Associates Limited
HOLLAND Mr.P., Lloyds Register of Shipping
HOLMES, Dr.J.E.R., United Kingdom Atomic Energy Authority, Winfrith AEE
HOLT, Mr. G., CEGB Generation Development and Construction Division
HUGGENBERG Mr. R., GNS
HUNTER, Mr.I.J., Nuclear Transport Limited

ISHII Mr.Sadayoshi, Kandenko Company Limited
IWASAKI Mr.Akihiro, Mitsui Company Limited
JOHNSTONE Mr.A., Electrowatt Engineering Limited
JOHNSON, Mr.S.F., ABB ATOM
JONES, Mr.E.W., Ministry of Defence (PE)
KAY Mr.A., Costain Engineering Limited
KINDLER, Mr.W., Reederei und Spedition "Braunkohle" GmbH
KNIPE Dr.A., United Kingdom Atomic Energy Authority, Winfrith AEE
KOBAYASHI Mr.Yoji, Utoku Express Company Limited
KOJIMA Mr.Katsuo, Ocean Cask Lease Company Limited
KOMORI Mr.Kiichiro, JDC Corporation
KOVACIC Mr.R., Nuklearna Elecrarna Krsko
KOVAN Mr.R., 'ATOM', United Kingdom Atomic Energy Authority HQ
KRIEGER Dipl.Ing. H., TüV Baden EV
KUTEA Mr.Damir, Nuklearna Elecrarna Krsko
LAKEY Professor J.R.A., Royal Naval College, Greenwich
LIBON, Mr.H., Transnubel SA
LINS, Dr.W., Kraftanlagen AG
LIVESEY, Mr.E., British Nuclear Fuels plc
LUCK, Mr. G.N., CEGB Generation Development and Construction Division
McCREESH, Dr.G.C., British Nuclear Fuels plc
McDONALD, Mr.J., Impell Corporation
McINERNEY Mr.P.T., UK Nirex Limited
McKENNA, Mr. M.L., Rolls-Royce & Associates Limited
McKURDY, Mr.B.J., United Kingdom Atomic Energy Authority, Harwell Labs.
McLAUGHLAN Mr.W., British Nuclear Fuels plc
McROBERTS Mr.D., UKAEA Dounreay Nuclear Power Development Establishment
MAMIYA, Mr.Shuichi, Kandenko Company Limited
MARLOW, Mr.B., UK Nirex Limited
MARSHALL Ms.P., 'NUCLEONICS WEEK'
MARTINEZ-LIZARRAGA Dr.M.A., Transnuclear SA
MATTHEWS, Mr.J., Ministry of Defence (PE)
MEREDITH Mr.P.V., United Kingdom Atomic Energy Authority, Winfrith AEE
MEEWEZEN, Mr.J.C.Williams Fairey Engineering Limited
MILES, Dr.J., Ove Arup & Partners
MISHIMA, Mr.Mikio, Hitachi Zosen Corporation
MORGAN, Ms.B., Technica Limited
MOULTON, Mr.R.J., United Kingdom Atomic Energy Authority, Harwell Labs.
NAGANO Mr.Isamu, Nuclear Transport Co. Limited
NAKAI Mr.Kunihiro, JGC Corporation
NARAI Mr.Toshihiro, Kobe Steel Limited
NICHOLSON, Mr.G., British Nuclear Fuels plc
NIELSEN, Mr.A., United Kingdom Atomic Energy Authority, Winfrith AEE
NISHI Mr.Akio, Japan Atomic Industrial Forum Inc.
NOURA Mr.Tsuyoshi, Transnuclear Limited
NUHAAN, Mr.F., Westinghouse International
OJIRO, Mr.Soken, Genshiryoku Daiko Company Limited
OLDROYD, Mr.G., Ove Arup & Partners
OMICHI, Mr.Teppei, Mitsubishi Metal Company Limited
OYINLOYE, Dr.J.O., Electrowatt (UK) Limited
PECKOVER Dr.R.S., United Kingdom Atomic Energy Authority, Winfrith AEE
PHAN-HOANG, Mr.L., Electricité de France
PINTO, Dr. S., Nuclear Assurance Corporation
PRICE, Mr.M.S.T., United Kingdom Atomic Energy Authority, Winfrith AEE
RAO, Mr.S., Impell Corporation
RESKE, Mr.H-J., Kraftanlagen AG
RIBBANS Mr.D.J., Ontario Hydro
RYCROFT Mr.C.I., E.C.Transport Limited
SAITO Mr.Daiju, Ishikawajima Harima Heavy Industries Company Limited
SAUL Mr.D., NEI International Research and Development Co. Ltd.
SCHWARZ Dr.G., GRS mbH

SHEPPARD, Mrs.J., Merseyside & North Wales Electricity Board
SIEVWRIGHT, Mr.B., UK Nirex Limited
SILLS Dr.R., British Nuclear Technology plc
SIMPSON Mr.M.A. British Nuclear Fuels plc
SMITH Mr.Lyall, Paul Scherrer Institute
SMITH, Mr.M.J.S., UK Nirex Limited
SMITH, Mr.P.J.A., Urenco Limited
SNAREY, Mr.V.L., Rolls-Royce & Associates Limited
SNOW Mr.M.W., Combustion Engineering Inc.
van SOELEN Mr.A., Covra N.V.
SPECKENHEUER, Mr. P., Thyssen Steel Limited
SPINK, Mr.H.E., British Nuclear Fuels plc
SPRAKE, Mr.V.M., Impell Corporation
STEGFELLNER, Mr. J., Voest Alpine
STEPHEN Mr.T.W., James Fisher & Sons plc
STRANG, Mr.C.R.H., Strachan & Henshaw
STREET, Mr.C.L., Ministry of Defence (Navy)
STUBBS, Dr.R.J., Nuclear Installations Inspectorate
SWINDLEHURST, Dr.W.E., British Nuclear Fuels plc
TAKENAKA Mr.Akira, Nuclear Fuel Transport Company Limited
TANGUY Mr.L., CEA CEN
TAPLEY Mr.C., Amersham International plc
TAYLOR, Mr.W.R., Atomic Energy of Canada Limited
THIEL Mr.Claude, Mory Technologis, Paris, France
TIMME, Mr.J. Drath & Schrader,
TOMACHEVSKY Mr.E., CEA IPSN
TYROR Mr.J.G., United Kingdom Atomic Energy Authority, SRD
URCH, Mr.K.C., UK Nirex Limited
UTSUMI Mr.Mitshuhiro, Nippon Shipping Company Limited
UYANNEH, Mrs.P.A., Institution of Nuclear Engineers
VANNESTE Mr.M., COGEMA SGM-CS
VARLEY, Mr.J., Nuclear Engineering International
VAUGHAN Dr.R., Croft Associates Limited
VERDIER, Mr.A., COGEMA
VILKAMO Mrs.S., Finnish Centre for Radiation and Nuclear Safety
WAKEFIELD, Dr.J.R., United Kingdom Atomic Energy Authority, Windscale
WALES, Mrs. J., Costain Engineering Limited
WALTON, Mr.D.G., University of Manchester
WARDEN, Mr.D. Amersham International plc
WESTON Mr.B., British Nuclear Fuels plc
WILCOCK, Mr.G., Rolls-Royce & Associates Limited
WILLCOCK, Mr.K., CEGB HQ
WILLIAMS, Mr.J., United Kingdom Atomic Energy Authority, Harwell Labs.
WILLIAMS, Mr.R.J., British Nuclear Fuels plc
WILLS, Mr.L.I. representing Smit International TA, Rotterdam
WILSON, Mr.P.W., British Nuclear Fuels plc
WOOD, Mr.I.A., GEC Energy Systems Limited